WAFER SCALE INTEGRATION, III

WAFER SCALE INTEGRATION, III

Proceedings of the Third IFIP WG 10.5 Workshop on
Wafer Scale Integration
Como, Italy, 6–8 June 1989

Edited by

MARIAGIOVANNA SAMI
Politecnico di Milano
Milano, Italy

FAUSTO DISTANTE
CSISEI, CNR, Politecnico di Milano
Milano, Italy

1990

NORTH-HOLLAND
AMSTERDAM · NEW YORK · OXFORD · TOKYO

ELSEVIER SCIENCE PUBLISHERS B.V.
Sara Burgerhartstraat 25
P.O. Box 211, 1000 AE Amsterdam, The Netherlands

Distributors for the United States and Canada:
ELSEVIER SCIENCE PUBLISHING COMPANY INC.
655 Avenue of the Americas
New York, N.Y. 10010, U.S.A.

Library of Congress Cataloging-in-Publication Data

IFIP WG 10.5 Workshop on Wafer Scale Integration (3rd : 1989 : Como, Italy)
Wafer scale integration, III : proceedings of the Third IFIP WG 10.5 Workshop on Wafer Scale Integration, Como, Italy, 6-8 June 1989 / edited by Mariagiovanna Sami, Fausto Distante.
p. cm.
Includes bibliographical references.
ISBN 0-444-88496-3
1. Integrated circuits--Wafer scale integration--Congresses.
I. Sami, Mariagiovanna. II. Distante, Fausto. III. Title.
TK7874.I329 1989
621.381'5--dc20 89-26671
CIP

ISBN: 0 444 88496 3

Printed in The Netherlands.

INTRODUCTION

In this third IFIP workshop on Wafer-Scale Integration, the increasing viability of the main theme is made evident by the mature treatment received by the subject areas which have already been discussed at the two previous meetings, as well as by new items that have been brought up for discussion, this time.

The attention given to **technologies** is notable. A relevant point is that technologies for restructuring are no longer the main item in this context – on the contrary, claims to good yield and low defect distributions are made – while actual production problem areas such as packaging, solutions for power distribution and health dissipation, even actual *structural engineering* (not to be overlooked for such large chips) are discussed in detail. At the same time, new technological solutions are envisioned; on the one hand the use of *optical* technologies for interconnections promises to overcome one of the main restrictions to architectures on a wafer, and on the other hand the extension of quick-prototyping solutions to the wafer dimension allows the introduction of wafer-scale systems in educational environments as well as in applications where a quick turnout and a small production would make traditional silicon solutions unacceptable.

Apart from technologies, most attention is paid to two main areas: architectures for wafer-scale and applications. Actually, it is obvious that these two areas t separate, since most architectures described are application-oriented ones.

Concerning **architectures**, three different lines of approach can be identified. *Evolutionary* solutions are proposed, mainly concerning array architectures and the restructuring techniques best suited to such structures. *Innovative* architectures are presented – it is interesting to note that, in particular, several papers deal with neural nets. Actually, architectures taken into account are again the ones that – through intrinsic regularity, characteristics of the interconnection network, foreseeable capacity of supporting defect-tolerance techniques – belong to the "classical" type suited to wafer-scale devices. An attractive point is constituted by the presence of architectures designed not just for experimental reasons but for *industrial production*: once again, this is a testimonial to the maturity reached by wafer-scale integration.

All in all, non-numerical applications are by far predominant: there are certainly architectural reasons for this preference – the regularity of image-processing architectures is obviously well suited to WSI – as well as reasons due to the peculiar requirements of the application area (a typical case is that of signal-processing devices), in particular when spaceborne applications, due to the simultaneous requirements of speed, compactness, and low power consumption are considered. On the other hand it may be a source of curiosity that, for example, high-performing floating-point arithmetic units or even vector units up to now have not been the subject of developments in this context: the numerical structures envisioned (e.g., an array of multipliers) are characteristic of the WSI area for their particular architecture rather than for application characteristics.

The Editors
Mariagiovanna Sami
Fausto Distante

CONTENTS

Introduction
M. Sami and F. Distante v

Neural Networks on Silicon
P. Treleaven and M. Vellasco 1

A Flexible Wafer Scale Neural Network
J. Ouali, G. Saucier, and J. Trilhe 11

Focused Electron and Ion Beam Personalized WSI Neural Networks
H.T. Lin, R. Philhower, and J.F. McDonald 25

ASP Modules: WSI Building-Blocks for Massively Parallel Computing
R.M. Lea 39

Hierarchical Redundancy and Self-Testing for a ULSI Poly-Processor System
N. Tsuda and T. Satoh 57

A Methodology for the Design of Fault-Tolerant, Testable, Wafer-Scale Processor Arrays Using Boundary Scan Architecture
N. Jarwala and Y. Zorian 73

Evaluation and Comparison of Selected WSI Reconfiguration Architectures in Terms of Yield and Yield per Area
M. Wesseling, U. Ramacher, and K. Goser 87

A Model for the Analysis, Design and Comparison of Fault-Tolerant WSI Architectures
V.S.S. Nair and J.A. Abraham 99

Testability and Design for Testability for WSI Systolic Hexagonal Arrays
D. Sciuto 109

On Testable Redundant PLAs for WSI Implementation and Manufacturing
Y.-N. Shen and F. Lombardi 121

Switches Activation and Desactivation Using the E.Beam
P. Girard and B. Pistoulet 137

Wafer Scale L–U Decomposition Array with a New Reciprocal Cell
V.K. Jain, D.L. Landis, and C.E. Alvarez 143

A Fault Tolerant Wafer Scale Architecture for Vision Processing
J.A. Trotter and W.R. Moore 155

Modern Aspects of Silicon Technology for Microelectronics
M. Domenici 163

Design of MOS Power Switches for Short-Circuit Protection in Large-Area Integrated Circuits
Y. Savaria, N.C. Rumin, B. Mandava, and S. Madhyastha 175

Organic Polymer Optical Interconnections for Photonics in WSI and WSHP
H.T. Lin, S. Dabral, and J.F. McDonald 187

The WASP Project: Bridging the Gap between Hypothetical Wafer-Scale Architectures and Engineering Reality
I.P. Jalowiecki, S.J. Hedge, K.D. Warren, and R.M. Lea 199

An Architecture for an Intelligent Multi-Mode WSI Memory
K.E. Grosspietsch 211

On Repairability/Unrepairability Detection and Yield Enhancement of WSI Redundant Memories
Y.-N. Shen and F. Lombardi 221

Westinghouse WSI Development Program
J.E. Brewer and M.E. Michael 237

Automatic Wiring Algorithm for WSI Interconnection Networks
J.M. Wills and V.K. Jain 247

An Architecture-Independent Model for Partitioning and Redundancy in Wafer-Scale Integrated Circuits
M.F. Beusekamp 257

Wafer Scale Integration for Signal Processing
E.E. Swartzlander, Jr. 269

A Modular Bit-Serial Convolver
L. Dadda and L. Breveglieri 279

WSI Implementation of a Video Rate 2 Dimensional FFT
R. Philhower and J.F. McDonald 291

The Reliability of Large Arrays for Matrix Multiplication with Algoriothm-Based Fault-Tolerance
W.G. Bliss and M.R. Lightner 305

A Practical Experiment in Defect Tolerance: The HYETI 16-Bit Microprocessor
R. Leveugle, M. Soueidan, and N. Wehn 317

Defect-Tolerant Chip Featuring Multiple Configuration and Built-In Test Approaches
U. Jagau, H.-J. Iden, and M. Grützner 333

Controlling Unit for a Pipelined Floating Point Hardwired Engine
S. Vassiliadis and E. Schwarz 343

A Configurable Single Wafer Bit-Sliced Convolver: Effect of the Interconnection Network on the Yield
S. Lasserre 357

Design of a Universal Switching Network for Reconfigurable 2-D Arrays
J.-L. Patry and G. Saucier 377

Low Temperature Fabrication to Enhance Discretionary WSI and WSHP Wiring Yield
S. Dabral and J.F. McDonald 393

Wafer Scale Integration, III
M. Sami and F. Distante (Editors)
Elsevier Science Publishers B.V. (North-Holland)

NEURAL NETWORKS ON SILICON

Philip Treleaven and Marley Vellasco

Department of Computer Science
University College London

Neurocomputers are currently being developed focusing either on increasing the performance, as with the Special-purpose approach, or on enhancing flexibility to emulate neural models, like the General-purpose approach. This paper gives a brief overview of the current work on neurocomputer developments and describes a neurocomputer, being implemented at University College London, that aims to optimise flexibility, performance and silicon area by using a silicon compiler. The neurocomputer is configured from identical chips, each containing a number of processing elements, and connected by broadcast busses. Central to the neurocomputer and its silicon compiler is the concept of a "generic" neuron model and a corrdesponding extensible VLSI architecture.

1. INTRODUCTION

The contrast in pattern processing ability between computers and humans led many research groups to study *Artificial Neural Networks.* These networks mimic the organization and functionality of the human brain and are based on mathematical models, or *neural models*.

Recent advances in neural network models have been largely supported by simulations on conventional computers[1]. However, the development of *neurocomputer* architectures, designed to optimise the computation of artificial neural networks, is necessary if these advances are to demonstrate their true value. Architecture developments are currently progressing along two avenues[2]:

- **General-Purpose Neurocomputers** - programmable neural computers for emulating a range of neural network models.

- **Special-Purpose Neurocomputers** - specialised neural network hardware implementations that are dedicated to a specific neural network model and therefore have potentially very high performance.

Special-purpose neurocomputers provide high performance and make optimum use of silicon area, but frequently are inflexible due to supporting a single model. On the other hand, general-purpose neurocomputers opt for generality, providing a framework to execute many different models, but have lower performance. In order to combine the advantages found in both approaches - *flexibility* and *performance* - we are developing a new proposal, based on *silicon compilation.* The approach is based on a "generic" neuron model, that integrates the basic features of neural networks, a parallel architecture of processing elements and a high level description language. A designer, using the description language, describes basically the network topology and the model's functions (including learning and recall), and then the silicon compiler generates the target architecture for that specific model. This method provides the performance of a specialised hardware and yet offers the flexibility of a general-purpose approach in the neural model specification.

This paper firstly gives a brief overview of the current work on neurocomputer developments and then, in the following sections, we describe the architectural features of the *Generic Neuron* approach.

2. NEUROCOMPUTERS

A neurocomputer, whether general-purpose or special-purpose, is essentially a parallel array of interconnected, primitive processing units that operates concurrently. When considering the set of possible architectures for the basis of a neurocomputer, the issues to be considered are: firstly the degree of programmability of the computer, secondly the number of physical processing units, and thirdly the complexity of an individual processor. These design issues, as shown by Figure 1, lead to systems ranging from dedicated hardware, analogous in complexity to traditional random access memories (i.e. RAMs), to conventional computers. Lowest in element complexity are RAMs, in which the "processing element" is about as small as usefully imaginable. Next are special-purpose neurocomputers, each with a neural network implemented directly in hardware and dedicated to one model. In the next zone are general-purpose neurocomputers, designed to support a range of neural network models. (Also shown for comparison are the two sub-classes of cellular arrays; computational arrays and systolic arrays.) The last two zones cover conventional computers that support neural networks through simulation programs. Popular simulation vehicles include the Intel iPSC and TRANSPUTER systems[3].

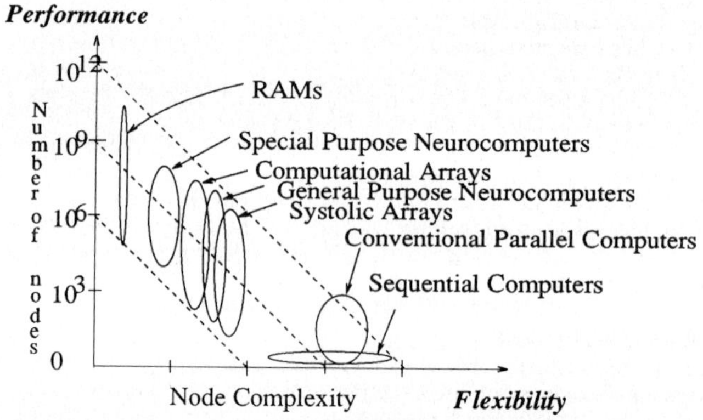

FIGURE 1: Spectrum of Neurocomputer Architectures

General-purpose neurocomputers are being developed in two categories: *co-processor boards*, such as the Neuro-07 from Japan, to accelerate the execution of neural networks, and *parallel processor arrays*, which aim to achieve real parallelism by having a large number of physical processing units. Many co-processor boards are already available commercially. Firstly the TRW MARK III[4] a co-processor for VAX computers which is able to support 65K virtual processing elements and 1M interconnections. Secondly, Hecht-Nielsen has developed the ANZA and ANZA Plus systems[5,6], coprocessor boards which interfaces to an IBM PC-AT. ANZA Plus has 1M processing elements and allows a total of 1M interconnections between all the PEs. Another important company is Scientific Applications International Corporation (SAIC) which produces the SIGMA-1 workstation[7] based around the DELTA accelerator board for the PC-AT. Performance characteristics are: 1M virtual PEs, 1M interconnections and update speed of 10M interconnections per second. Lastly, Texas Instruments have announced their ODYSSEY co-processor that plugs into a TI Explorer Workstation. The ODYSSEY comprises four TMS32020 digital signal processing chips and 1M bytes of memory.

Many research projects are also developing processor arrays. For example, the IBM Palo Alto Scientific Center has implemented a complete system consisting of the Network Emulation Processor (NEP) hardware and the Computation Network Environment (CONE) software[8]. Another interesting processor array based on Inmos TRANSPUTERS has been developed at the NASA Johnson Space Center in Houston. Neural Network Environment Transputer System (NNETS)[9] comprises forty, 32-bit/10-MIPS TRANSPUTERS each with 256K bytes of local memory and interconnected via four 10M-bits full duplex serial links to other TRANSPUTERS. NNETS has a potential performance of 400 MIPS with 12M bytes of memory. Garth of Texas Instruments, working with Cambridge University, has developed NETSIM[10], a 3-D array of processing elements, each based on specially designed chips plus a 80188 microprocessor.

Developments in **Special-Purpose** neurocomputer architectures can be categorised by the implementation technology into: silicon[1,11,12], optical[6,13,14] and molecular[15,16]. Development of dedicated silicon hardware implementations (using digital or analog circuits) of neural networks are underway at a number of locations[16,17]. AT&T have develop several prototype neural network chips. Example chip designs[11] include: a hybrid digital/analog programmable connection matrix, an analog connection matrix with adjustable connection strengths, and a digital pipelined best-match chip. Bell Communications Research (Bellcore Labs.) have built a first analog neural network with adaptive model synapses[18]. At the California Institute of Technology (CALTECH), a team led by Carver Mead, has implemented neural network chips[12]. Their initial prototype contained 22 active elements and a full interconnection matrix of 462 elements, implemented in 4 micron NMOS technology. One of the latest was a 289 neuron chip in CMOS. Mead is investigating neural chips for the early vision and auditory processing functions, using the intrinsic nonlinearities of MOS transistors in a subthreshold regime. Aleksander at Imperial College London has developed a series of systems called Wisard[19]. Wisard II is organised as a hierarchical network of PEs, with each PE being constructed from commercial RAM. The RAMs address inputs are used to detect binary patterns, with the input field for an image (comprising 512 x 512 binary pixels) being connected to the first layer of PEs.

The current developments in general-purpose and special-purpose neurocomputers have shown their effectiveness, either by providing a good degree of flexibility to experiment with different models, or by implementing small scale neural network applications directly into hardware to achieve the best result in terms of performance. Nevertheless, no optimal solution for neurocomputer architecture has still been found. To develop an "ideal" neurocomputer, we believe one should look for a compromise between these two contrasting issues: flexibility and performance.

At University College London we are looking for this trade-off by means of a silicon compiler. Central to the neurocomputer and its silicon compiler is the concept of a *generic neuron model* and a corresponding extensible *VLSI architectural framework.*

3. GENERIC NEURON

In order to achieve both high performance and flexibility a more general neuron model should be devised. The *generic neuron* model should be flexible enough to cope with different topologies and learning/recall algorithms and, at the same time, should be a very simple design to attain high performance.

The generic neuron model we have developed is based on the main features of neural network models, comprising the basic variations of learning and recall procedures. The model is shown in Figure 2. The generic neuron model consists of a set of state inputs **S**, that receives the state inputs from other PEs, and a state output **s** that sends the processing element's own state to other PEs. A group of error inputs **E** and an error output **e** are also provided to cope with models that use backward flow of data, such as Back Propagation[20]. Internally, the generic neuron includes a table of weights **W** (associated with both groups of inputs) and three different functions: the activation function **f1**, to evaluate the neuron's state; the weight updating function **f2**; and the error calculation function **f3**.

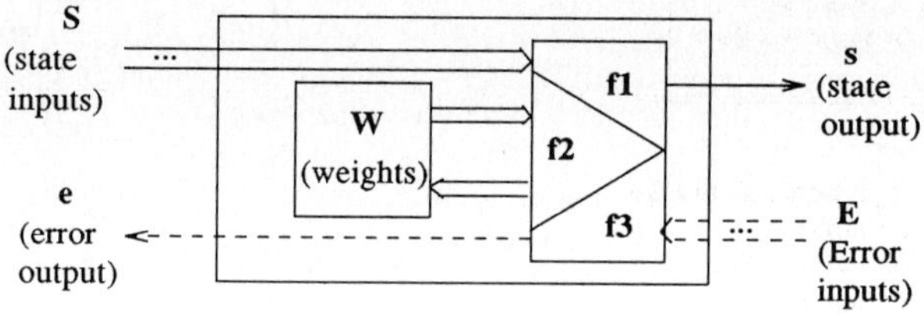

- Activation Function - s = f1 (S, W)
- Weight Updating - ΔW = f2 (S, E, W)
- Error Calculation - e = f3 (s, E, W)

FIGURE 2: Generic Neuron

The activation function **f1** is performed in two steps: the propagation rule, that defines how the state inputs are modified by the associated weights; and the threshold function that actually uses the result of the propagaion rule to obtain the state output. So, **f1** is generally a function of **S** and **W**:

$$s = f1\ (S, W) \tag{1}$$

The weight updating function **f2**, executed during learning phase, is typically a function of the old weights **W**, the state inputs **S** and the neuron's error output **e**:

$$\Delta W = f2\ (S, W, e) \tag{2}$$

The generic neuron's error value **e**, applied to the weight updating function, is evaluated by the error calculation function **f3**. It is typically a function of the input errors **E**, the neuron's state **s** and sometimes the weight values **W**:

$$e = f3\ (s, E, W) \tag{3}$$

By describing these three functions and the network topology, it is possible to specify a large range of neural models. Figure 3 summarises the description of three neural models in terms of their functions and the topology.

Based on the structural design of the generic neuron, we defined the architectural framework in terms of the interconnection scheme and the processing element structure.

Neural Network Model	Network Topology	Recall/Learning Phase			
		f1		f2	f3
		Propagation Rule	*Threshold Function*	*Weight Updating*	*Error Calculation*
Hopfield/ Kohonen	single-layer with feedback	net = $\sum S.W$	hard limiter	$\Delta w_{ij} = s_i . s_j$	----
Perceptron	single-layer feed-forward	net = $\sum S.W$	hard limiter	$\Delta w_{ij} = \eta . s_i . e_j$	$e_j = t_j - s_j$
Back Propagation	multi-layer bi-directional links	net = $\sum S.W$	sigmoid	$\Delta w_{ij} = \eta . s_i . e_j$	output layer: $e_{jo} = T'(net).(t_j - s_j)$ hidden layers: $e_{jh} = T'(net).\sum E.W$

Notation:

w_{ij} = weight from PE_i to PE_j
s_i = state from PE_i
e_j = output error of PE_j
t_j = target output of PE_j
η = constant less than 1
$T'(net)$ = derivative of the threshold function

FIGURE 3: Example of Neural Models' Descriptions

4. ARCHITECTURAL FRAMEWORK

For attaining a good degree of flexibility and performance the interconnection strategy and the processing element should be defined considering some fundamental requirements like expandability, modularity and minimum silicon area.

4.1. Interconnection Structure

The interconnetion strategy is one of the main problems in neural networks. In terms of interconnection structure, three main requirements were established: *regularity*, *scalability* and *expandability*.

The huge amount of connections inside a neural network and the considerable number of different interconnection patterns among the models, imply an interconnection strategy that provides the necessary flexibility. The system should be able to form complex topologies and to handle all the connections with minimum overhead in communication protocols and routing algorithms.

Design scalability is essential to reduce the cost per interconnection, when implementing large networks. The interconnection scheme should be regular to cope with the improvements in the circuit and packing technologies, without affecting the chip pin count when a larger number of PEs is packed onto one chip.

The number of PEs in a neural network is application/implementation dependent, demanding an expandable system. The user should be able to develop neural networks with any number of processing elements, without prohibitively increasing the cost of the interconnection structure.

Based on the above criteria, we opted for the bus interconnection strategy shown in Figure 4. The communication is accomplished through common broadcast busses, and each processing element gains access to the bus sequentially, through the command of a central controller.

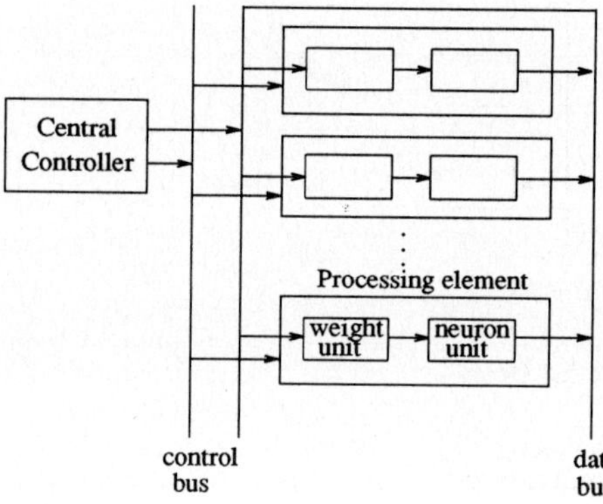

FIGURE 4: Architecture Interconnection Strategy

The bus strategy provides the necessary flexibility, as it can effectively realise any topology desired, and its regular structure simplifies the interconnection of a large number of PEs. Moreover, its regularity contributes to the ability to pack more PEs onto one chip, without affecting the total number of pins. This feature is illustrated in Figure 5.

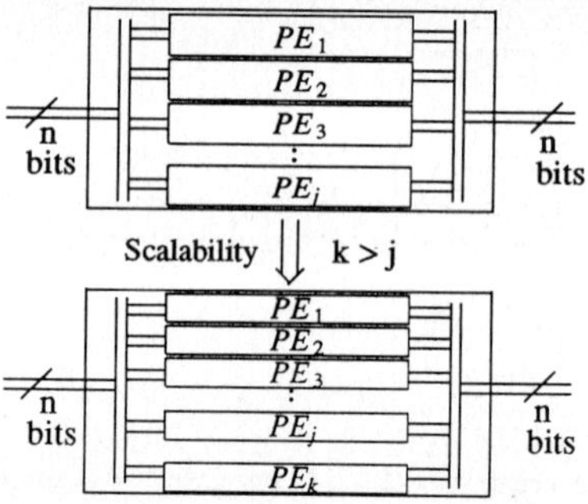

FIGURE 5: The Scalability Property

One can argue that the performance in communication through a bus is not very good when compared to dedicated interconnection structures, like the hypercube. Nevertheless, because of constraints in pin count, complex interconnection structures are normally restricted to narrow or even serial links off chip, dropping their performance considerably[21].

In addition, in order to improve the bus performance, the system can still be expandable into many busses, clustering PEs that receive the same input into the same bus (for instance, the PEs of the same layer). This extension is shown in Figure 6.

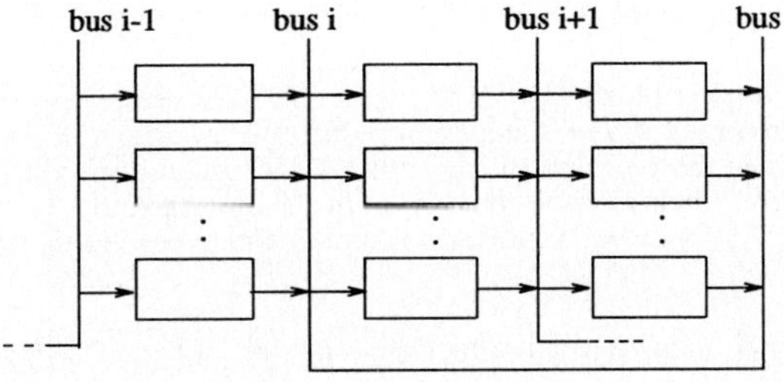

FIGURE 6: Multi-Bus Structure

4.2. Processing Element Design

The design of the processing element observed two main issues: simplicity, to attain low silicon area; and modularity to reduce communication between modules and to facilitate replicability.

Using the general properties of the generic neuron described in section 3, we designed the architectural framework of Figure 7. The generic neuron, as shown in Figure 2, can be divided in two logical blocks: a block that calculates the neuron's outputs and another that deals with the weight updating.

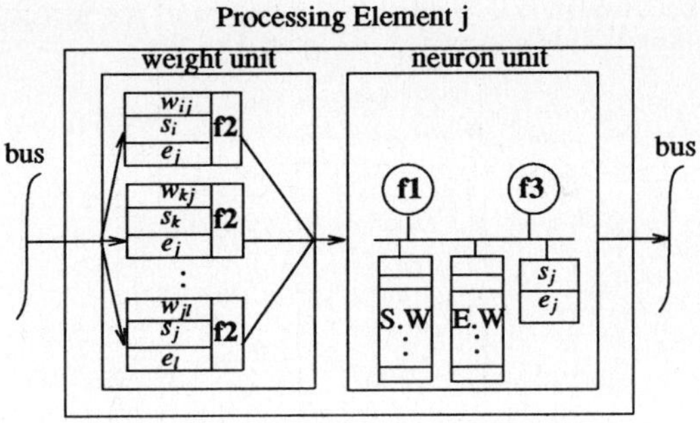

FIGURE 7: Processing Element Architecture

So, the processing element's structure is divided in two basic units namely ***weight unit***, which accomplishes the synaptic function, and ***neuron unit*** that carries out the neuron functions. The weight unit executes the updating of the weights and the neuron unit performs basically the calculation of the processing element's new state and error value.

The weight unit, illustrated in Figure 7, comprises the weight updating function f2 and a number of memory blocks, representing each connection, with the three necessary parameters of equation 2: the weight, the state and the error associated with the connection. As shown in Figure 2, the generic neuron has two groups of inputs: the state inputs associated with the forward flow of data; and the error inputs related to the backward flow of data. Since both groups can have associated weights, the weight unit has two different types of memory blocks: ***forward blocks***, related to the state inputs' weights, and ***backward***

blocks, associated with the error inputs' weights. The forward block parameters are the input state of the connection, the associated weight and the PE's error calculated in the neuron unit. Correspondently, the backward block comprises the error input of the connection, the associated weight and the PE's state evaluated in the neuron unit.

When the flow of data is bi-directional (as with the Back Propagation model), both processing elements involved in the connection require the weight value to accomplish some calculations. To avoid the problem of transmitting the updated weight between processing elements, the weight block is duplicated in both PEs interconnected. So, for each backward block in one PE there is a replicated forward block in the other PE involved in the connection.

The neuron unit deals with the activation function **f1** and error calculation function **f3**. According to equation 1, to calculate the PE's new state, the neuron unit needs to store the input states modified by the associated weights (S.W). Similarly, according to equation 3, to evaluate the PE's error value (e_j), the neuron unit requires to save the PE's state (s_j) and the input errors (sometimes also modified by the associated weights).

The processing element structure is very simple and modular, comprising the necessary memory and functions to execute the neural model. This approach provides high performance and allows the easy replication of a large number of PEs, in order to compose a complete system.

5. INTERNAL ARCHITECTURE

The internal architecture of the two units inside the processing element must establish the best compromise between performance and minimum silicon area. Therefore, the VLSI properties should be adequately exploited in the design of the processing elements. The simplified internal architecture is depicted in Figure 8.

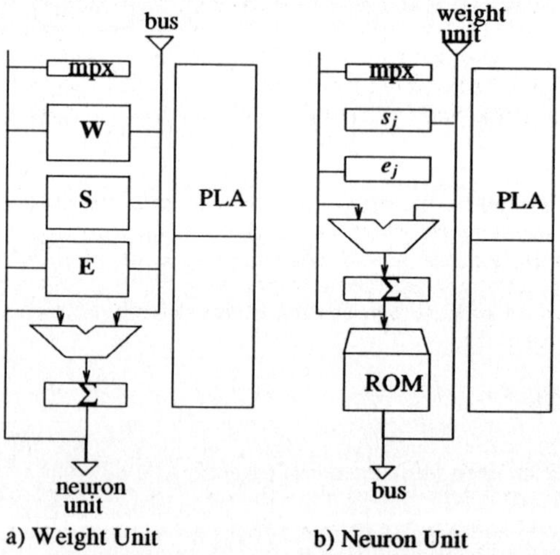

FIGURE 8: Simplified Datapaths

The weight unit contains basically: the memory blocks to store the weights (W), the states (S) and the errors (E); an arithmetic-logic unit (ALU); and the control structure (PLA) to accomplish the weight updating function. The silicon area necessary to realise this unit is greatly dependent on the amount of memory required to store the function's parameters, which is expressed by the number of interconnected PEs.

The neuron unit is essentially composed of: registers to store the PE's state (s_j) and error (e_j); an ALU; a PLA to control the execution of the state and error calculation functions; and a ROM memory to perform the threshold function. The silicon compiler, using the designer's specification of the threshold function, digitise it and implement it in a form of look-up table. Simulation measure marks using 256 and 512 table inputs, have proved the advantage of using such approach in benefit of greatly minimising silicon area. Both units contain a *mpx* register to help in one of the most common calculation: multiplication.

The use of PLAs as the control structure, for both the weight and neuron units, provides the required generality to support the great variety of neural models. The neural model's functions will be described in the high level language and the silicon compiler will translate it into finite state machines, producing the PLAs. Consequently, the PLAs' micro-code is the basic change when implementing different neural models. This approach also copes with the development of new neural algorithms since a previous knowledge of the possible neural functions is not required, as usually happens when using a library of hardware implemented functions.

We have an operational simulator for the neurocomputer and are currently implementing a prototype chip in 2μm CMOS technology, which should be finished by the end of 1989. The silicon compiler is also being developed. The aim is to produce a series of chips to identify the optimum structures primarily for the processing unit.

6. CONCLUSIONS

Many different neurocomputer architectures have been proposed for neural networks, although no optimal solution has been found. Special-purpose neurocomputers optimise performance and silicon area, at the expense of flexibility. In contrast, general-purpose neurocomputers optimise flexibility for performance. We presented in this paper a neurocomputer architecture, embodied in a silicon compiler, that attempts to combine flexibility with performance.

Through a high-level language, the user describes the neural network model in terms of the connectivity, the functions to be performed and the desired precision for the weights and states. Using this description, the silicon compiler produces the processing elements, with the internal functions specific to the neural model described. In addition, the silicon compiler packs a number of processing elements onto one chip, largely according to their necessary amount of memory.

ACKNOWLEDGMENT

The authors gratefully acknowledge the help of Carlo de Oliveira for his suggestions in the VLSI design and Marco Pacheco for his valuable insights from his own project of a general-purpose neurocomputer. We finally thank the CNPq of Brazil and ESPRIT PYGMALION Project for the financial support.

References

1. IEEE COMPUTER, "Special Issue on Artificial Neural Systems," *IEEE COMPUTER*, vol. 21, no. 3, March 1988.
2. P.C. Treleaven, "Neurocomputers," *Department of Computer Science, University College London*, 1989.
3. P.C. Treleaven and et al, "Computer architectures for artificial intelligence," *Springer Verlag*, vol. 272, pp. 416-492, 1987.
4. R.M. Kuczewski, M.H. Myers, and W.J. Crawford, "Neurocomputer Workstations and Processors: Approaches and Applications," *Proc. IEEE First Int. Conf. on Neural Networks*, pp. III-487-500, June 1987.
5. R. Hecht-Nielsen, "Neurocomputing: picking the human brain," *IEEE SPECTRUM*, vol. 25, no. 3, pp. 36-41., 1988.
6. R. Hecht-Nielsen, "Performance Limits of Optical, Electro-Optical, and Electronic Neurocomputers," *Optical and Hybrid Computing SPIE*, vol. 634, pp. 277-306, 1986.
7. SAIC, "DELTA/SIGMA/ANSim, editorial," *Neurocomputers*, vol. 2, no. 1, 1988.
8. W.A. Hanson and et al, "CONE - Computational Network Environment," *Proc. IEEE First Int. Conf. on Neural Networks*, pp. III-531-538., June 1987.
9. R.T. Savely, "The Implementation of Neural Network Technology," *Proc. IEEE First Int. Conf. on Neural Networks*, pp. IV-477-484, June 1987.
10. S.C.J. Garth, "A Chipset for High Speed Simulation of Neural Network Systems," *Proceedings of the IEEE First Annual International Conference on Neural Networks*, pp. III-443-452, June 1987.
11. H.P. Graf, W. Hubbard, L.D. Jackel, and P.G.N. deVegvar, "A CMOS Associative Memory Chip," *Proc. IEEE First Int. Conf. on Neural Networks*, pp. III-461-468, June 1987.
12. M.A. Sivilotti and et al, "VLSI architectures for implementation of neural networks," *AIP conference proceedings 151, ed.*, 1986.
13. Y.S. Abu-Mostafa and D. Psaltis, "Optical Neural Computers," *Scientific American*, vol. 255, no. 3, pp. 88-95, March 1987.
14. D. Psaltis and et al, "Learning in Optical Neural Computers," *Proc. IEEE First Int. Conf. on Neural Networks*, pp. III-549-555, June 1987.
15. M. Conrad, "The Lure of Molecular Computing," *IEEE Spectrum*, vol. 23, no. 10, pp. 55-60, October 1986.
16. New Science Associates Inc., "Neural Networks: Prospects for Commercial Use," *New Science Associates, South Norwalk, CT*, Spring 1988.
17. O. Port and J. Wilson, "They're Here: Computers That Think," *Business Week*, pp. 84-85, January 1987.
18. J.A. Alspector and et al, "Stochastic learning networks and their electronic implementation," *IEEE Conf. on Neural Information Processing Systems - natural and synthetic*, November 1987.
19. I. Aleksander and et al, "WISARD - A radical step forward in image recognition," *Sensor Review*, pp. 120-124, July 1984.
20. D.E. Rumelhart and J.L. McClelland, "Parallel distributed processing: explorations in themicrostructure of cognition," *MIT Press, Cambridge, Mass.*, vol. 1 & 2, 1986.
21. Tse-yun Feng, "A Survey of Interconnection Networks," *Computer*, pp. 12-27, December 1981.

Wafer Scale Integration, III
M. Sami and F. Distante (Editors)
Elsevier Science Publishers B.V. (North-Holland)

A Flexible Wafer Scale Neural Network

J. Ouali, G. Saucier * J. Trilhe

Institut National Polytechniquo de Grenoble / CSI
46, Avenue Félix Viallet
38031 Grenoble - FRANCE

* SGS Thomson
Avenue des Martyrs
38019 Grenoble - FRANCE

1. Introduction

Neural computing relying on non algorithmic information processing has led in the recent last years to a tremendous research effort all over the world [1..6]. It appears clearly that neural computing will be practical and efficient if dedicated, fast architectures exist. Presently, neural computing is simulated on existing Von Neumann or parallel computers (CRAY2, GAPP, Transputer). Attempts have been made to build dedicated architectures with commonly available components and some other studies concentrate on the design of either analog or digital specific components [7..16].

This paper focusses on the design of a flexible wafer scale neural architecture. The goal is to provide an implementation frame for a large variety of architectures (Hopfield, layered network) including both learning and relaxation phases. Some other parameters such as precision are also taken into account as design parameters. A silicon compilation approach allows to provide architectures with a given number of bits for state and coefficient encoding without new design efforts.

Since there is an urgent demand for large arrays, extension to wafer scale architectures is studied. Neural networks should be a good candidate for Wafer Scale Integration because of their inherent defect tolerance. Therefore, a study will point out to what extend end of manufacturing defects are assimilated to defects tolerated by neuro-computing and what defect tolerance has to be provided for the other defects (fatal defects for instance) as in a classical wafer scale approach [17..19].

2. Basic choices for the architecture

It is obvious that neuro-computing is not a mature field at the moment and that a lot of different architectural proposals exist. One of the first requirements for our architecture is to implement both Hopfield and multilayered networks (figure 1a, 1b) where the number of layers will be specified by the user. Therefore, the same design has to support both architectures. We shall see that both software programmable switches and silicon compilation techniques will be used to meet this requirement.

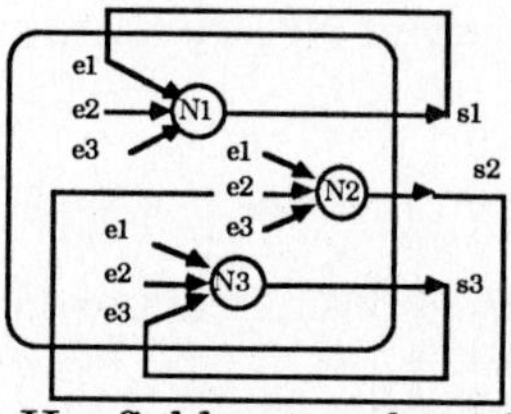

Figure 1a : Hopfield network with 3 neurons

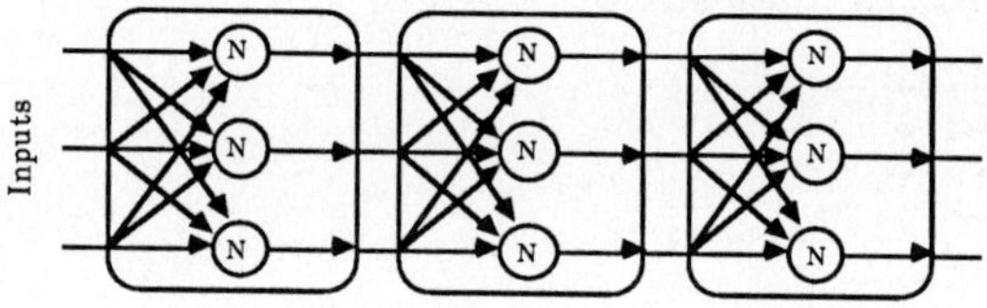

Figure 1b : A 3-layer network

The second requirement is that both learning and relaxation phases should be executable on the same architecture. Therefore, a "neural processor" containing a controller will be associated with each neuron. The controller will stored the two types of operations and will be extendable and modifiable easily.

The third choice is a flexible precision for both the state encoding of the neuron and the coefficients. This will be achieved using a bit-sliced data path and a silicon compiler approach will allow to meet exactly the user requirements.

The implementation of physically well-identified neurons will also allow to benefit from the defect tolerance of neuro-computing. Defective neurons identified attend of manufacturing will be disconnected by laser techniques and the chip will be usable. This will increase the yield of such chips allows to go to large chips.

Finally, let us mention that digital circuitry only is able to fulfill our requirements ; therefore, CMOS technology will be used.
The previous choices are summarized in figure 2.

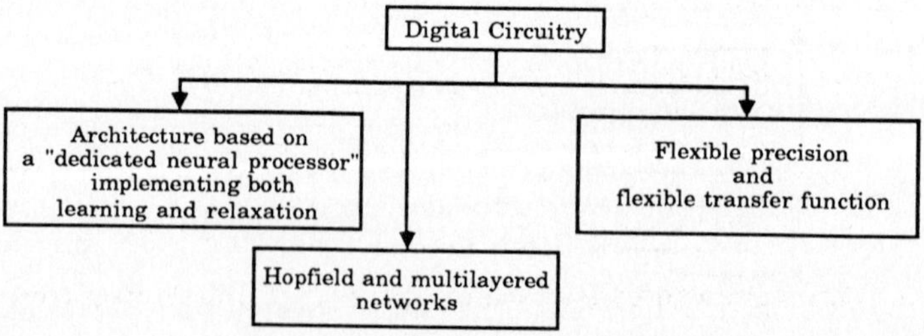

Figure 2 : Architectural choices

3. The basic architectural choices

3.1. The neural processor

As previously mentioned, a physical entity is associated with each neuron. It is an application-specific processor, the controller of which implements both learning and relaxation phases. We suppose that the neural network is made up of one layer (Hopfield network) or several layers (multilayered network). Each neuron is connected to all the neurons of the previous layer in the layered network or to all other neurons in a Hopfield network. Symbolically, during the relaxation phase, the neural processor implements the classical operations shown in figure 3.

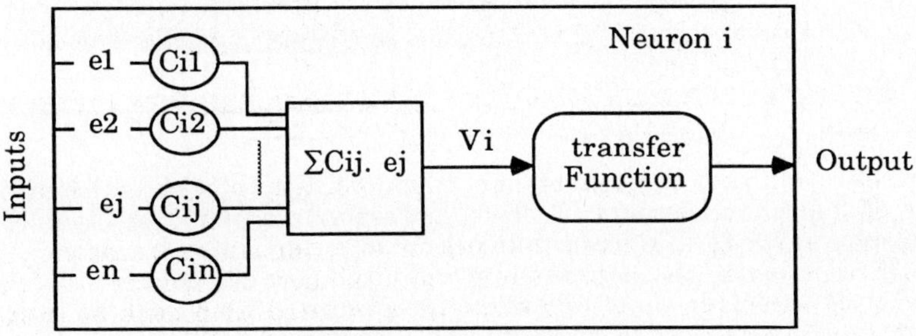

Figure 3 : Function performed by the neural processor during the relaxation phase.

3.2. Connections between neurons

It has been pointed out frequently, that a large number of connections between the neurons is a severe drawback for their implementation on silicon. Practically, virtual connections are implemented to replace spatial connections [20]. This means that if the neuron cannot be connected to a large number of neighbours, the state of these neighbours will circulate in a shift register and the neuron will pick up at the right instant the values of the states of its predecessors.

This principle is illustrated in figure 4a which shows a 3-node network. The 3 neurons are connected through a shift register which propagates the state values among the 3 neurons. The computation of the potentials for the 3 neurons is illustrated in figure 4b.

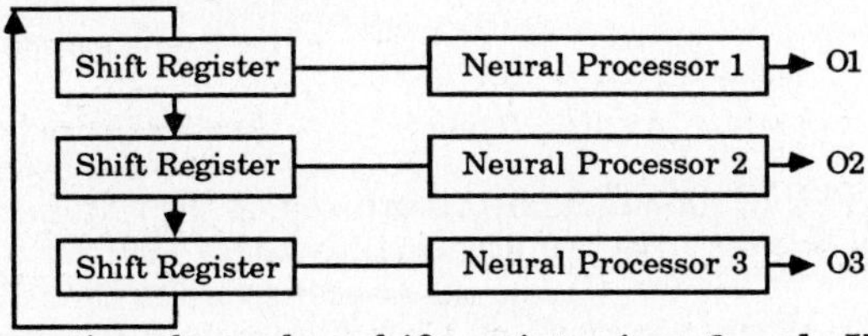

Figure 4a : Connection through a shift register in a 3-node Hopfield network

$$V1 = C_{11}.e_1 + C_{12}.e_2 + C_{13}.e_3$$
$$V2 = C_{21}.e_1 + C_{22}.e_2 + C_{23}.e_3$$
$$V3 = C_{31}.e_1 + C_{32}.e_2 + C_{33}.e_3$$

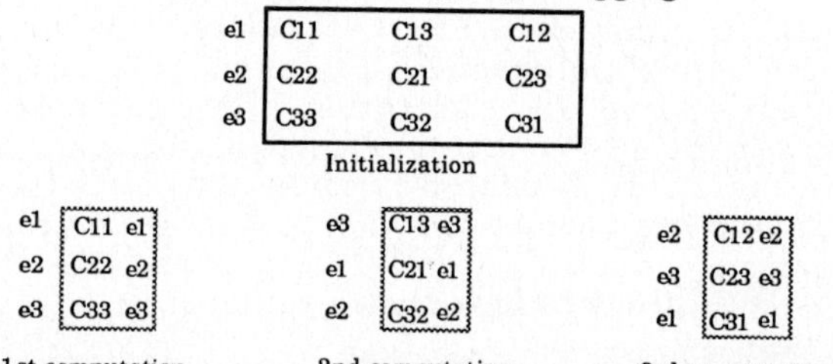

Figure 4b : Computation of the potentials in the previous Hopfield network.

3.3. Global architecture : Elementary blocks and data flow for potential computation

The global architecture is made up of elementary blocks which are 2 D arrays of neuron processors. A block corresponds either to a layer or to a portion of a layer in a layered network or to a portion or a whole Hopfield network. In a block, the neurons are organized in rows between two buses. These buses are made up of bus segments (figure 5) which can be connected by soft switches (software programmable switches) when required during the phases of neuro-computing.

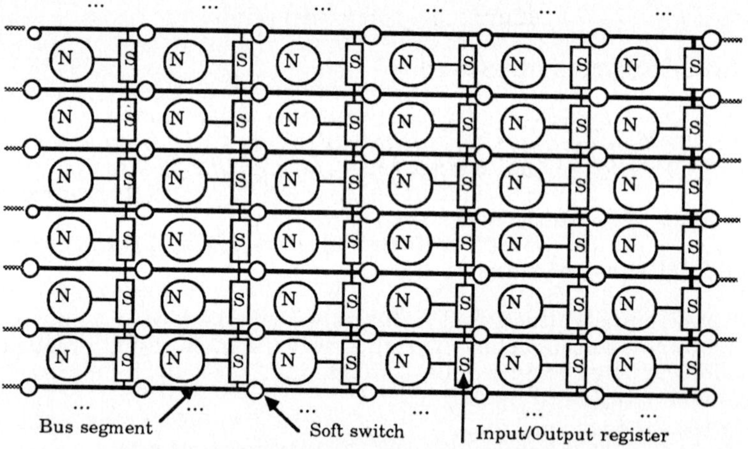

Figure 5 : Global architecture (partial view)

We will detail this global architecture using the example of 32 neurons realized on a 8 x 4 array. Complete architectures will be built later on by assembling such basic blocks. These 32 neurons either constitute a layer in a multilayered network, or a portion of a whole Hopfield network.

First, let us show how the computation of potentials is performed ; the equations of potentials Vi are given in figure 6.

$$V1 = C1,1e1 + C1,2e2 + C1,3e3 + \ldots\ldots + C1,31e31 + C1,32e32$$

$$V2 = C2,1e1 + C2,2e2 + C2,3e3 + \ldots\ldots + C2,31e31 + C2,32e32$$

$$\vdots$$

$$V31 = C31,1e1 + C31,2e2 + C31,3e3 + \ldots\ldots + C31,31e31 + C31,32e32$$

$$V32 = C32,1e1 + C32,2e2 + C32,3e3 + \ldots\ldots + C32,31e31 + C32,32e32$$

Figure 6 : operations required for evaluating the potentials of the neurons

In a first computation step, bus segments are connected as shown in figure 7. The switches of the second column are open for rows 2 to 8. The state of neuron N1 is available on the bus segment of the second row for both neurons N3 and N4, the state of neuron N3 is available similarly on the bus segment of the third row and the same for neurons N5, N7, ..., N29. The computations performed during this step are shown in figure 7a.

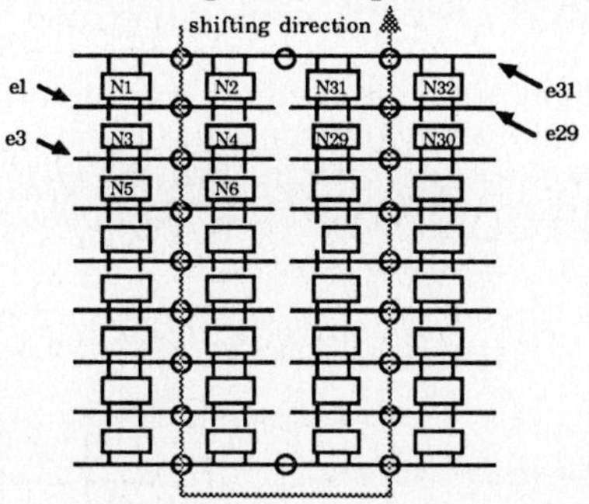

Figure 7 : A block of 32 neurons.

$$V1 = \boxed{C1,1e1} + C1,2e2 + C1,3e3 + \ldots\ldots + C1,31e31 + C1,32e32$$

$$V2 = \boxed{C2,1e1} + C2,2e2 + C2,3e3 + \ldots\ldots + C2,31e31 + C2,32e32$$

$$V3 = C3,1e1 + C3,2e2 + \boxed{C3,3e3} + \ldots\ldots + C3,31e31 + C3,32e32$$

$$V4 = C4,1e1 + C4,2e2 + \boxed{C4,3e3} + \ldots\ldots + C4,31e31 + C4,32e32$$

$$\vdots$$

$$V31 = C31,1e1 + C31,2e2 + C31,3e3 + \ldots\ldots + \boxed{C31,31e31} + C31,32e32$$

$$V32 = C32,1e1 + C32,2e2 + C32,3e3 + \ldots\ldots + \boxed{C32,31e31} + C32,32e32$$

Figure 7a : operations performed in the first computation step

In the second step of the computation, a circular shifting is performed in the first layer as indicated in figure 7b. The state of neuron N1 has been shifted on the bus segments of the third bus row and is now available for both neurons N5 and N6.

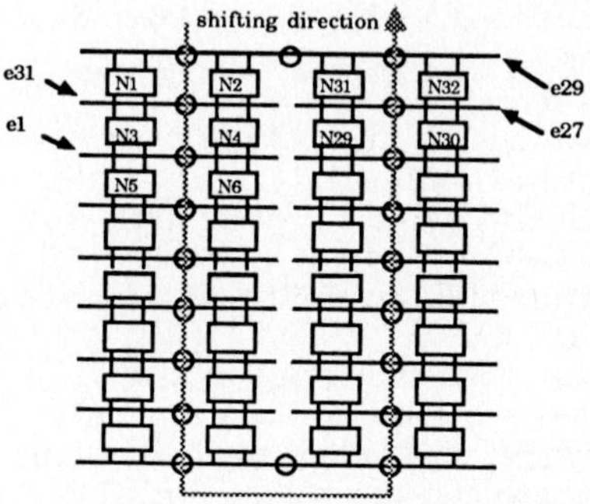

Figure 7b : Second computation step

It is clear that 16 elementary shiftings - a complete circular shifting - will be necessary to perform half of the complete computation. Two complete circular shiftings (32 elementary shiftings) will be required to end up the computation. Figure 8 gives the operations performed during the 16th. shifting.

Once this computation is over, the controller initializes the computation of the transfer function.

V1	=	C1,1e1	+	C1,2e2	+	C1,3e3	+.........+	C1,31e31	+	C1,32e32
V2	=	C2,1e1	+	C2,2e2	+	C2,3e3	+.........+	C2,31e31	+	C2,32e32
V3	=	C3,1e1	+	C3,2e2	+	C3,3e3	+.........+	C3,31e31	+	C3,32e32
⁞				⁞						
V31	=	C31,1e1	+	C31,2e2	+	C31,3e3	+.........+	C31,31e31	+	C31,32e32
V32	=	C32,1e1	+	C32,2e2	+	C32,3e3	+.........+	C32,31e31	+	C32,32e32

previously computed — currently in computation

Figure 8 : operations performed after the 16 th. shifting

4. Detailed architecture :

4. 1. Neuron structure :

The neural processor (Figure 9) is made up of :

1) four identification registers :

- N1 containing the number of bits encoding the state
- N2 containing the number of rows in the block
- N3 containing the number of coefficients or number of predecessors of the neuron
- N4 containing the shifting direction of the inputs in the layer.

for example : if N4=0 the shifting of the inputs is performed down
if N4=1 the shifting of the inputs is performed up

2) a local memory which stores the coefficients {Cij} and other informations such as the parameters used to approximate the transfer function

3) an A.L.U able to perform mutiplications and additions/subtractions

4) a controller which controls the potential, the state computation of the

neuron and the data transfer with the other neurons
5) an input/output register
6) three counters C1, C2, C3
C1 is initially loaded with the value of the register N1 and is decremented during the multiplication
C2 is initially loaded with the value of the register N2 and is decremented during the shifting
C3 is initially loaded with the value of the register N3 and is decremented after each read cycle in the memory

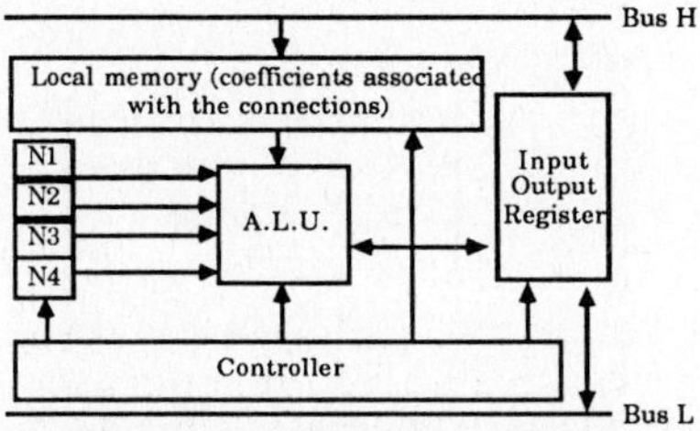

Figure 9 : Neural processor

4. 2. Algorithm used for the potential computation

The A.L.U. used in the relaxation phase (figure 10), computes the potential of the neuron $\sum Cij.ej$ where the Cij are the coefficients and the ej are the input states. Both of the Cij and ej are represented in 2's complement and encoded on N1 bits. The A.L.U. is made up of 2 operators, the multiplier and the adder subtractor.

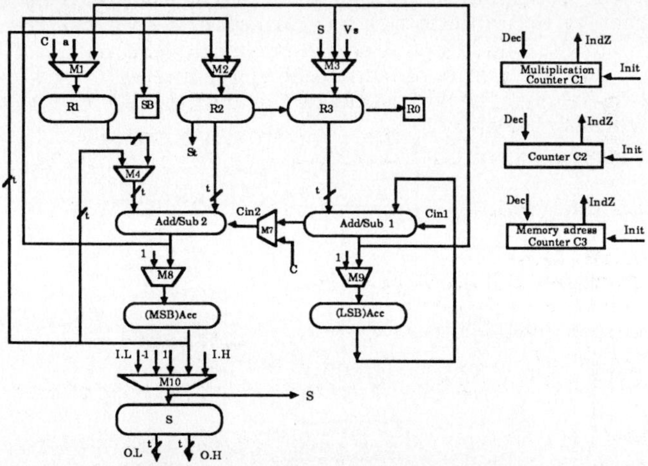

Figure 10 : The A.L.U structure

The multiplication performs a Booth's algorithm for 2's complement fixed point representation ; it is based on addition subtraction and shifting according to figure 14a.

```
C2 <------- N2
While C2 ≠ 0     Do
  Begin
     make "OPR"
     make "Shift"
     Decrement C2
  End
```

Figure 10a : Algorithm used for the multiplication

The operation (OPR) will be generated from the LSB bits of register R3 and register R0 as follows :

(LSB)R3	R0	Operation
0	0	No operation
0	1	addition
1	0	Substraction
1	1	No operation

The multiplier uses :
- an 8 - bit adder subtractor (Add/Sub 2),
- three 8 - bit registers R1 (Multiplicand), R2 (Buffer), R3 (Multiplier)
- a one bit register R0
- a multiplication counter (C1), initially loaded with the value of the register N1

The multiplication result encoded on 16 bits will be stored in R2 & R3 registers. After each multiplication, we add the contents of R2 & R3 to the contents of the accumulator ((MSB)Acc, (LSB)Acc encoded on 8 bits each one). In order to accumulate the partial potential, we use the two adder subtractor operators (Add/Sub 2 and Add/Sub 1). The result of the partial sum of the potential and of the final potential will be stored in the accumulator. We give the algorithm of the multiplication with the partial sum of the potential (figure 11).

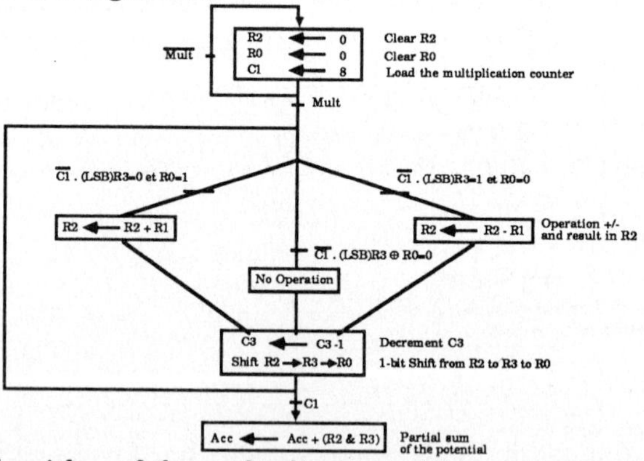

Figure 11 : algorithm of the multiplication with computation of the partial sum of the potential.

Once the potential is computed, the neuron calculates its state using the transfer function.

4.3. Algorithm used for the transfer function

The transfer function is a sigmoidal transfer function (figure 12a) which can be either stored by a table of values or hardwired.

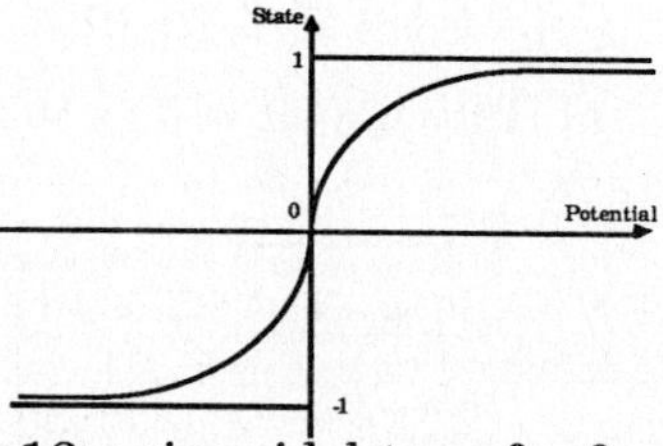

Figure 12a: sigmoidal transfer function

The first solution is simple and straightforward but the memory size becomes important if a high precision is required. The second solution is more interesting. Our study is based on the choice of the function which approximates the sigmoidal transfer function.

The two classical approximation methods are the interpolation and the least squares approximation.The drawback of the polynomial approximation is the cost of the computation which increases when the degree of the polynomial is high. In the particular case of the least squares, we impose certain constraints to the polynomial in order to approximate the positive non-linear part of the sigmoidal function limited by the interval $[0, Vs]$. When the potential value is greater than Vs the state will be equal to one and the neuron is saturated (figure 12b). The transfer function is symmetrical to the origin, therefore ; for the negative potential, the state will be computed using the absolute value of the potential and reversed to get the final result.

$$
\begin{aligned}
O(V) &= a(Vs - V)^2 + 1 && \text{if } V<Vs \\
O(V) &= +1 && \text{if } V>Vs \\
O(-V) &= -1 && \text{if } V>-Vs \\
O(-V) &= -O(|V|) && \text{if } V<-Vs
\end{aligned}
$$

where a is a negative real and Vs the saturation potential

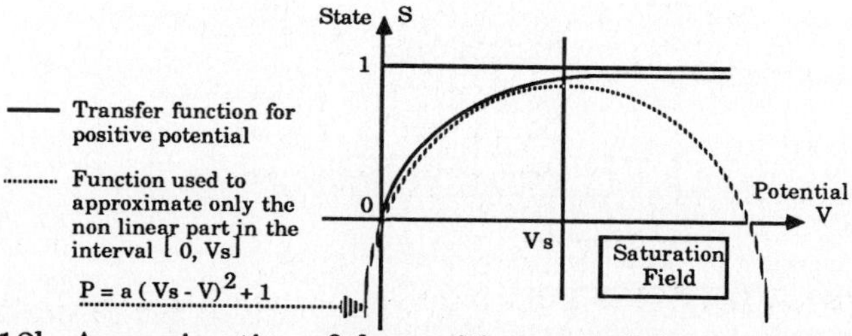

Figurel2b: Approximation of the positive part of the transfer function

As already mentioned, the coefficients and the states are represented in fixed point 2's complement. t1 bits encode the integer part of the state, t2 bits encode the integer part of the coefficient and t3 bits encode the integer part of the saturation potential. The integer part of the neural potential will be encoded on (t1 + t2) bits. The test of the saturation will be done as follows on the t (t = t1+t2-t3) MSB of the potential :
- if the exclusive-or of the t bits is equal to zero (St = 0), then the potential is between -Vs and Vs and the state will be computed (figure 12c).
- if the exclusive-or of the t bits is equal to one (St = 1), then the potential V is out of the interval [-Vs , Vs]. The sign bit will be stored in the register SB.

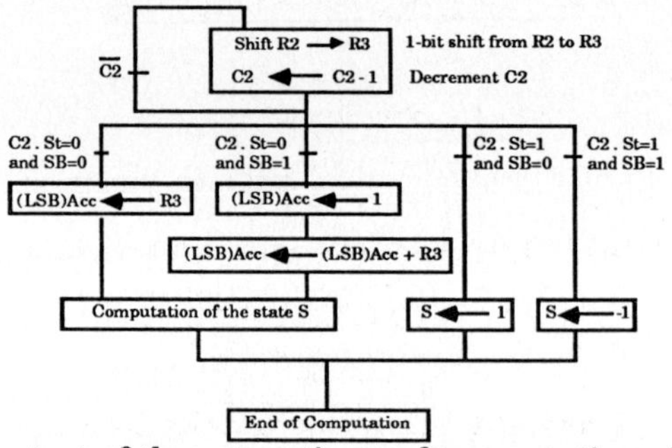

figure 12c : test of the saturation and computation of the state.

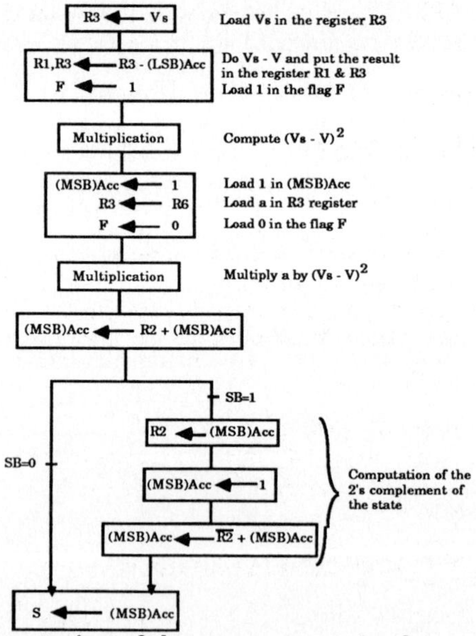

Figure 12d : computation of the neuron state in the non saturation case

4.. 4. Global flowchart for the relaxation phase

In the relaxation phase, each neuron computes its potential and its state as shown in (figure 13).

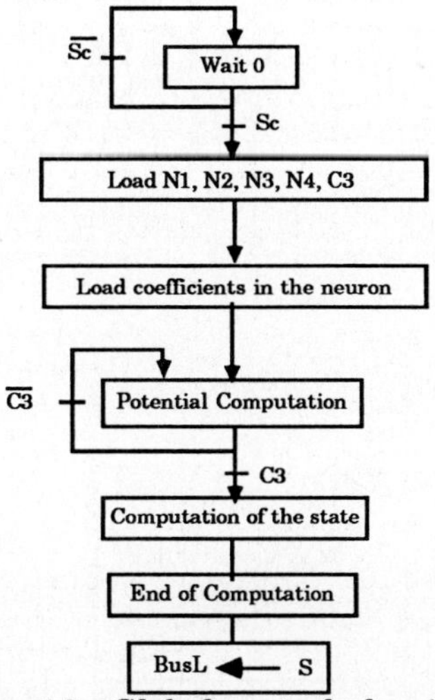

Figure 13 : Global control algorithm

If one column of neurons is selected, a first initialization step loads the values of N1, N2, N3, N4, C3. A second initialization step loads the coefficients in the memory.

5. Wafer scale integration

5.1. Motivations and basic choices

Wafer scale integration is a suitable target for neural networks, first because the structure is very repetitive and second because inherent defect tolerance can be expected from neural computation. Nevertheless some adequate facilities have to be implemented. First an efficient test or self test of the neurons must be provided ; the defective neurons will be disconnected by laser fuses like in the other WSI demonstrators realized in the ESPRIT projects [17..19].

Robust defect tolerant power, clock and basic control signals, also previously defined, will be used. Since a conventional step and repeat manufacturing process must be used, manufacturing will be based on a reticle-based approach.

5.2. Size estimation of wafer scale neural networks

The number of neurons which may be implemented on a wafer is shown in figure 18 according to the precision of the coefficients (encoded on k bits) and of the precision of the state of the neuron encoded on k1 bits. It appears that about 400 to 800 neurons can be implemented on a classical 4 inches Wafer.

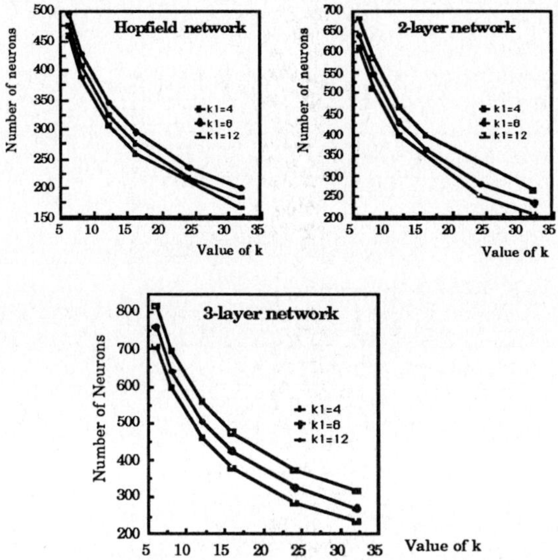

Figure 14: Number of neurons on a 4 inches wafer

Conclusion

A highly flexible neural architecture on silicon is proposed. It implements different target architectures and is customized by using soft switches and a silicon compilation approach. The processor associated with a neuron allows to implement both learning or relaxation phases and an extension to a wafer scale product is studied.

Acknowledgements

The authors are gratefull to L. Personnaz, G. Dreyfus from ESPCI (Paris) and M.Weinfeld from Ecole Polytechnique (Paris) for helpful discussions.

References

[1] "Disordered systems and biological organization"
Ed. by E. Bienenstock et al., Springer, 1986.

[2] "Neural networks for computing"
Ed. by J. Denker, American Institute of physics, 1986.

[3] "Parallel distributed processing"
Ed. by D.E. Rumelhartand, I.L. McClelland, MIT press, 1986.

[4] "Neural Computers"
Ed. by R. Eckmiller and C. von der Malsburg, Springer 1988.
[5] "Neural information processing systems"
Ed. by D.Z. Anderson, American Institute of Physics, 1988.
[6] "Neural networks : from models to applications"
Ed. by L. Personnaz and G. Dreyfus, IDSET, Paris, 1989.
[7] "Stochastic learning networks and their electronic implementation"
J. Alspector, R.B. Allen et al., Neural information processing systems, Ed. by D.Z. Anderson, American Institute of Physics, 1988.
[8] "A programmable binary synaptic matrix chip for electronic neural networks"A. Moopenn, H. Langenbacher et al., Neural information processing systems, Ed. by D.Z. Anderson, American Institute of Physics, 1988.
[9] "Bit-serial neural networks"
A.F. Murray, V.W. Smith, Z.F. Butler, Neural information processing systems, Ed. by D.Z. Anderson, American Institute of Physics, 1988.
[10] "Dynamics of microfabricated electronic neural networks"
D.B. Schwartz, W. Hubblard et al., Appl. Phys. Lett. 50, 1110, 1987.
[11] "A novel associative memory implemented using collective computation"
M. Sivilotti, M.R. Emerling, C. Mead, Proc. Chapel Hill Conf on VLSI. Ed. by Henru Fuchs, Computer science proceding 1985
[12] "Analysis and design of a limited-interconnect, feedforward neural array" M.R. Walker, L.A. Akers, Neural networks : from models to applications, Ed. by L. Personnaz and G. Dreyfus, IDSET, Paris 1989.
[13] "VLSI neural networks for fast pattern matching"
H.P. Graf, W. Hubbard, Neural networks : from models to applications, Ed. by L. Personnaz and G. Dreyfus, IDSET, Paris 1989.
[14] "A digital VLSI module for neural networks"
M. Duranton, J. Gobert, N. Mauduit, Neural networks : from models to applications, Ed. by L. Personnaz and G. Dreyfus, IDSET, Paris 1989.
[15] "A VLSI asynchronous cellular architecture dedicated to multilayer neural networks" B. Faure, G. Mazaré, Neural networks : from models to applications, Ed. by L. Personnaz and G. Dreyfus, IDSET, Paris 1989.
[16] "A new VLSI architecture for neural associative memories"
M. Verleysen, B. Sirletti, P. Jespers, Neural networks : from models to applications, Ed. by L. Personnaz and G. Dreyfus, IDSET, Paris 1989.
[17] "Defect tolerance in the wafer scale array for image processing"
J.L. Patry, E.F.Kouka , G. Saucier , Int. Workshop on Defect and Fault Tolerance in VLSI Systems, Sprinfield Massachusetts, October, 1988
[18] "Reconfigurable 4.5 MBit SRAM WSI Memory"
D. Marron, J Trilhe, B. Nasreddine, E.F.Kouka,Y. Wang, Int. Workshop on Defect and Fault Tolerance in VLSI Systems, Massachusetts, Oct, 88
[19] "Defect Tolerance in a 16 bit Microprocessor"
R.Leveugle, M.Soueidan and N. When, Int. Workshop on Defect and Fault Tolerance in VLSI Systems, Massachusetts, October, 1988
[20] L. Personnaz, A. Johannet et al., Neural networks : from models to applications, Ed. by L. Personnaz and G. Dreyfus, IDSET, Paris 1989.

Wafer Scale Integration, III
M. Sami and F. Distante (Editors)
Elsevier Science Publishers B.V. (North-Holland)

Focused Electron and Ion Beam Personalized WSI Neural Networks

H. T. Lin, R. Philhower and J. F. McDonald

Center for Integrated Electronics
Rensselaer Polytechnic Institute
Troy, NY, 12181, USA

Neural networks offer the promise for parallel computation on a scale never before attempted. Sixth generation architectures will make extensive uses of such networks for cognitive systems which will also exhibit extensive fault tolerance, and a high degree of restructurability. In this paper we will explore highly restructurable wafer integrated neural networks (WINN). Fault tolerance will be implemented using a variety of strategies, including the exploitation of the network restructurability, by providing discretionary personalized wiring using electron beam lithography, and using personalized focused ion beam repair.

1. Introduction

The possibility of electrical circuit implementations of neural networks based on analogies to biological models have intrigued computer designers for many years. The earliest work of this variety is the well known research of McCulloch and Pitts [1], Hebb [2], and leading to Rosenblatt's single layer perceptron [3] followed by related studies by Minsky and Papert [4]. Essentially one can regard these networks as analog versions of Threshold Logic as discussed, for example, by Hu [5], Lewis and Coates [6]. Dertouzos [7], Sheng [8], and Muroga [9] in which linear discriminate functions are thresholded with one or more nonlinear mapping functions.

Recently there has been a resurgence of interest in neural networks partially stimulated by the appearance of various commercial products such as the Hecht-Nielsen ANZA Neurocomputer, the TRW MARK III-1 Neurocomputing Workstation, and the SAIC delta-Sigma Neuro-accelerator for the IBM PC [10]. Hence, with the appearance of some tangible products have come the requisite commercial applications necessary for the economic viability of this technology.

In addition to these early commercial ventures, there has been a surprising persistence in the development of the theoretical framework for neurocomputing. Typical of this effort have been the evolution of the very influential presentations of Hopfield [12], McElice [13], Kohonen [14.15,16] and Hamming [17]. Some of these models can be analog, while others are explicitly binary and digital. Generally these models have more explicit use of feedback, giving the appearance of what today for binary signals is called a programmable logic sequencer (PLS). The PLS is a canonic form well known for its ability to realize an arbitrary algorithmic automaton in the form of a finite state machine as discussed by Arbib [18-19]. In its simplest form, inputs to the PLS are combined with feedback signals in a combinatorial circuit to produce a state transition sequence. Many theorems exist for decomposing such machines into simpler cascade or parallel realizations such as those described by Hartmannis and Stearns [20]. The combinational circuit can be synthesized with with modern spectral based threshold logic techniques as discussed by Edwards [21], and others [22] which frequently reduces the device count in the circuts when compared with more traditional Boolean operators. More importantly, the use of weights in the thresholding devices provides a means for making the devices adaptive, or at least adjustable. Even in this context of a binary sequential circuit the threshold logic element relies on analog weights and thresholds for its operation. Most of the neural networks, especially the Hopfield model are simply analog extensions of this concept, although the analog context can provide an environment linking to certain mathematical control concepts such as stability and convergence issues [23, 24, 25].

2. "Analog" vs. "Digital" Neural Systems

Great system density is thought to be possible using analog neural networks because the threshold device weights are realizable with resistances of small physical size. As a result, work by Mead [26] has precipitated a great deal of interest in analog neural nets. However, one problem which plagued early logic designers using threshold logic was the error introduced due to difficulties in specifying these weights with precision. Failure to use accurate weights can lead to the threshold device triggering to large values when only small errors arise in the weights. The weights can be selected to make the network maximally tolerant to this source of error, and, in fact, many neural "algorithms" can be designed to be somewhat fault tolerant. One can argue that the human brain is essentially analog and suffers very little from such noise as discussed by Winograd and Cowan [27]. Nevertheless, there is a limit which will ultimately be exceeded (most likely in large systems) when the numbers of weights or their dynamic range becomes excessive. This limit may also involve excessive fan in or fan out, which the sigmoid threshold devices may not accommodate. Finally, and most importantly, switching noise in large systems operating at high speed may be introduced into the signals. Switching noise is not much of a problem in the real biological system, but it is most certainly a problem in extremely large, dense digital systems. At this point the advantages of the exclusively analog approach might be compromised, and at least a hybrid if not an all digital approach would be safest.

Moreover, analog neural networks employing fixed resistance values don't lend themselves to "learning" by adjusting these resistor values. Attempts have been made to apply varying voltages to the gate of a FET as a means for controlling the transconductance of these transistors thereby providing a variable weight. The control voltage has been successfully applied with charge in a CCD array cell, or MNOS charge storage. However, there are limits to this approach in large networks due to slow charge leakage, dynamic range limitations, dark current sensitivity, and nonlinearity, age degradation and drift in the devices. On the other hand digital neural networks can more readily exhibit the flexibility of variation of weights with great precision and reliability (even to extent of using floating point computations if the needs exist). Furthermore digital processing is, and probably always will be, more available and more adaptable to shrinking lithography and advances in device technology than analog simply because it is more robust as the extreme nonlinearity of these devices constantly "launders" the digital signals, minimizing the effect of switching noise in large systems.

Analog circuits involving series connections of large resistances present another problem in limiting system speed. These circuits combine the two worst enemies of system speed; resistors and capacitors. Since these RC circuits are in feedback paths for the neural networks their slow charge and discharging behavior impose severe impediments to electrical realization of neural networks. We must keep in mind that real neurons use very different physics from our simplistic electrical replicas, so clinging to an analog form because it resembles the biological system may be misleading.

Another key limitation of the analog neural network is the difficulty of handling large (e.g. $> 10,000$) fan in or fan out situations which arise in applications such as image cognition. That is, each neuron could possibly need connections to all the others. On the other hand, binary digital circuits which can compute a linear discriminant to almost any number of terms and precision are trivial to construct. These can simply accumulate the result of adding weighted terms almost without limitation assuming they employ sufficient bits of precision. Mathematically,

$$X(i,j) = F\lfloor\sum_{k=1}^{N}\alpha(i,k)X(k,j-1)\rfloor \tag{1}$$

where $X(i,j)$ is the i-th component of the vector of neural values as it passes through the j-th iteration of the systolic array, $\alpha(i,k)$ is the matrix of synaptic weights, $F(\cdot)$ is the threshold function, k indexes the connections between the $(j-1)$ iteration of the neural X values and each of the the threshold elements, and N is the total number of neurons.

Parallel processing of these sums can be conducted with duplicate multiply and accumulate units as was discussed by Blayo and Hurat [28, 29]] at the first US Wafer Scale Integration Conference in San Francisco in January of 1989, and earlier during the previous summer in the Oxford University Connectionist Architecture Workshop. Similar systolic arrangements have been discussed by Kung [30]. In the Blayo and Hurat approach a multiply accumulate unit is used for each neuron or time shares computations for several neurons. Each processor makes a pass in systolic fashion over the data while propagating this data around to other units.

This process of passing the data around completely avoids fan out and fan in problems very easily. At the end of a recursive iteration for the neural net the new data is substituted for the old data in a very straightforward manner. Such a method of organization is only possible with digital systems. Speed can be further enhanced by advancing the technology employed in construction of these adders as shall be discussed further in the next section of the paper. Also, the systolic method of mesh interconnection lends itself to increasing the amount of parallelism almost indefinitely, so the architecture scales nicely to larger systems.

Eventually analog techniques could be mixed with the proposed approach. However, in this paper only the all digital neural networks are considered because other strengths of this approach are emphasized. Specifically, the ability of the system to adapt weights, thresholds, and even interconnections is explored. Eventually the all digital approach, is expected to benefit most directly from the advances in lithography in the semiconductor industry, which are now entering the submicron domain. When this technology approaches 0.25 micron design rules it is not clear whether the analog approach will offer any density advantage over the digital approach. We note, however, that the impact on the digital system of faults is quite different from that of the all analog system. The all digital system is eminently fabricatable, and exceptionally robust to manufacturing and environmental variability.

3. The Programmable Gate Array (PGA) - Highly Restructurable Logic

It is desired to demonstrate in quick-turn around, timely fashion a highly restructurable WSI implementation of large neural networks. The WSI approach is attractive primarily because of the improved reliability likely to accrue from a "connectorless" system. Other attractive features include reduced heat dissipation, lower average current requirements, lower capacitance, lower switching noise, and of course, improved packaging density [31]. The high density interconnections (HDI's), which is another term for wafer scale interconnections (WSI's) which can be superposed on the WSI platform can provide more than adequate support for any remaining fan in and fan out required by the all digital neural network, as well as any discretionary wiring. In addition, another problem that becomes progressively more vexing as circuit density increases is the pin-out problem for each die. In WSI the reduced size of interconnect can accommodate far more pin-out sites per die. The current Xilinx parts (which are discussed here below) require 177 pin packages. This can be expected to grow to 256 and eventually 512 pads per die in the SEMATECH limit. In WSI these pads are chemically bonded to the site to site interconnections.

A suitable host for the kinds of processors suggested by Blayo and Hurat is a wafer consisting of dies containing sufficiently flexible logic clusters with restructurable general interconnections which can be "soft" programmed. Soft programming of interconnections can be accomplished with a sparing application of pass transistors or transmission gates in CMOS. The delay imposed by these switches can be minimized by providing sufficient resources in the form of long metal runs which can be connected to sources and sinks of logic signals. By providing for moderate quantities of these soft programming switches interstitially distributed throughout the die in various wiring channels it is possible to nearly completely restructure these connections dynamically. In this way it is possible to realize nearly any logic function desired without having to alter masks for metallization or devices.

Such a soft repair technology is available in its infancy today for logic circuits, although it has not been widely recognized in this light. Instead the industry has termed this kind of part a programmable gate array (PGA). Pioneered by Xilinx, and now second sourced by AMD/MMI the PGA is a more flexible solution to user programmed logic than the ubiquitous programmable logic array (PLA), and more recently the programmable logic sequencer (PLS). Essentially random logic can be realized by the PGA [32, 33]. Naturally, inclusion of enough soft switches to make the entire circuit mapping arbitrary wastes some circuit resources. However, the Xilinx packing of switches is so ingenious that a 10,000 gate equivalent PGA (the 3090) fits neatly into a space less than one square centimeter using 1.25 micron technology! Fabricated on 6 inch wafers containing approximately 160 dies the a single integrated system of nearly 1,600,000 gate equivalents is potentially feasible, if all the wafer sites could be made usable, something which we might almost be able to approach with highly repairable PGA sites.

The technique for providing programmable connections consists of providing a "hidden" or "shadow" switch control memory shown here in (Figure 1). This memory can be loaded on powerup, or anytime, for that matter. The time to load the memory for the current Xylinx parts is on the order of 10 milliseconds for a parallel "byte" wide serial load path, or about 100 milliseconds if the load is "bit" serial. It is this ability to restructure itself continuously that is unique about PGA based hardware design. For some period the system can be configured as a neural net, and 10 milliseconds later it can be reconfigured as a signal processor. Then 10 milliseconds later it can be reconfigured as a different neural net. From our perspective this extremely fine grained restructurability lends itself also to yield enhancement.

We note that the PGA chip is an array of standard configurable logic blocks (CLB's). These blocks consist, in the current product offerings, of 4-5 input memories which can be loaded on powerup from the "shadow" memory, and then behave as extremely fast ROM's (Read Only Memories) thereafter. These memories then can be used to store a primitive truth table for some logic building block. In addition to this fast ROM the CLB contains some flip flops, and multiplexers. Figure 2 shows the internal structure of one of the CLB's.

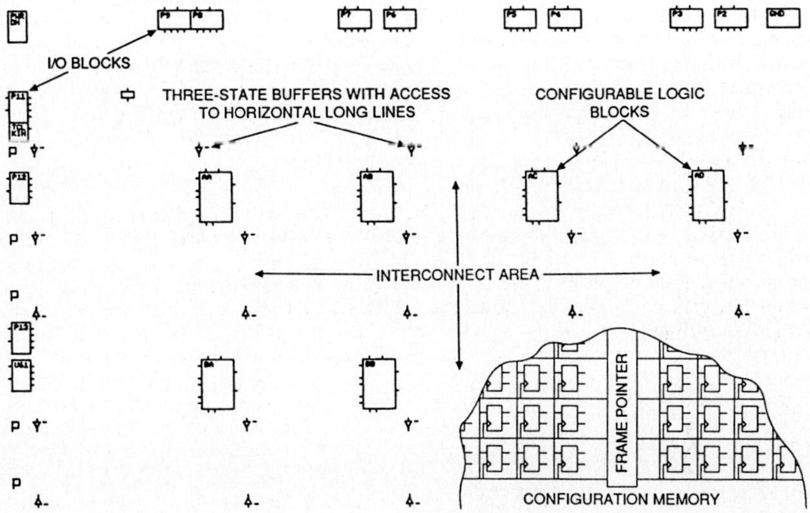

Fig. 1: Programmable Gate Array "Uncovered" to Reveal the "Shadow" Control Memory for the Programmable Soft Switches

It should be noted that the approach we advocate here is a an enhancement of the approach suggested by Blayo and Hurat which employs soft programming switches only to route peripherally around faulty die sites on a wafer. This is accomplished by supplying each die only with NESW (North East South West) soft routing. No other restructurability is provided. Since full die yield can be as low as 15 to 30 % in reasonably advanced CMOS processing on large dies, the implementation suggested by Blayo and Hurat might be very undense. We propose to use soft programming switches much more aggressively inside each die site (as exemplified by the PGA), and not just around the periphery of each site. In this manner even a site with a few faults inside of it might be made to accommodate a useful function. For example, if a few of the CLB's don't work correctly we could map our desired interconnections around the faulty site, bu reloading the shadow memory. We note that this is the logic equivalent of site repair in successful DRAM production. In DRAM's each site is internally restructured to get a working memory out of it. With high probability this repair is successful. Only certain fatal flaws such as a direct short from power to ground in a site prevent this strategy from succeeding.

This is where the flexibility of the PGA approach becomes significant. All those sites which do not contain an absolutely fatal flaw (e.g. direct short between power and ground, lithography shift destroying all contacts, open in power or ground ring) could be made usable by simply mapping desired logic functions onto the host site in such a way as to avoid the faults. Of course, extremely powerful CAD tools, which place and route much more rapidly than the conventional offering from the vendor are needed for this mapping. Conceivably each die site might have a different interconnection pattern (intra-chip) within it to avoid the faults that occur there. Little or no manual

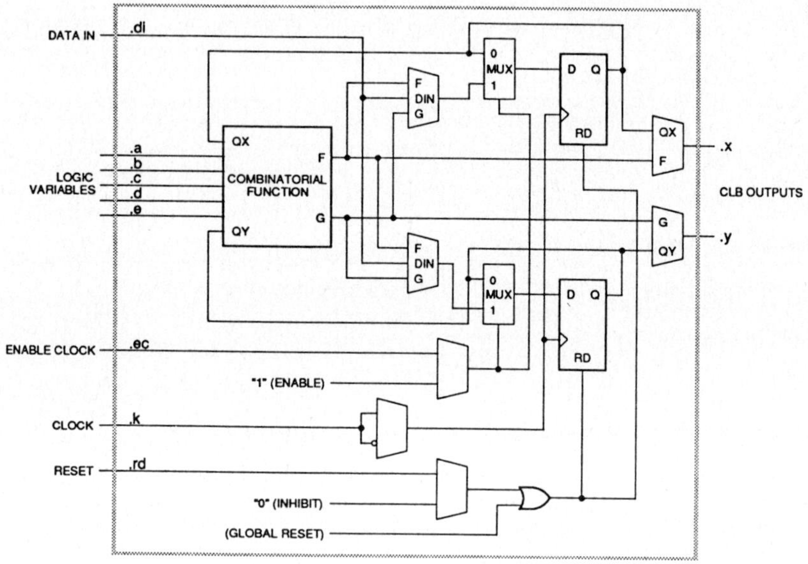

Fig. 2: Structure of the Configurable Logic Block (CLB) for the AMD/XILINX Series 3000 PGA Showing Truth Table Memory

intervention would be possible, in spite of the need for care in routing critical paths.

As we have noted this is the soft switch logic equivalent to row and column test and repair for large DRAM circuits. Naturally, the cell to cell wiring might still be different from wafer to wafer if some sites contain the absolutely fatal flaws. This might still require the chip to chip (inter-chip) wiring to be different on different wafers, however, one could anticipate harvesting a larger fraction of sites and making them usable by this process that was implied by the Blayo and Hurat paper. In effect they assumed that all die sites which contained a fault would be omitted. This can result in too large a fraction of the die sites being sacrificed, a problem which could be anticipated to grow worse with shrinking design rules as yields suffer.

Here for once, is a general building block which is so totally restructurable that the wafer scale approach actually permits use of faulty cells and could in principle raise the wafer yield potentially to nearly 100%! This approach is not limited to neural network applications. In fact, combinations of neural and non neural logic circuits are readily possible, and dynamic changes of the logic structure become quite simple. This is the quintessential environment for the study of neural networks!

The testing of the cells on the wafer is already an accomplished facet of this exciting technology. Partially due to the regularity and flexibility of the PGA structure the dies are highly testable during wafer probe. Furthermore, the possibility that additional testing might be required before supplying die to die interconnect will not incur a high cost, since an IBM PC based emulator/tester pod set is available along with a pioneering die test probe which is being developed by Tektronix. Supplying an interface between these two should not be difficult. Debugging of cell mappings will be readily conducted using conventional packaged parts using this same PC based system which includes all the CAD tools to design the soft programming switches. The CAD prototyping workbench system is another designers dream. No longer is it necessary to wait for chips to return from foundry for testing. The chips are right there waiting to be tested and the tester is ready to go!

Preliminary examination indicates that these cells can readily accommodate the system proposed by Blayo and Hurat, and that this approach is amenable to image processing and signal processing applications at attractive (video) signal bandwidths. One can readily see how the configurable memory component of a CLB will accommodate a full adder, and how the MUX circuits can be used in a carry select implementation. If the X_j values for equation (1) are binary (the thresholding device is an ideal clipper), then the products in the sum become simple AND operations, and the additions in the sum are readily mapped onto the CLB building blocks in the PGA's.

4. A Five Year PGA Technology Projection -

A clearly defined track of predictable lithographic shrinkages will take the aforementioned technology approach to an extremely high level of integration without any substantial changes in the system approach taken. Using currently available six inch diameter CMOS wafers, 1.25 microns for the die lithography, and 8 microns design rules for the interdie wiring neural networks can be embedded in systems containing 1.6 million gate equivalents operating at maximum clock rates of from 35 - 70 MHz. Networks of this size and speed are arguably suitable for image recognition applications. Furthermore, as has already been noted all of this technology is suitable for an extremely fast demonstration of the basic concepts. However, as this technology evolves through the straightforward lithographic extensions projected by the national SEMATECH project, this approach can be expected to accommodate systems of more than 100 million gate equivalents using 0.35 micron CMOS circuits employing 4 level intradie interconnect and operating at clock rates above 100 MHz (approaching 200 MHz) on eight inch wafers. For such systems conventional packaging would predictably be inadequate, and the high density controlled impedance, interconnections afforded by wafer scale integration will be needed.

Finally, certain features lacking in the current PGA's will be addressed as early as the next generation of the technology. Already the 4000 series will include memory islands which can be written as well as read, thus providing distributed logic-in-memory. This same technology which is expected in the 4th quarter of 1989 will already be at 0.8 micron design rules leading to nearly 2.5 times as much logic per wafer, and 100 MHz flip-flops. This technology growth path will undoubtedly accommodate much of the scale up to larger sized neural architectures. The packaging of these wafer scale networks will permit much more efficient use of pin-out, since according to Rent's rule such growth is at a rate only at about the one half power of the number of gate equivalents. Hence, 100 million gate systems could be expected to use only about 3000 total external pins, and probably much less due to the systolic nature of the neural networks we propose here. Normal separate die packaging of the entire system would still need these 3000 external connections plus roughly 100,000 additional internal connections.

5. "Gentle" Discretionary Wiring Process

In order to post process wafers after wafer probe test one needs to employ very "gentle" processing. This problem is also encountered in large DRAM processing where tested and repaired cells are subjected to a relatively high temperature Silicon Nitride deposition, and exposed to some radiation damage due to sputterer metal deposition, and reactive ion etch plasma radiation. Nevertheless, the process works so well that Innova and other WSI manufacturers can employ the exactly identical interconnection patterns to each wafer to configure a WSI sized memory. The difficulty with that strategy is that a single unrepairable die could prevent the use of this identical interconnection pattern, at the complete loss of the wafer. To avoid this problem, some minimal alteration of the post wafer probe test wiring is provided. The occasionally irreparable dies which would contain "poison pill" faults (if you incorporate them the wafer expires) still result in aborting the entire wafer. For the Xilinx wafers this would result in incurring a cost of many thousands of dollars per aborted run. Consequently a provision for major restructuring of the interconnections and power/ground connections is essential, and discretionary wiring is needed which in effect personalizes the wafer to avoid the "poison pill" faults.

The low temperature processing might accomplished with a novel Vapor Depositable Polymerization (VDP) process [34]. Using this technique a special polymer (aliphatic tetrafluorinated poly para xylylene - AF4 Parylene) is deposited on a the wafer while it is held at room temperature. This polymer has an unusually low dielectric constant between 2.14 and 2.36 (depending on the method of deposition). Hence it is nearly as attractive as Teflon as an insulator, but it does not suffer from cold creep. AF4 also has no pin-holes, and is strikingly resistant to penetration by contaminating solvents. Its water take up is essentially zero, and only hydrogen and helium seem to diffuse through it (and there is surprisingly very little of that). This is the key step, and it replaces the Silicon Nitride layer used by DRAM manufacturers. It also replaces polyimide which has been used by some manufacturers in wafer scale hybrid multichip packaging as an insulator.

After deposition of the AF4 another parylene is deposited on top of this until the layer is planarized. This sacrificial layer is etched back until only the flat AF4 remains. Then the layer is patterned by RIE in a method suitable for lift-off, and the trenches etched into the AF4 by this step are metallized. This step requires use of a low temperature, low ionizing radiation metal deposition technology. This is the partially ionized beam (PIB) or ionized cluster beam (ICB) deposition method [35]. Excess metal is removed by selective dissolution of a release layer as is traditional for the lift off technique. Following this is an optional application of plasma amorphous carbon or diamond like carbon coating in a thin layer for additional encapsulation [36].

The aforementioned processing is repeated until all pillar vias and metal layers are in place [37,38]. Special CAD tools for deriving the personalized interconnect are already developed [39] and a technique for using electron beam lithography to pattern the discretionary wiring is already in place [40]. A backup procedure for repairing damaged WSI interconnect is also available [41]. There are already so many companies working on WSI that it appears to be possible to have the wafers mounted in an appropriate pin grid array style package.

6. Electron Beam Discretionary Personalization

The fine grained fault tolerance afforded by the restructurability of interconnections in PGA's will not accommodate severe "poison pill" faults which could drag down the wafer scale system. For example, failure of the "shadow" memory to load properly could prevent rerouting through or around the chip to circumvent the faults in that die site. Inability to do this would probably be caught at the wafer probe test. Therefore, a method for surgical "work around" is to pattern a special set of interconnections suitable for mapping the desired system onto the usable sites on the wafer. Since the number of totally irretrievable sites is presumably minimal, the amount of discretionary modifications from wafer to wafer could be minimized. To allow for all contingencies we have elected to personalize our wafers by electron beam lithography direct write patterning of the aforementioned high yield "gentle" wiring process.

The direct write ebeam system shown in Figure 3 is illustrative of the technology which has been available for nearly 20 years, but only recently at wafer throughput rates that begin to look attractive (a rate of one wafer exposure every few minutes would be extremely attractive). The term direct write refers to the process by which the electron beam exposes the resist directly on a wafer. The immediate advantage is the ability to pattern interconnections without having to create a mask plate set which would be unique to one and only one wafer. Additionally, the electron beam direct write system is similar to a step and repeat system which focuses on every section of the wafer before writing, yet each writing field can be different and unique to the wafer being written. Figure 4 shows a section of about 8 such writing fields containing discretionary personalization wiring patterns exposed in PMMA resist by the ebeam. The abutment of these writing fields (which are 4mm x 4mm for the IBM EL-2 and 10mm x 10mm for the IBM EL-3) must be exact since lines must proceed from one field to the next with perfect "stitching" even when the wafers are extremely warped due to stress from high temperature processing encountered during the early fabrication steps (diffusion or post implant annealing). Even the later WSI wire fabrication steps if they are not chosen properly can introduce large amounts of stress which can further warp the wafer.

Fig. 3: IBM EL-2 direct write electron beam lithography system

Note in Figure 4 the regular array of registration marks (dots) on the 4mm x 4mm centers. Wiring cannot go through these regions since the ebeam must be able to scan

Fig.4 : Sample EL-2 direct write electron beam discretionary personalization wire pattern exposures. Note the regular array of white dots on 4mm x 4mm centers which are exposed areas around the registration marks.

these marks. The marks consist of 1 micron deep trenches 5 microns x 100 microns in width and length. The sharpness of the trench wall is one of the secrets to a good registration process. Therefore no processing should cover up the marks or lessen the edge definition of these trenches. The ebeam scans over these marks and receives a distinct signature as the beam "steps" up or down on one of these edges [42].

7. Focused Ion Beam Repair

As the WSI fabrication process nears its completion a large amount of money has been invested in each wafer. A fault introduced somewhere near the end of that expensive process would extinguish any hope of selling the item as product, and consequently retrieving any of the investment. The problem is a good deal more worrisome to managers than any other difficulty encountered in the industry. It is taken as a given assumption that every effort is made to minimize such faults. Yet they can occur, and some method for repairing the damage is desirable. This last step can constitute an additional form of personalization, and to be successful may require inclusion of spare or redundant long runs of wire, because ion beam write rates in the presently available systems, are fairly slow.

Lasers have been successfully employed to restructure WSI. However, laser personalization requires the preplacement of fuses for cutting and welding, and these fuses can not be located everywhere. Consequently if a fault occurs in an inaccessible location, or one not anticipated by the fuse placement designer, the focused ion beam (FIB) offers an interesting alternative. The FIB can cut or deposit small amounts of metal or dielectric anywhere on the wafer. It is not limited to selected fuse locations for such operations, and no special fuse materials are required.

A typical focused ion beam system is shown in Figure 5, with a cutaway drawing in Figure 6 showing key components in the system. The specific model shown is the VG

IBL 100S. A Liquid Metal Ion Source (LMIS) is located at the top of the ion column, where a metal or alloy is heated and permitted to wet a 1 micron radius of curvature needle tip. An extractor voltage applied on a nearby aperture causes the liquid to form a sharp 50 Angstrom cusp on the needle tip. It is from this extremely small point that the ions are emitted in the conventional source design.

Fig.5: The VG IBL 100S Focused Ion Beam System

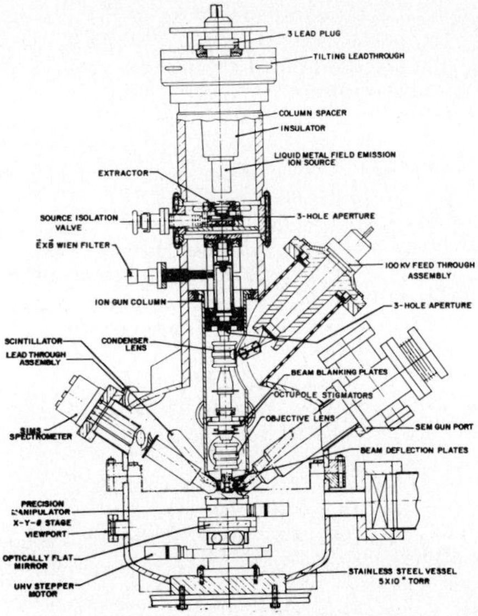

Fig.6: Cutaway View of the VG IBL 100S FIB

The ions are further accelerated by floating the heater, and extractor voltages on a high voltage supply (0-100Kv) toward the wafer table at the bottom of the column which is near ground potential. A series of two lenses (objective and condenser lenses) brings the beam to focus on the wafer. Ion beam systems have extremely large depth of focus (fraction of a centimeter). So wafer warp is almost not a problem except for extremely submicron features. An E X B mass filter selects the desired ion species in the case of an alloy source or in the presence of isotopes. For extremely fine isotope separation there is an additional aperture located near the bottom of the column.

Ions at sufficient energy begin to mill or sputter material off of a surface. This can be used to remove dielectric coating or cut metal wiring. Indium has been found to be a uniquely stable ion source with long life, and good milling power. In addition to this milling or cutting ability, the FIB can deposit materials from a gas source. For example a tiny jet of metal containing gas such as trimethyl aluminum or gold can be used to inject these metal containing molecules at toward the impact point on the wafer surface. Under the right conditions of gas flow and accelerating voltage, metals can be deposited on the wafer.

FIB repair is unique in that it can operate equally well on large WSI level interconnections, and on the submicron wiring in the dense fine line interconnections one could expect to be fabricating for WSI in the not too distant future. Should the defect occurring toward the end of the WSI fab process involve a crack or other defect on the die site wiring levels then submicron focused ion beam milling or deposition might be the only path to repair the wafer.

Acknowledgement

The assistance and support provided by Eastman Kodak, IBM, Raytheon and DEC over the years is gratefully acknowledged.

References

[1] W. Pitts, and W.S. McCulloch, “How We Know Universals: The Perception of Auditory and Visual Forms,” Bull. of Math. Biophys. 9:127-147, (1947).

[2] D. O. Hebb, “The Organization of Behavior,” New York, J. Wiley, (1949).

[3] F. Rosenblatt, “The Perceptron: A Probabilistic Model for Information Storage and Organization in the Brain,” Psycholo. Rev/ 65:386-408, (1958).

[4] M. Minsky, and Seymour Papert, “Perceptrons,” MIT Press,(1969).

[5] S.-T. Hu, “Threshold Logic,” U. California Press, (1965).

[6] P. M. Lewis, and Coates, C. L., “Threshold Logic,” Wiley, New York, (1967).

[7] M. L. Dertouzos, “Threshold Logic, a Synthesis Approach,” MIT Research Monograph #32, MIT Press, Cambridge, MA, (1965).

[8] C. L. Sheng, “Threshold Logic,” Academic Press, New York, (1969).

[9] S. Muroga, “Threshold Logic and Its Applications,” J. Wiley, (1971).

[10] B. Soucek and M. Soucek, “Neural and Massively Parallel Computers - The Sixth Generation,” Wiley-Interscience, (1988).

[11] J. S. Denker, Ed., "Neural Networks for Computing," AIP Conference Proceedings, Snowbird, UT, 1986.

[12] D. W. Tank and J. J. Hopfield, "Simple "Neural" Optimization Networks: an A/D Converter, Signal Decision Circuit, and a Linear Programming circuit," I.E.E.E. Trans. on Circuits and Systems, CAS-33(#5), 533, (1986).

[13] R.J. McEliece, E. Posner, E.R. Roderick, and S.S. Venkatesh, "The Capacity of the Hopfield Associative Memory," I.E.E.E. Trans. on Infor. Theo., IT-33(#4), 461, (1987).

[14] T. Kohonen, "Content Addressable Memories," Springer-Verlag, New York, (1980).

[15] T. Kohonen, "Self-Organizing and Associative Memory," Springer-Verlag, New York, (1984, 1988).

[16] T. Kohonen, "Self-Organized Formation of Topologically Correct Feature Maps," Biolog. Cyber., 43, 59, (1982).

[17] R. P. Lippmann, "An Introduction to Computing with Neural Nets," I.E.E.E. ASSP Mag., 4 (1987).

[18] M. A. Arbib, "Theories of Abstract Automata," Prentice Hall, (1969).

[19] M. A. Arbib, "Brains, Machines, and Mathematics," Springer-Verlag, (1987).

[20] J. Hartmannis, and R.E. Stearns, "Algebraic Structure Theory of Sequential Machines," Prentice Hall, New Jersey, (1966).

[21] C. R. Edwards, "The Application of Rademacher-Walsh Transforms To Boolean Function Classification and Threshold Logic Synthesis," I.E.E.E. Trans. on Computers, Vol. C-24, 48, (1975).

[22] S. L. Hurst, D. M. Miller, and J. C. Muzio, "Spectral Techniques in Digital Logic," Academic Press, (1985).

[23] M. J. S. Smith and C. L. Portmann, "Practical Design and Analysis of a Simple "Neural" Optimization Circuit," I.E.E.E. Trans. on Cir. and Syst., CAS-36(#1), 42, (1989).

[24] L. O. Chua, and L. Yang, "Cellular Neural Networks: Theory," I.E.E.E. Trans. on Cir. and Syst., CAS-35(#10), 1257 (1988).

[25] L. O. Chua, and L. Yang, "Cellular Neural Networks: Applications," I.E.E.E. Trans. on Circ. and Syst., CAS-35, 1275, (1988).

[26] C. Mead, "Analog and Neural Networks," Addison Wesley, (1989).

[27] S. Winograd, and J. D. Cowan, "Reliable Computation in the Presence of Noise," MIT Press Tech. Monograph #22, (1965).

[28] F. Blayo, and P. Hurat, "A Reconfigurable WSI Neural Network," Proc. 1st Annual International Conf. on WSI, (E. Swartzlander and J. Brewer), San Francisco, 141 (1989).

[29] W. Moore, "Connectionist Architectures," Adam Hilger, Proceedings of the Workshop at Oxford, (1989).

[30] S. Y. Kung, "VLSI Array Architectures," Prentice Hall, (1988).

[31] J. F. McDonald, E. H. Rogers, K. Rose, A. J. Steckl, "The Trials of Wafer Scale Integration," I.E.E.E. Spectrum, 32, (1984).

[32] "The Programmable Gate Array Data Book," Xilinx, Corp., 2069 Hamilton Avenue, San Jose, CA.

[33] "The LCA Applications Handbook," AMD Corp., Sunnyvale, CA.

[34] N. Majid, S. Dabral, and J. F. McDonald, "The Parylene- Aluminum System for Multilevel Interconnections in WSI and WSHP," I.E.E.E. Journal of Elec. Materials, JEM-18(#2), 201, (1989).

[35] R. Selvaraj, S.N. Yang, J. F. McDonald, and T.-M. Lu, "Ion Cluster Beam Metallized Interconnections for Wafer Scale Integration," Proc. VLSI Multilevel Interconn. Conf., 440, (1987).

[36] S. Dabral, J. F. McDonald, and T.-M. Lu, "Plasma Deposition of Amorphous Carbon for Encapsulation of Wafer Scale Interconnections," Proc. VLSI Multilevel Interconn. Conf., (1989).

[37] Lt. B. J. Donlan, J.F. McDonald, G.F. Taylor, R.H. Steinvorth, and A. S. Bergendahl, "CAD and FAB for WSI," VLSI Design, 6(#4), 34, (1985).

[38] Capt. B. J. Donlan, J. F. McDonald, R. H. Steinvorth, M.K. Dhodi, G.F. Taylor, and A.S. Bergendahl, "The Wafer Transmission Module (WTM)," VLSI Systems Design, 7(#1), 54, (1986).

[39] Capt. B. J. Donlan, and J. F. McDonald, "A Placement and Routing Program for WSI," Proc. I.E.E.E. ICCAD, 462 (1986).

[40] J. F. McDonald, M. Stanton, R. Rajapakse, H. Lin, R. Selvaraj, N. King, M. Haslam, "Adaptive Discretionary Wiring for WSI Using Electron Beam Lithography," SPIE, Vol. 773, "Electron-Beam, X-Ray, and Ion Beam Lithographies," 140, (1987).

[41] J. F. McDonald, R.U. Rajapakse, H.T. Lin, R. Selvaraj, J.C. Corelli, H.S. Jin, S. Balakrishnan, A. J. Steckl, "Optimized Focused Ion BEAM (FIB) for Inspection and Repair of wafer Scale Interconnections"" SPIE, Vol. 773, "Electron-Beam, X-Ray, and Ion Beam Lithographies," 206, (1987).

[42] H.T. Lin, S.Dabral, and J.F. McDonald, "High Yield In-Situ Fabrication of Multilevel Interconnections for WSI," in Wafer Scale Integration, E. Swartzlander, Chapter 8, Kluwer Publishing, (1989).

[43] H. T. Lin, J. F. McDonald, J.C. Corelli, S. Balakrishnan, and N. King, "Focused Electron and Ion Beam Repair Strategies for Wafer-Scale Interconnections in Thin Film Packaging," J. Thin Solid Films, Vol. 166, 121-130, (1988).

[44] J. F. McDonald, and H.T. Lin, "Application of Focused Ion Beam for Testing, Repair, and Fabrication of Metal and Optical Interconnections in Wafer Scale Integration," Proc. I.E.E.E. 1st International Conf. on Wafer Scale Integration, E. Swartzlander, and J. Brewer, Eds., 161-172, Jan. 3-5, (1989).

Wafer Scale Integration, III
M. Sami and F. Distante (Editors)
Elsevier Science Publishers B.V. (North-Holland)
IFIP, 1990

ASP MODULES: WSI BUILDING-BLOCKS FOR MASSIVELY PARALLEL COMPUTING

R. M. Lea

Aspex Microsystems Ltd.
Brunel University
Uxbridge, United Kingdom.

1 INTRODUCTION

For nearly three decades, the elusive target of an extremely powerful super-computer has stimulated the development and evaluation of parallel computer architectures. In the 60s and 70s, pioneer machines (e.g. Unger, Solomon, Illiac IV, Staran, CLIP, (ICL) DAP) demonstrated evolving parallel processing concepts and in the 1980s first-generation multiprocessors (e.g. Alliant FX8, Sequent Balance, BB&N Butterfly, iPSC and Transputer systems), array processors (e.g. MPP, (AMT) DAP and GAPP) and hypercube architectures (e.g. NCube and Connection Machine) have become commercially available; each claiming impressive performance for mainly ad hoc benchmarks. Indeed, many pundits are now proclaiming the age of very powerful compute-, graphics- and database-servers, based on Massively Parallel Processors (MPPs), for networked CAD/CAM, military, scientific, medical and office applications.

However, although first-generation MPPs have achieved significant increases in processing performance, they have not yet demonstrated a sufficiently significant breakthrough in application flexibility and cost-effectiveness to be commercially successful.

First-generation MPPs are highly machine-oriented, each being dominated (in cost and performance) by the particular inter-processor communication network (e.g. pipeline, mesh, hypercube, tree, pyramid, Omega, Banyan etc.) favoured by the machine's architect; which may have had curiosity or, even, status value, but may have borne little relation to real (problem-solving) application requirements.

Thus, second-generation MPPs will not only compete for the TOPS challenge trophy nor just represent an upgrade in parallel computer implementation technology. Instead, it is likely that second-generation MPPs will target more realistic user/application-oriented commercial opportunities, thereby allowing freedom from force-fitting applications on to mismatching inter-processor communication networks.

1.1 Real-time Massively Parallel Processors (MPPs)

In contrast to the first-generation parallel computer experience, an important application class for second-generation MPPs will include real-time signal and data processing systems operating in real-world locations. Achieving performance levels up to 1 TOPS (Tera (1E12) Operations Per Second) and supporting continuous data input, such general-purpose (problem-solving) computing systems will attempt the rapid detection, analysis and decision-making (based on recognition or understanding) of object-related information and determination of some appropriate response, as indicated in Figure 1.

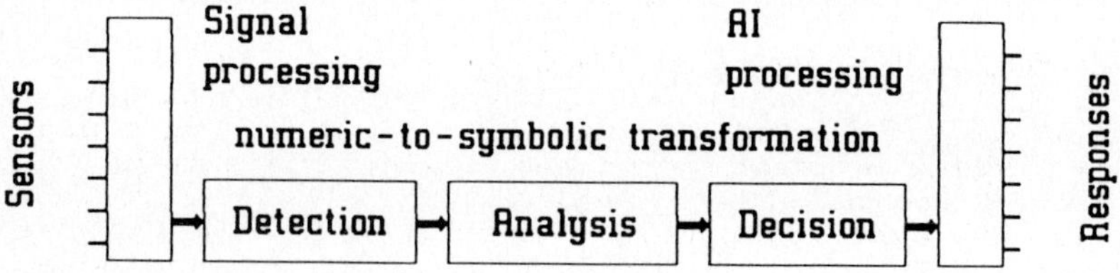

Figure 1. Real-time signal and data processing system

Sensors may include any form of acoustic (e.g. sonar and speech), visual, thermal, mechanical, electromagnetic (e.g. radio and radar) or nuclear radiation energy and applications include remote sensing, surveillance and tracking in aerospace, avionics, seismological and increasingly in biomedical and automotive environments. In particular, real-time computer vision systems for robotics suggest ideal MPP application examples, with a processor allocated to each pixel of the image to be processed.

Typical detection tasks range from signal reconstruction or restoration (to compensate for sensor non-uniformities, ageing and eventual failure), through filtering (to improve signal-to-noise ratios and enhance particular signal features) to correlation (to discriminate relevant signal information from background clutter).

Typical analysis tasks range from grouping data associated with meaningful objects, through specific transforms for feature extraction and edge or contour following (to establish feature connectivity) to object quantification (for length or area measurement or centre-of-mass coordination) and listing of relevant object properties.

Typical decision tasks range from pattern matching (correlation of invariant object descriptors with model descriptions) and navigation of semantic networks through hypothesis analysis to interpretation of object-related information to facilitate decision-making and response determination.

Real-time signal and data processing applications are characterised by the need to provide a rapid response to information derived from a high-speed continuous data stream. For example, real-time computer vision with non-interlaced (60 Hz) raster-scanned image frames requires the completion of all processing tasks (including data input and output) within a frame-time of 16.7 msec, suggesting processing rates of 10 - 1000 GOPS (GigaOPS (e.g. 1E9 8-bit adds per second) and input data rates approaching 1 Gbytes/sec. Clearly, such high-performance can only be achieved with massively parallel processing.

However, such end-to-end processing (viz. problem-solving) systems encompass a sequence of tasks, from signal processing through some form of numeric-to-symbolic data conversion (usually achieving considerable data reduction) to AI (Artificial Intelligence) processing. Indeed, processing tasks are not solely confined to the low-level arithmetic operations on 1D or 2D arrays typical of first-generation parallel computing, but the wide variety of tasks also includes higher-level operations on more complex data structures (e.g. sets, tables, trees or graphs). Moreover, the nature of these application environments ensures dynamically changing functionality requirements as sensor technology and application understanding improve.

Consequently, such real-time signal and data processing systems present the most formidable technical challenge. Indeed, not only will rapid response to object-related information, derived from high-speed continuous data streams, require a step-function advance in high-throughput computer architecture, but cost-effective application will also push state-of-the-art limits of computer implementation technology. Indeed, second-generation MPPs must satisfy significantly different application/design requirements to their first-generation predecessors.

1.2 Engineering reality

Second-generation MPP applications are likely dictate an engineering approach (as opposed to a computer science approach) to parallel computer design. For example, real-time parallel computer systems must operate reliably within environmental restrictions under maintenance schedules and cost limitations. Such engineering issues are discussed below.

(1) Environmental restrictions

Certain MPP application environments (e.g. space, avionics, underwater and automotive) present the most formidable design challenge by severely restricting size and weight, assembly and packaging techniques and power supply and dissipation, as discussed below.

(a) Size and weight

Typical cavities for the above application environments measure significantly less than 1 cubic-foot and weight reduction can be a key issue. Rather less demanding, but nonetheless important, are the size and weight requirements for workstations, which must fit acceptable (e.g. under desk) locations of, say, under 5 cubic-feet. Nevertheless, 1 TOPS/ft^3 and 1 TOPS/lb seem reasonable design targets for second-generation MPPs.

In contrast, the 65,536-processor Connection Machine, delivering 2.7 GOPS (in terms of 12-bit additions) and representing the state-of-the-art in first-generation MPP compaction, occupies 75.4 cubic-feet. Clearly, second-generation MPP applications will require a step-function (i.e. 4 orders-of-magnitude) increase in the density of microelectronics assembly and packaging to achieve the required reduction in the size and weight of first-generation MPPs.

Consequently, the trend towards Wafer-Scale Integration (see (6) below) is of significant interest for size and weight reduction.

(b) Assembly and packaging

The application environments mentioned above often present a most formidable assembly and packaging challenge, due to harsh operational conditions (e.g. extreme temperature range, radiation, electrical noise, mechanical vibration, humidity or even a corrosive atmosphere), which further stimulates interest in Wafer-Scale Integration (see (6) below).

(c) Power supply and dissipation

The application environments mentioned above typically restrict supply and thermal design with power limits of less than 1KW. Thus, 1 GOPS/W seems a reasonable design target for second-generation MPPs.

In contrast, the CM-1 and CM-2 versions of the 65,536-processor Connection Machine, dissipate 12KW and 28KW respectively. Thus, even in an office environment, the power dissipation of such an MPP workstation would be unacceptable. Clearly, second-generation MPP applications will also require a step-function (i.e. 3 to 4 orders-of-magnitude) reduction in the power consumption of first-generation MPPs.

(2) Reliability and maintenance

The reliability and ease of maintenance of relatively expensive machines, such as MPPs, will undoubtedly become key marketing issues. Consequently, maximising MTBF (Mean Time Between Failures) and minimising MTRF (Mean Time to Repair Faults) will be important requirements for second-generation MPPs.

In particular, for the reasons given below, highly-compact fault-tolerant Wafer-Scale modules, are recommended for the implementation of highly-reliable and remotely-maintainable second-generation MPPs.

(a) Fault-tolerance

Second-generation MPPs would offer improved system reliability and ease of maintenance, by adding redundancy to the processor ensemble and achieving in-service fault-tolerance with self-testing and electronic reconfiguration to isolate faulty processors. Moreover, fault-tolerant second-generation MPPs could support graceful degradation or even fail-safe systems.

(b) Modular systems

With service lifetimes extending beyond 10 years, ease of maintenance will be a key issue for second-generation MPPs, which will stimulating a trend towards modular parallel computing systems. Indeed, ease of maintenance would be greatly improved by the use of line replaceable modules, since simple plug-in module replacement would provide the opportunity to minimise down-time and maximise availability.

(c) Monolithic integration

Increasing micro-miniaturisation also offers improvements in reliability, due to the monolithic integration of chip-to-board-to-chip interconnections, the consequent reduction of failure-prone board connectors. Consequently, the trend towards Wafer-Scale implementation (see (6) below) is reinforced by this advantage.

(3) Cost-lifetime

The history of maxi, mini and microcomputer development suggests that first-generation MPPs will rapidly become obsolete as inevitable advances in microelectronics technology (e.g. increasing chip size and reducing feature size) allow more exciting products to enter the market place. Thus, second-generation MPP customers are likely to be wary of technological change. Indeed, wise customers will be looking for security and growth potential in 'future-proof' investments in MPP technology.

With service lifetimes extending beyond 10 years, architectural and software durability under functional upgrade (as application requirements change) and technological upgrade (without modification of existing software) will be very important requirements for second-generation MPPs. Indeed, the trend towards modular systems (see (2) above) could provide some degree of system (design) stability and technology independence, with simple replacement of MPP modules.

(4) Cost-performance

Emerging from the first-generation MPP experience is the realisation that the cost-performance of second-generation MPPs will not be dominated by the exciting increase in processing power, but by the much more mundane details of implementation cost of the processor ensemble and, especially, the inter-processor communication network plus the peripheral costs of the global memory, control unit and data interfaces.

Indeed, significant increases in processing performance have enabled first-generation MPPs to stimulate commercial interest in parallel computing; applications exist, parallel algorithms are being developed and architectural principles have been established. However, first-generation MPPs are unlikely to break through cost-barrier currently impeding commercial exploitation. For example, first-generation MPPs claim up to 1MOPS/$1K, which represents only O[1] improvement in cost-performance for O[N] parallelism.

In contrast, second-generation MPPs must aim at 2 - 3 orders-of-magnitude improvement, in order to provide an adequate return on the high capital investment required to install new hardware, for programmer retraining and, especially, for new software development. Hence, 1 MOPS/$ represents a realistic cost-performance target.

In view of the dominating influence of implementation cost and the inevitability of processor redundancy (see (2) above and, especially, (5) below), it follows that second-generation MPPs should be specifically designed to exploit the economic opportunities of microelectronics technology.

(5) Silicon-efficient chip architecture

Implementing MPP architectures in Silicon introduces a different set of cost-performance trade-offs to those previously encountered by first-generation MPP designers; these being rather more cost-reduction than performance-enhancement oriented. In contrast to first-generation MPP developments, top-down design of MPP architectures (designed without detailed regard to microelectronic opportunities and constraints) is likely to be less cost-effective than bottom-up design of MPP chips (without regard to architectural requirements).

An important lesson can be learned from the history of computer memory technology. Random-access memories incorporate massive redundancy; since, except for the addressed storage location, all other locations are inactive. Consequently, early computer memories, based on expensive implementation technologies, were kept small to maintain cost-efficiency. However, as technology improved, minimising memory implementation cost, memory redundancy became less of a problem. Indeed, memory is now so cheap that even modest computers can boast a large main store. Clearly, advances in microelectronics technology have reduced the memory redundancy problem to insignificance. Thus, the challenge is to apply a similar technological trend to improve the cost-effectiveness of the processor ensemble.

With this challenge in mind, it is significant that although memory locations have been greatly reduced in size, memory redundancy remains constantly close to maximum. In contrast, the much lower level of processor redundancy tends to compensate for higher processor complexity.

A concerted attempt to render second-generation MPPs as cost-effective as random-access memories, would be driven by the following design attempts to achieve Silicon-efficient MPP chip architectures.

(a) Maximise the packing density of the processor ensemble

In order to reduce the cost of inevitable processor redundancy to a negligible level, this objective requires the maximisation of the number of processors which can be packed on an economical Silicon die; which entails minimisation of processor layout area to benefit from a high intrinsic chip yield.

(b) Incorporate the inter-processor communication network on chip

This objective entails the minimisation of the geometrical layout of the inter-processor communication network and supporting global routing for the processor array. As layout feature sizes continue to reduce and VLSI chip sizes increase, the inter-processor communication network progressively dominates the device architecture, such that placement of individual processors is becoming a secondary consideration. Thus, integrating the network within the processor ensemble is of major importance for the reduction of MPP implementation cost.

(c) Minimise input-output pinning overheads

This objective recognises that the cost microelectronic assembly is dominated by interconnection technology.

(6) Microelectronics and packaging technologies

The relentless SSI-to-MSI-to-LSI-to-VLSI technological trend has been driven by decreases in layout feature size and increases in chip area. Reductions in feature size are continuing, but at an ever increasing price in terms of investment in wafer processing equipment. Chip areas are increasing, in step with improved wafer processing technology, but such progress is severely limited by defects in the processed Silicon. Larger wafers increase the probability of finding working chips, but, as chip size increases overall chip yield drops rapidly. Consequently, it is timely to question the cost-effectiveness of seeking further increases in chip complexity by such traditional routes.

An alternative approach to increased levels of monolithic integration is evident in the inevitable trend towards Wafer-Scale Integration (WSI), which provides the opportunity to satisfy the second-generation MPP requirements discussed in (5) above by integrating a major sub-system on a single (undiced) Silicon wafer.

There are two competing forms of WSI, these being

monolithic WSI: integrating redundant (viz. standby) logic blocks and associated reconfiguration circuitry with functional logic, within the same Silicon wafer, to allow otherwise fatal defects to be isolated

hybrid WSI: mounting independently tested and unpackaged VLSI chips (diced from different wafers) on the Silicon wafer, which supports a number of isolated metallisation layers, to form a HDMI (High-Density Multi-layer Interconnect) - VLSI multi-chip (hybrid) module.

The two forms compete in size, with the monolithic overheads of defect-tolerance being offset by the chip interconnection overheads of the hybrid approach. Monolithic WSI gains the advantage in performance, weight, power consumption and dissipation (due to output pad drivers), reliability (due to fewer chip-to-chip interconnections) and ease of maintenance (if the defect-tolerance can also support fault-tolerance) and eventually in cost. Hybrid WSI eliminates the need for defect-tolerance since all circuitry is functional (but lacks the fault-tolerant capability) and its fabrication technology is currently better established. Of course, the VLSI chips of hybrid WSI could incorporate fault-tolerant circuitry, but then monolithic WSI would gain the extra advantage in size. At present, industrial practice favours hybrid WSI, since it complements existing technology, and, it is not clear that the (speculative) advantages of monolithic WSI will change this attitude.

Both forms of WSI can benefit from emerging highly-compact packaging techniques, based on new materials such as Aluminium Nitride, Silicon Carbide and Carbon fibre (with thermal characteristics closely matched to those of Silicon). However, this is an accident-prone and fast-changing field and it will be essential align WSI and packaging developments.

In summary, engineering barriers in size, weight, reliability, power consumption and cost are currently impeding progress towards the commercial exploitation of second-generation MPPs for real-time parallel computing. Indeed, it is clear that, in order to achieve a breakthrough, it is not just TOPS performance that is required, but second-generation MPPs must also achieve step-function improvements in TOPS/ft^3, TOPS/lb, TOPS/failure, GOPS/W and MOPS/$. For example, assuming that weight and power problems can be resolved, it is generally recognised that a low-cost highly-compact fault-tolerant real-time computer vision system could launch high-volume products in the robotics market.

Current trends in microelectronics and packaging technologies towards highly-compact plug-in replaceable HDMI-VLSI multi-chip (hybrid) modules augur the desired breakthrough and, at the same time, simplify problems of tolerating harsh environmental conditions. Moreover, simple module replacement satisfies the need for ease of maintenance and provides a convenient means for both functional and technological upgrade during operational lifetime.

1.3 Second-generation MPPs

In terms of the engineering issues discussed above, the development of computer storage has been a great success. Indeed, as the ultimately cost-effective computer component, RAM storage represents an ideal model for the development of second-generation MPPs.

Consequently, since traditional computer storage is organised with a string topology (i.e. all application data structures are mapped to a string of (groups of) words in RAM or bytes on disk), it seems logical prospect for second-generation MPPs to be organised as a string of processors.

Moreover, further consideration of the introduction presented above suggests that a fault-tolerant homogeneous string of MIMD (Multiple-Instruction control of Multiple-Data streams) processors, would satisfy both the architectural and engineering requirements of second-generation MPPs.

According to application requirements, a string of identical MIMD processing modules would be plugged into the control bus and Data Communications Network of an appropriate modular MPP racking system, as indicated in Figure 2.

Each MIMD modules would comprise a simply scalable and fault-tolerant string of inter-communicating SIMD (Single-Instruction control of Multiple-Data streams) processors, each incorporating an independent Data Interface (DI), as shown in Figure 3. Edge-connection constraints suggest 2 (32-bit) Data Interfaces (DIs) and a single Control Unit (CU) per SIMD sub-module.

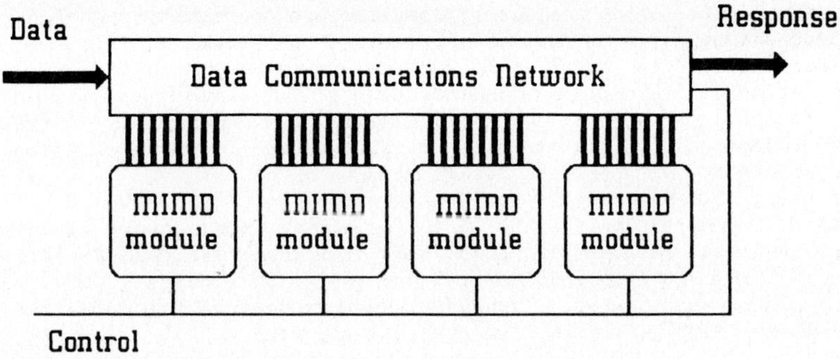

Figure 2. Modular second-generation MPP

Similarly, at the next level down in this homogeneous parallel computing hierarchy, each SIMD processor would comprise a simply scalable and fault-tolerant string of a large number of identical fine-grain processors and a reconfigurable inter-processor communication network. As with RAM storage, the SIMD processors would be internally organised to exploit the most recent advances in state-of-the-art microelectronics.

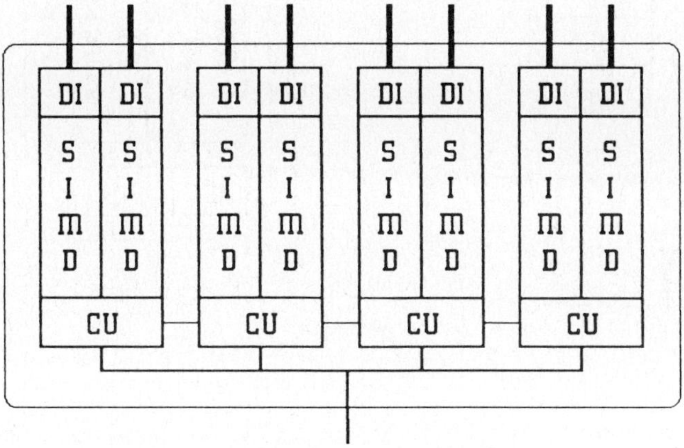

Figure 3. MIMD module

2 ASP MODULE ARCHITECTURE

Based on the encouraging results emerging from research into parallel computing technology at Brunel University and being developed by Aspex Microsystems Ltd., ASP technology comprises highly versatile hardware and software modules for the construction of second-generation MPPs capable of TOPS performance [1,2].

ASP (Associative String Processor) architecture and support software provide the base technology for the development of versatile parallel processing building-blocks (i.e. modules of the form shown in Figure 3) for the simple construction of modular second-generation MPP systems, satisfying both the architectural and engineering requirements discussed in the introduction above. Indeed, it is claimed that the ASP is the first parallel computing architecture to achieve this design goal.

An ASP module comprises a MIMD parallel processing structure of intercommunicating ASP substrings, each supported with an ASP Data Buffer (ADB) and an ASP Control Unit (ACU), as shown in Figures 3 and 4. The ASP module also incorporates a single control interface and a multiple Data Interfaces (DIs), to connect to the system rack.

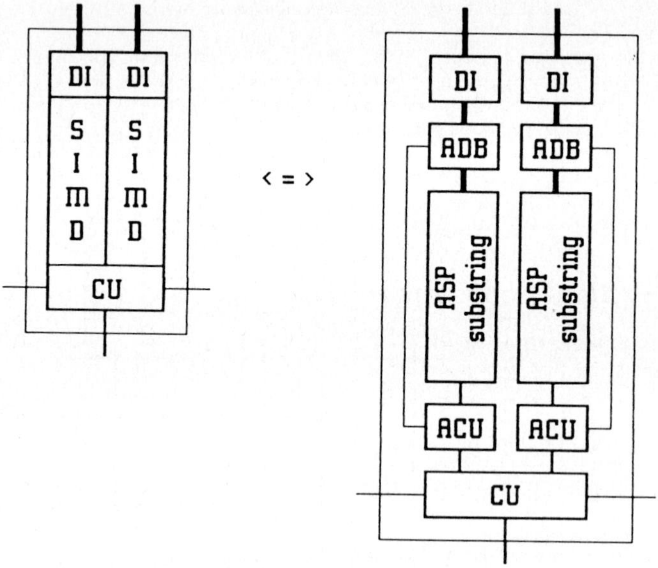

Figure 4. ASP (SIMD) sub-module

Based on a fully-programmable, simply scalable, reconfigurable and fault-tolerant homogeneous parallel processing architecture, ASP modules offer considerable application flexibility, maintaining high computational efficiency over a particularly wide range of signal and data processing applications, due to

(1) simple configuration of ASP modules to simplify the development of MPP systems which are well matched to functional application requirements

(2) pipelining and overlapping input-output data transfers (via the Data Communications Network) with parallel processing (within ASP substrings), by separating input-output from processing with the ASP Data Buffers (ADBs) and Vector Data Buffers

(3) overlapping of sequential (scalar) processing (in the ASP Control Units (ACUs)) with parallel processing (in ASP substrings)

(4) mapping different application data structures to a common string representation within ASP substrings (supporting content-addressing, parallel processing and a reconfigurable inter-processor communication network).

Furthermore ASP modules not only offer highly-compact implementation the MIMD/SIMD configuration discussed above, but also they exploit the following benefits of associative processing.

(1) elimination of processor (location) addressing, for the purposes of

(a) achieving unlimited architectural scalability

(b) implementing cost-effective fault-tolerance with simple (hierarchical) bypassing of faulty processor substrings (see below), which, until failure occurs, are available for parallel processing

(c) further minimisation of sequential processing overheads

(2) minimisation of inter-processor data movement, with high-speed activity transfer between processor subsets and in-situ processing.

In particular, the ASP module incorporates no specifically redundant processors; all processors are available for useful work. In operation, testing and reconfiguration of a modular ASP system would yield a working set of ASP substrings for subsequent parallel processing.

Thus, ASP modules provide an unprecedented opportunity for second-generation MPPs which require dynamic allocation of processors for the flexible and progressive trade-off of parallel processing power for increasing fault-tolerance. Indeed, fault-tolerant configurations of ASP modules could be assembled for application as highly-reliable, self-repairable and gracefully-degradable second-generation MPPs. For example, data input could be regularly suspended, for say 10ms in every minute, to allow testing and string reconfiguration. Moreover, multiple-buffering ASP configurations could avoid such interruption of data input.

2.1 ASP substrings

Each ASP substring comprises a string of identical APEs (Associative Processing Elements), which are connected to an Inter-APE Communications Network (running in parallel with the APE string), as shown in Figure 5. All APEs share common bit-parallel Data, Activity and Control Busses and a single feedback line (Match Reply, MR) maintained by the ASP Control Unit (ACU), which also maintains the LinK Left and LinK Right ports (LKL and LKR) of the Inter-APE Communications Network [1,2].

The Inter-APE Communications Network implements a globally-controlled and dynamically-reconfigurable tightly-coupled APE interconnection strategy, which supports cost-effective emulation of common network topologies (viz.

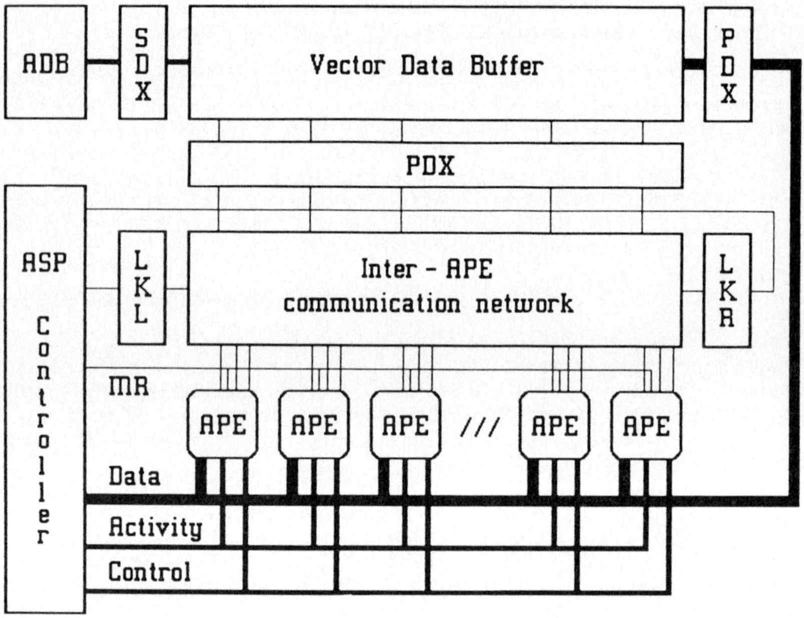

Figure 5. ASP substring

arrays (e.g. vector, matrix and binary n-cube), tree networks, graph (e.g. semantic) networks, address permutation (e.g. exchange, shuffle and butterfly) networks and shifting networks). Most significantly, the APE interconnection strategy supports simple modular network extension, to enable tailoring of parallel processing power to match user requirements.

At an abstract level, and assuming a programmable connection between the LKL and LKR ports shown in Figure 5, the Inter-APE Communications Network can be considered as a hierarchical chordal-ring structure, with the chords bypassing APE blocks (and groups of APE blocks), as indicated in Figure 6, to

accelerate inter-APE communication signals : APE blocks not including destination APEs are automatically bypassed for both circuit-switched and (if required) packet-switched modes of inter-APE communication and, if appropriate for the former mode, activated in a single-step

provide APE block defect/fault-tolerance: APE blocks failing a test routine, either in manufacture or service, can be switched out of the string, such that defective or faulty blocks would be simply bypassed.

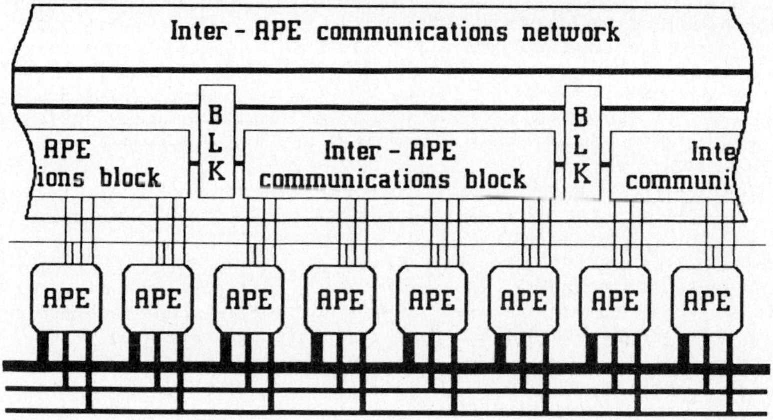

Figure 6. APE block bypassing

As shown in Figure 7, each APE incorporates a 64-bit Data Register and a 6-bit Activity Register, a 70-bit parallel Comparator, a single-bit full-

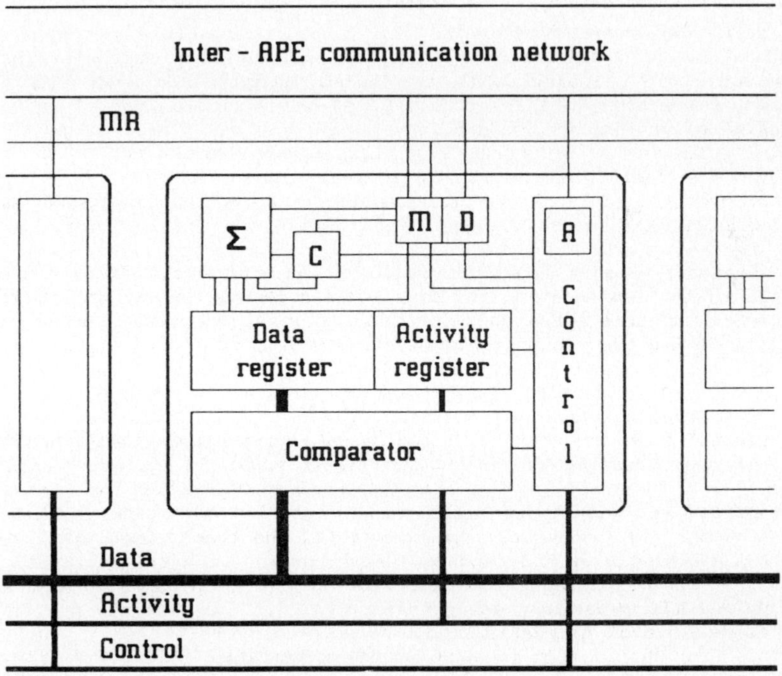

Figure 7. Associative Processing Element (APE)

adder, 4 status flags (viz. C to represent arithmetic Carry, M and D to tag Matching and Destination APEs and A to activate selected APEs) and control logic for local processing and communication with other APEs.

In operation, each ASP substring supports a form of set processing, in which the sub-set of active APEs (i.e. those which match broadcast data and activity values) support scalar-vector and vector-vector operations. Matching APEs are either directly activated or source inter-APE communications to indirectly activate other APEs. The Match Reply (MR) line indicates whether none or some APEs match.

The Vector Data Buffer (shown in Figure 5) optimises the efficiency of input-output data transfer, by supporting a dual-port exchange of vector data (i.e. output dumped and input loaded in a single step) with alternating Primary Data eXchange (PDX) and Secondary Data eXchange (SDX).

2.2 WSI ASP module implementation

In practice, ASP module form-factors will depend on the packaging standards adopted for the modular construction systems of particular application environments. Of the possible packaging standards being considered for aerospace applications in the 1990s, a light-weight SEM-E compatible module (measuring 6.4" x 5.88" x 0.6" [3]) is assumed for this paper.

A priori consideration suggests that SEM-E compatible packages could accommodate 4 ASP (SIMD) sub-modules, each mounted on a suitable substrate which would provide electrical and thermal support for 2 WSI ASP (WASP) devices, 2 (32-bit) Data Interfaces (DIs) and a single Control Unit (CU), as indicated in Figures 4 and 8.

WASP devices could achieve similar SIMD processor packing densities with either monolithic or hybrid WSI technologies [4-6], as discussed in sub-section 1.2.(6).

As indicated in Figure 8, a WASP device is composed from 3 different "chip" reticle-masks; implementing the Data Routers (DRs), the ASP substrings and the Control Routers (CRs).

Each of these "chips" integrate a defect/fault tolerant design; such that APE blocks (see Figure 6), individual ASP "chips" and entire rows of ASP "chips" can be bypassed. In addition, the DR and CR reticles incorporate defect/fault-tolerant routing to implement a single Data Interface (DI) and a single connection to the control interface shown in Figures 3 and 8.

The WASP project was started at Brunel University in October 1983 and has been funded since 1985 under a UK Alvey (VLSI) contract, involving Plessey, GEC, ICL and Middlesex Polytechnic, which has included the fabrication and evaluation of defect/fault-tolerant test chips and the WASP 1, WASP 2a and WASP 2b WSI ASP technology demonstrators [4,5]. In a separate project, the WASP 3 application demonstrator for iconic-to-symbolic image processing, scheduled for fabrication in 1990, is being developed by Aspex Microsystems under a US SDI IS&T contract. In summary, research strategy is to approach the ideal WASP target, as described above, with a stepping-stone sequence of experimental prototypes (i.e. WASP 1/2/3/4 ...); analysis of each

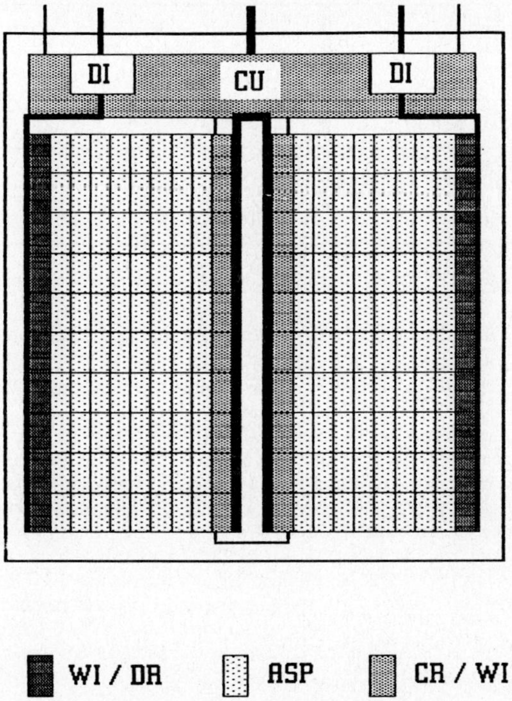

Figure 8. WSI ASP (WASP) substrate

prototype helping to redefine WASP potential and to specify the experimental objectives of the next step.

Extrapolation, from the results of current experimental progress, suggests that, with standard 1.2um CMOS (double-layer metal) fabrication technology, (nominal) monolithic WSI ASP harvests of 4,096, 8,192 and 16,384 APEs can be confidently forecast for 4", 5" and 6" wafers respectively. Moreover, WSI ASP economics determine that, with die aspect ratios of 2:1, a 6" wafer could (nominally) harvest 2 x 8,192-APE monolithic WSI ASP devices. Consequently, a 65,536-APE WSI ASP module with 8 (32-bit) data channels can be, conservatively, forecast [2].

3 CONCLUSIONS

The progress of WSI ASP hardware and software research heralds the development of highly-versatile and fault-tolerant ASP modules for the construction of second-generation MPPs; with significant benefits, compared with contemporary MPP implementations, in terms of the engineering reality discussed in sub-section 1.2. Indeed, the following design targets summarise the potential of WSI ASP modules in these terms.

Configuration	1, 2 or 4 MIMD processors
Performance	100 GOPS (12-bit adds at 40 MHz)
Input-output bandwidth	320 Mbytes/sec (at 20 MHz)
Number of processors	65,536
Package size	6.4" x 5.88" x 0.6" (SEM-E)
Power dissipation	< 100 W
Market cost	< $100,000

Being based on extrapolation of experimental results, the first 4 design targets are fairly conservative. Since, a performance of 1 TOPS (i.e. 1000 GOPS) could be achieved with only 10 ASP modules within a 6" cube, the TOPS/ft^3 target seems achievable. Moreover, the quoted cost includes ample provision for marketing overheads, software support and profit, thereby indicating that the 1 MOPS/$ target could be exceeded with higher volume marketing. In fact, the most speculative prediction is that for power dissipation, aiming to exceed the 1 GOPS/W target with a compromise between the 50 Watts quoted for SEM-E packaging technology [3] and the desire to maximise performance. Hence, current research is addressing this issue not only in terms of minimising power dissipation in WSI circuit design, but also in improvement of the thermal characteristics of appropriate assembly and packaging materials.

Of course, such performance forecasts refer only to the low-level arithmetic operations of numeric-to-symbolic data conversion process of real-time signal and data processing. Indeed, a more realistic performance indication can be gained from real-time signal and data processing benchmarks, which include all processing from data input to the output of the required response. To this end, the ASP architecture has been evaluated in terms of the 2 US DARPA Image Understanding benchmarks set in 1986 and 1988 and the European CERN LAA (particle Physics track analysis) benchmark in 1989. Although beyond the scope of this paper, the results [7,8], nevertheless, indicate significant performance advantages for ASP modules compared with contemporary parallel computers.

In summary, ASP modules offer considerable promise for second-generation parallel computer applications. Perhaps, more significantly, ASP modules offer simple scalability and, indeed, the prospect of breaking through the cost-barrier currently impeding the wider commercial exploitation of massively parallel computing systems.

ACKNOWLEDGEMENTS

The author gratefully acknowledges enthusiastic contributions to ASP-based projects from past and present members of the Parallel Computing Technology and Aspex Microsystems Ltd. at Brunel University, the support of the UK Alvey initiative and US SDI IS&T funding.

REFERENCES

[1] R.M. Lea, "The ASP: a cost-effective parallel microcomputer", IEEE Micro, October 1988, pp. 10-29.

[2] R.M. Lea, "ASP modules: cost-effective building blocks for real-time DSP systems", Journal of VLSI Signal Processing, 1989, Vol. 1, pp. 61-76

[3] R. Morgan and M. Soraya, "Future military avionics applications of wafer-scale technology", Proc. IEEE Int. Conf. on Wafer Scale Integration (eds. Swartzlander and Brewer), IEEE Computer Society Press, 1989, pp. 1 -12.

[4] R.M. Lea, "A WSI image processor", Wafer Scale Integration (ed. Swartzlander), Kluwer Academic Publishers, Chapter 5, 1988.

[5] I.P. Jalowiecki, K.D. Warren and R.M. Lea, "WASP: A WSI Associative String Processor", Proc. IEEE Int. Conf. on Wafer Scale Integration (eds. Swartzlander and Brewer), IEEE Computer Society Press, 1989, pp. 83 - 93.

[6] K.D. Warren, J.H. Reche, W.J. Jacobi and R.M. Lea, "A 3D HDI ASP: a cost-effective alternative to WSI signal processors", Proc. IEEE Int. Conf. on Wafer Scale Integration (eds. Swartzlander and Brewer), IEEE Computer Society Press, 1989, pp. 267 - 276.

[7] A. Krikelis and R.M. Lea, "Performance of the ASP on the DARPA architecture benchmark", Proc. Frontiers 88, 2nd. Symp. on the Frontiers of Massively Parallel Computation, Fairfax Virginia, October 1988.

[8] A. Krikelis, I. Kossioris and R.M. Lea, "Performance of the ASP on the DARPA architecture benchmark II", Proc. DARPA Image Understanding Benchmark Workshop., Avon Connecticut, October 1988.

Wafer Scale Integration, III
M. Sami and F. Distante (Editors)
Elsevier Science Publishers B.V. (North-Holland)

HIERARCHICAL REDUNDANCY AND SELF-TESTING FOR A ULSI POLY-PROCESSOR SYSTEM

Nobuo TSUDA and Tetsuji SATOH

NTT Communications and Information Processing Laboratories
1-2356 Take Yokosuka-shi Kanagawa 238-03, Japan

A design method employing 3-level hierarchical redundancy reconfigured by self-testing is proposed for defect tolerance of a large area poly-processor ULSI/WSI chip. A prototype 16-bit processing element chip with micro-program control has been designed and tested employing 3-micron CMOS technology, to ensure the first level redundancy and related testing applied in the element. Yield estimation based on newly-proposed effective active area considerations and the processing element model indicates that a ULSI/WSI chip on a scale 30 to 100 times larger than a general non-redundancy chip may be realized by employment of 3-level hierarchical redundancy, and by employment of sub-micron device technology.

1. INTRODUCTION

Large scale integration of a poly-processor system is a means of realizing compact, high-throughput machines for special purpose highly repetitive operations.

The authors have been studying defect tolerance technologies based on redundancy configurations for the wafer scale integration of memories and cellular structured logic circuits[1-7]. As the first step in ensuring a redundancy configuration for the poly-processor system, they have developed a micro-program controlled 16-bit processing element chip with redundancy employing 3-micron CMOS technology[5]. This chip can be applied in the assembly of a poly-processor system for highly-parallel pattern matching operations in handwritten Japanese Kanji-Kana character recognition.

This paper describes the design of a redundancy configuration and related testing using the micro-program control applied in the developed processing element chip. Also, it covers the estimated performance of redundancy configurations based on effective active area considerations, as well as the estimated feasibility of future ULSI/WSI poly-processor systems by sub-micron technology employing the developed processing element model and hierarchical redundancy.

2. HIERARCHICAL REDUNDANCY FOR A POLY-PROCESSOR SYSTEM

A ULSI/WSI poly-processor system model employing 3-level hierarchical redundancy is shown in Fig. 1. The basic structure of the system is a one-dimensional processor-array with serial and parallel connection. The developed prototype chip corresponds to a processing element (PE) in this figure. The system configuration in this type can support both SIMD and MIMD operations, and can simulate operations on another two-dimensional structure by arrangements of interface circuits and switches.

The "K out of K+R" redundancy reconfigured by self-testing is applied in cellular parts in each PE, each element group and a set of element groups, respectively. Replacements by spares on each level are achieved in a system initializing operation resulting from built-in testing and external testing by a host processor. Each PE can be accessed from the host processor, using its own physical address in a testing mode.

A PE in which the number of non-defective circuit blocks is not satisfied is treated as a defective PE. Also, an element group in which the number of complete PEs is not satisfied is treated as a defective group, which allows for construction of a complete two-dimensional PE-array by reconfiguration. Defective PEs and excess spare PEs in each element group are separated from the parallel connection, and are also bypassed in the serial connection by providing inhibit flags. After the testing, new serial logical addresses are assigned to the PEs of the complete array.

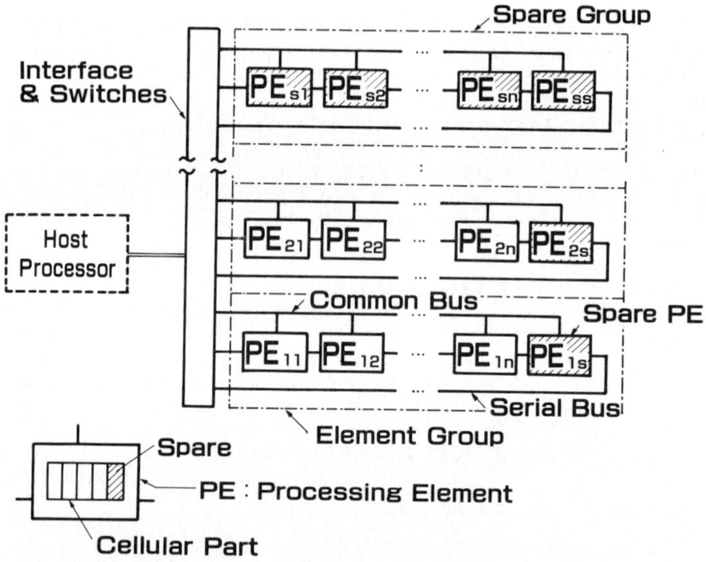

FIGURE 1 ULSI/WSI Poly-processor System Model Employing 3-level Hierarchical Redundancy

Hierarchical redundancy in this way has a great advantage, in that almost all circuits of the array can be effectively covered with replacements by spares. This allows for construction of a large area chip on which a large number of PEs are integrated.

The redundancy configuration shown in Fig. 1 can be expanded and applied to a ULSI/WSI poly-processor system employing another structure in PE-connections.

3. DESIGN OF A PROTOTYPE PROCESSING ELEMENT

3.1 Basic Functions for Pattern Matching

The first application of the developed prototype PE chip is in highly-parallel pattern matching for the recognition of hand-written Japanese Kanji-Kana characters, as shown in Fig. 2.

In a recognition system, the image pattern of an input unknown character is converted into a feature vector with several hundreds of vector components in a pre-processing stage. Reference vectors corresponding to about three thousands of character species have been stored in the dictionary RAMs of the PEs.

The parallel matching operation of the PE-array is achieved in a SIMD-like manner. Every PE performs calculation of the weighted Euclidean distances under successive broadcasting of the PE-commands and input feature vector components through the common bus.

After the distance calculations, distance values smaller than a certain threshold value are tagged with their character codes, and are then loaded into the cascade-connected sorter units. Through the sorting operation, the character code with the smallest distance value is extracted and output as the first candidate character.

As for total computational capability, it is estimated that a pattern matching processor composed of 1000 PEs can execute matching operations with a 3,200-character set employing 512-component feature vectors in a recognition throughput of 130 characters per second.

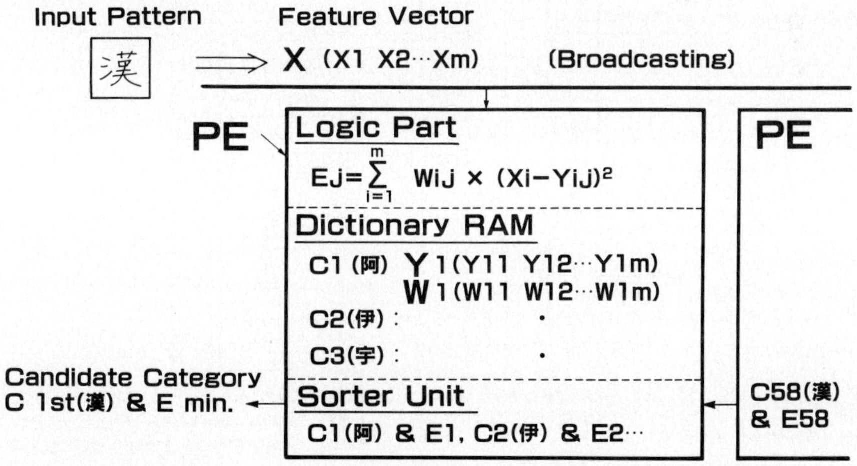

FIGURE 2 Parallel Pattern Matching Operation on PE-array for Hand-written Japanese Kanji-Kana Characters

3.2 Chip Implementation Employing 3-micron CMOS Cells

The prototype PE chip is basically composed of a 16-bit width logic part designed for digital signal processing with a 3-bus architecture, a 61-bit x 512-word micro-program ROM, a 16-bit x 2K-word dictionary RAM, and a 16-bit width sorter unit, as shown in Fig. 3.

This configuration allows for the following simultaneous actions in an instruction step: calculations in the logic part, reading or writing in the RAM, and subsequent address setting to the micro-program ROM. The sorter unit can work independently of the main body of the PE.

Calculations of Euclidean distance values are mainly achieved by the ALU, multiplier (MUL) and register-file (RF), with a throughput of 13 steps per vector component under the pipelined reading of reference feature vector components from the RAM. The cycle time of one step is 300 nano-seconds in case of the prototype PE chip.

The cascade connection of the sorter units can execute candidate extraction based on an odd-even sorting algorithm in 4N steps, the number of which is 4 times the number of the cascade-connected units.

A total of thirty-seven commands have been prepared for the micro-program controlled operations. Seventeen are for 4 types of successive distance calculations, and twenty are for simple operations, including testing and replacements.

A photograph of the PE chip is shown in Fig. 4. About 460K transistors, including those for redundancy, have been implemented on a 380-square-millimeter chip, employing 3-micron P-well CMOS technology. Main features of the PE are summarized in Table 1.

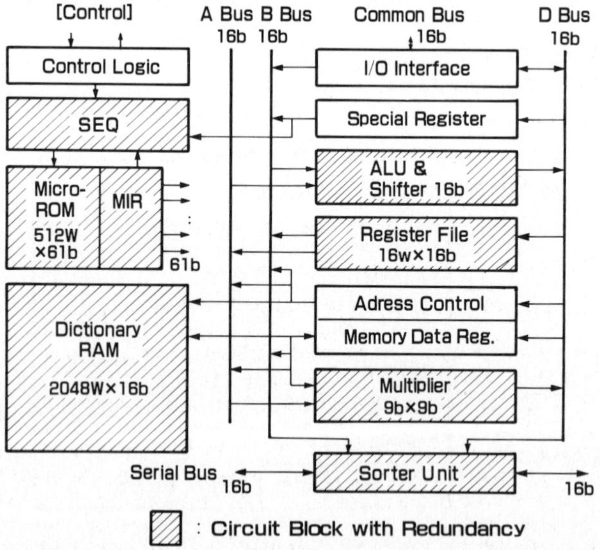

FIGURE 3 Basic Circuit Block Diagram of Prototype PE

FIGURE 4 Photomicrograph of Prototype PE Chip
(Chip Size: 18.5 x 20.5 mm , 3-micron CMOS Technology)

TABLE 1 Main Features of Prototype PE

- **Arechitecture : 16–bit DSP with 4KB RAM and Sorter**
- **Major Function : Weighted Euclidean Distance calculation**
- **Clock Cycle : 300ns**
- **Power Dissipation : 1W (5V)**
- **Total I/O Terminals : 109**
- **Device Technologies : 3μ m P-Well CMOS, Double Al Layers**
- **Total Device Count : 460K Transistors**
- **Chip Size : 18.5×20.5mm² (380mm²)**

4. REDUNDANCY CONFIGURATION IN PROCESSING ELEMENT

The prototype PE chip employs "K out of K+R" redundancy or duplication in major circuit blocks. The device count of each block and applied redundancy configurations are summarized in Table 2.

In the logic part, "4 out of 5" or "8 out of 9" redundancy is applied in the micro-instruction register (MIR), ALU, register-file (RF) and sorting unit, since they have bit-sliced circuit structures. Duplication by a simple "1 out of 2" redundancy is applied in the sequencer (SEQ) and multiplier (MUL), since the former is composed of random logic circuits and the latter has a large number of cell connections.

The micro-program ROM, a mask-programming type, is designed for fail-safe capability by means of duplication in every bit. The dictionary RAM occupies almost half of the chip area. Hence, two-level hierarchical redundancy is applied therein to attain high yield performance.

Approximately 85 % of the circuits in a PE have defect tolerance capability via redundancy. The total amount of additional circuits for spares and testing is 35 % of the on-chip circuits, as summarized in Table 3.

TABLE 2 Redundancy Configuration of Prototype PE

Circuit Block	Device Count	Redundancy
• Logic Part	44.3K (32.8K)	
· SEQ	5.2K	Duplication
· MIR	6.0K	8 out of 9
· ALU	3.5K	4 out of 5
· RF	4.3K	8 out of 9
· MUL	7.2K	Duplication
· SORT	6.6K	8 out of 9
· Core Part	11.5K	
• Micro-ROM	87.5K (39.8K)	Dup. & Fail-safe
• Dictionary RAM	325.6K (208.8K)	2-lev. Redundancy
Total of PE	457.4K (281.K)	

*( ) : Values for Basic Circuits Without Redundancy.

TABLE 3 Design Results of Redundancy and Testing

- Total Active Area of PE : 250mm^2 (163mm^2)
  - · Logic Part : 93.7mm^2 (70.0mm^2)
  - · Micro-ROM : 35.0mm^2 (15.9mm^2)
  - · Dictionary RAM : 120.7mm^2 (77.4mm^2)
- Total additional Area : 35% (Spares : 33%, Testing : 2%)
- Redundancy Coverage : 85% of Total Basic Active Area
- Built-in Testing Coverage : 80% of Total Active Area

*( ) : Values for Basic Circuits Without Redundancy.

5. SELF-TESTING FOR RECONFIGURATION

5.1 Outline of Testing and Reconfiguration

The testing and reconfiguration of a PE-array are achieved in a system initializing operation in an on-board manner. Concerning testing of the PE, a closed circuit composed of the micro-program ROM, sequencer (SEQ) and micro-instruction register (MIR) is automatically tested and reconfigured in a built-in manner in the first stage.

Subsequently, the ALU and other circuits in the logic part are tested under micro-program control by the reconfigured ROM and sequencer, with support of an external host processor. The sorter unit can be tested individually in each PE in a bypassed mode. It is then tested in a cascade-connected manner via array-level function testing. The replacements by spares in the ALU and other components are achieved as a result of flag setting by the host processor.

The testing and reconfiguration of the dictionary RAM is also automatically achieved by using the reconfigured logic part in a built-in manner. The testing procedures conducted in this way are summarized in Table 4.

TABLE 4 Testing Procedures for Prototype PE

Order	Operations	Testing Type	Checker
1	Initial Reset		
2	Micro-ROM Bit-line Test	Built-in	Bit-line Checker
3	Sequencer and MIR Test	Built-in	Comp. & Parity checker
4	I/O, ALU and RF Test	External	Host Processor
5	MUL and Sorter Test	External	Host Processor
6	RAM Bit-line Test	Built-in	Bit-line Checker
7	RAM Read/Write Test	Built-in	Word-line Checker & ALU
8	PE Function Test	External	Host Processor

*Replacements are achieved at the end of each testing.

Eighty percent of on-chip circuits can be tested through built-in testing procedures, and others can be tested in external testing procedures. Total testing time for the reconfiguration in a PE is about 100 milliseconds, and about 95 % of the procedures performed within the testing time can be achieved in a parallel manner on the PE-array.

5.2 Built-in Testing and Reconfiguration of the ROM and Sequencer

The micro-program ROM is composed of duplicated ROM units and simple fail-safe logic circuits which are prepared for every bit of the read data, as shown in Fig. 5.

The fail-safe capability is realized on the basis of a circuit characteristic whereby major defects in ROM cells cause "1" stacked errors in the outputs from ROM units. Hence, in cases wherein two outputs from the duplicated ROM units differ from each other, the output in "1" is regarded as an error, and "0" is output from the fail-safe logic circuit as correct read data of the micro-ROM. At the same time, defects which mainly cause "0" stacked errors can be detected in initial bit-line checking and run-time word-line checking, which are the same methods employed in RAM units. In this case, the error outputs from the ROM units are forced into "1", so as to be corrected by the fail-safe logic circuits. The details of error detection and correction in the micro-ROM are summarized in Table 5.

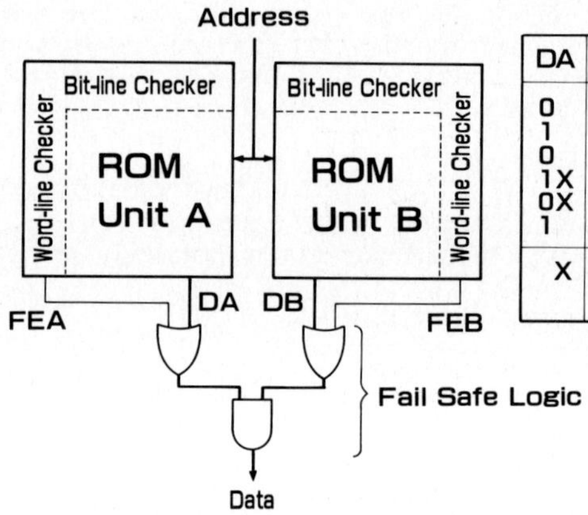

DA	DB	EFA	FEB	Data	
0	0	0	0	0	Correct
1	1	0	0	1	
0	1X	0	0	0	
1X	0	0	0	0	
0X	1	1	0	1	
1	0X	0	1	1	
X	X	0	0	X	Error
		1	1	X	

X : Error Output

FIGURE 5 Micro Program ROM with Fail-safe Capability by Duplication

TABLE 5 Error Detection and Correction in ROM Unit

Defect modes	Error outputs	Countermeasures
Bit-line to Cell Non-contact	'1'	Fail-safe
Bit-line Cut Off	'1'	Fail-safe
Bit-line to Bit-line Short	'0'	Bit-line Check
Bit-line to Vss-line Short	'0'	Bit-line Check
Vss-line Cut Off	'1'	Fail-safe
Word-line Non-select	'1'	Word-line Check
Word-line Double-select	'1'/'0'	Word-line Check
Word-line to Bit-line Short	'1'/'0'	Bit-line Check
Cell Transistor Through	'0'	Bit-line Check
Cell Transistor Cut Off	'1'	Fail-safe

The sequencer (SEQ) is also composed of duplicated blocks, each of which can run independently on sequences specified by a common micro-program, as shown in Fig. 6. In the testing procedure, normality in 16-way test sequences is checked by a parity checker and a comparator in each block, and non-defective one of the two blocks is selected automatically. The micro-instruction register (MIR) is also tested and reconfigured through this procedure.

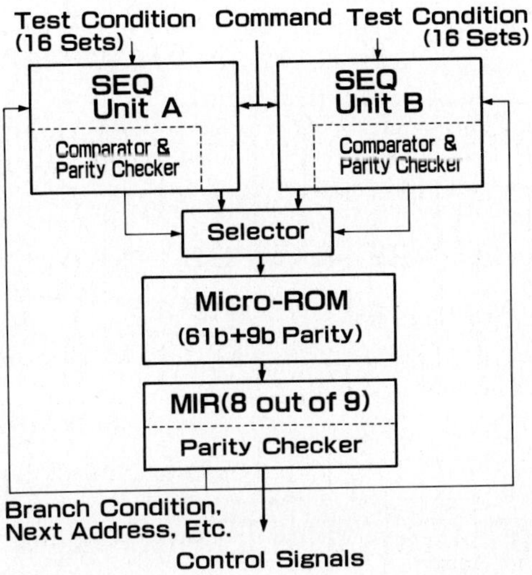

FIGURE 6 Testing and Selection of Duplicated Sequencer

5.3 Built-in Testing and Reconfiguration of Dictionary RAM

The dictionary RAM is composed of four basic units and a spare unit with 8K-bit net capacity, as shown in Fig. 7. Each of the units is of a 6-transistor static type, and has both spare bit-line blocks and spare word-line blocks, as shown in Figs. 8a and 8b. The redundancy configuration is "16 out of 17" for bit-line blocks, and "8 out of 9" for every quarter of word-line blocks.

Testing and reconfiguration are achieved in the following three stages. In the first stage, defects which cause short or open of bit-lines are checked by a read-out of test data from special ROM cells, and defective bit-line blocks are then replaced by spares.

In the second stage, RAM cells in every word-line block are tested on the basis of writing and reading procedures, and defective word-line blocks are replaced by spares. Here, the test data is prepared in the ALU, and checking is also achieved via the ALU. Defects which cause double selection or non-selection of word-lines are checked out with the

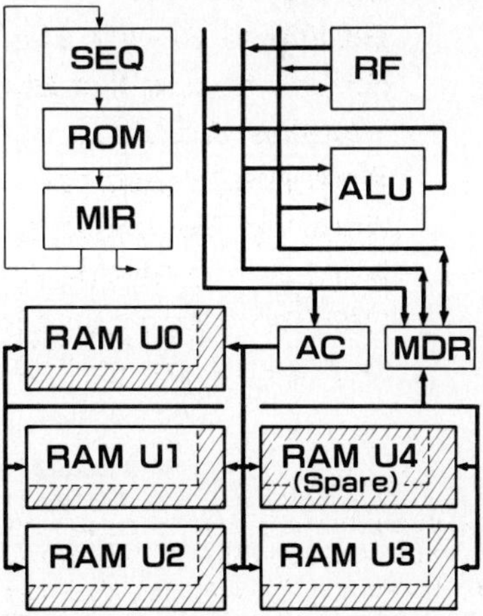

FIGURE 7 Dictionary RAM Test Circuit

help of the word-line checkers composed of special ROM cells and an exclusive-OR circuit. Replacements on the word-line block level and the RAM unit level are achieved at the end of second stage testing.

Finally, the reconfigured RAM is tested, also on the basis of writing and reading procedures, to indicate whether or not the RAM is complete.

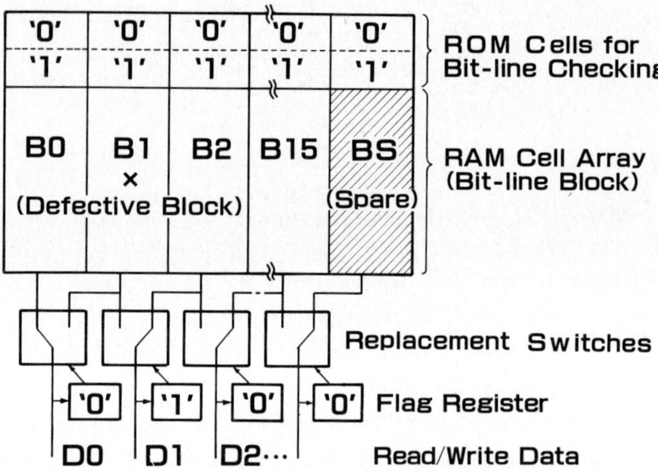

FIGURE 8a Redundancy Configuration of RAM Unit Bit-line Blocks

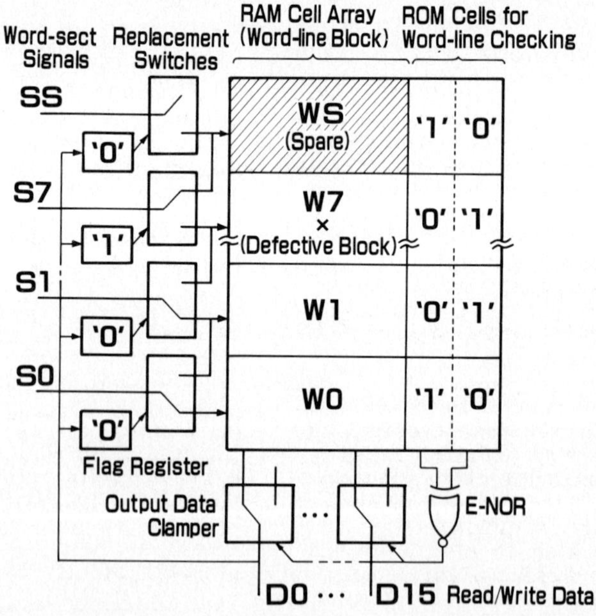

FIGURE 8b Redundancy Configuration of RAM Unit Word-line Blocks

6. YIELD ESTIMATION BASED ON EFFECTIVE ACTIVE AREA CONSIDERATION

6.1 Computation of Effective Active Areas Reduced by Redundancy

The yield enhancement effect resulting from redundancy can be estimated by use of an effective active area consideration induced from reverse calculation of the yield equation, assuming a single equivalent area, as follows.

Yield of an unit circuit (Yunt), implemented in active area A and fabricated under defect density D conditions, is given by the following equation, based on Poisson distribution.

$$\mathrm{Yunt} = \mathrm{EXP}\,(-A * D)$$

The yield of a circuit (Ycir) employing "K out of K+R" redundancy and using unit circuits is given by the following combination function. Here, Ycor is yield of a core part including switches for replacements.

$$\mathrm{Ycir} = \mathrm{Ycor} * \left(\sum_{i=K}^{K+R} {}_{(K+R)}C_i * \mathrm{Yunt}\hat{\ }i * (1-\mathrm{Yunt})\hat{\ }(K+R-i) \right)$$

Yield of an LSI chip (Ylsi) which employs a complicated redundancy configuration is given as a function of (Ycir)s.

$$\mathrm{Ylsi} = F\,(\mathrm{Ycir1}, \mathrm{Ycir2}, \cdots)$$

Assuming that Ylsi is given by a single circuit on effective active area (Aeff), Aeff can be induced by reverse calculation.

- **Effective Active Area** : $\mathrm{Aeff} = (-1/D) * \mathrm{LOG}\ e(\mathrm{Ylsi})$

Here, Aeff is not so strictly related to defect density D, since the term (LOG e(Ylsi)) is approximately a linear function of D in a nearly-fixed D condition. Hence, it can be used as an absolute performance factor which indicates whether the LSI chip employing redundancy is feasible or not in comparison to general chip size without redundancy.

By using Aeff, the following area Reduction Ratio (RR) and Figure of Merit (FM) can be defined as relative performance factors of an applied redundancy configuration.

- **Area Reduction Ratio** : $\mathrm{RR} = \mathrm{Aeff}/\mathrm{Abasic}$
- **Figure of Merit** : $\mathrm{FM} = \mathrm{RR} * ((\mathrm{Aadd} + \mathrm{Abasic}) / \mathrm{Abasic})$

Here, Abasic is the active area of the basic circuit, and Aadd is the one of the additional redundancy circuits. Area Reduction Ratio (RR) represents efficiency of redundancy directly; while, the Figure of Merit (FM) represents the balance of the area Reduction Ratio (RR) and the area Increase Ratio of redundancy.

6.2 Yield Enhancement Effect of the Processing Element

Concerning the prototype PE chip employing the 3-micron design rule, a 163-square-millimeter Abasic can be reduced to a 54-square-millimeter Aeff by an 87-square-millimeter Aadd for redundancy, as shown in Fig. 9.

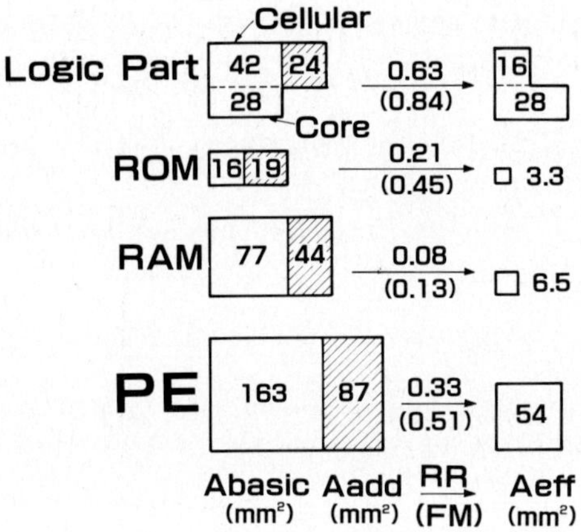

FIGURE 9 Effective Active area Reduction of Prototype PE and its Components (3-micron CMOS)

The area Reduction Ratio (RR) and Figure of Merit (FM) are 0.33 and 0.51, respectively. The computation of these factors is achieved employing a standard D condition wherein about 30 % yield can be attained in case of 50-square-millimeter non-redundancy chips. Hence, the factors mean that a prototype PE chip implemented on a large area chip can be fabricated with an adequate yield similar to that of general small-size chips.

The area Reduction Ratio (RR) of the logic part is 0.63, and those of the ROM and RAM are 0.21 and 0.08, respectively. These values are dependent on the block size for the replacements, total spare amounts, and coverage of the redundancy in each component. The values of the ROM and RAM indicate that the areas of the cell-arrays can be reduced sufficiently. However, that of the logic part is rather heavy, due to the existence of a core part.

The trend of the RRs employing 3 to 0.3-micron design rules are shown in Fig. 10. The values are almost constant except for the RAM, since the areas of the core part and the additional switches are dominant. At the same time, the RAM perfectly covered by 2-level hierarchical redundancy has a potential with a considerably higher grade of effective active area reduction followed by block size reduction in the sub-micron region.

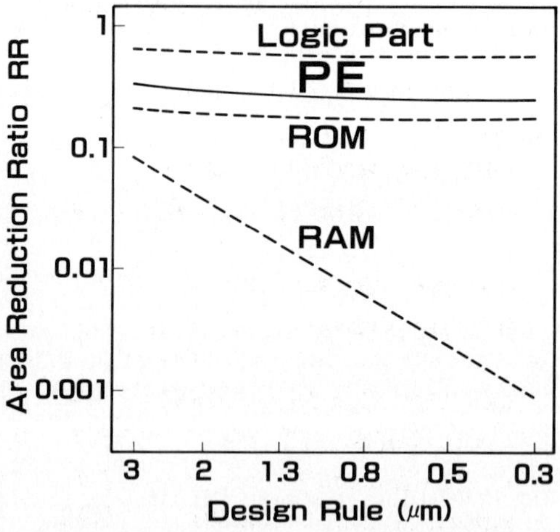

FIGURE 10 Area Reduction Ratio (RR) Estimation of PE and its Components under Finer Design Rules

7. PERFORMANCE ESTIMATION OF THE ULSI/WSI POLY-PROCESSOR SYSTEM

The feasibility of future ULSI/WSI poly-processor systems employing hierarchical redundancy can be estimated on the basis of the evaluated Aeff of the PE. In general, reduction of the design rule has the advantage of integration scale expansion. Furthermore, the employment of hierarchical redundancy makes it possible to integrate a larger number of PEs than in the case of non-redundancy. The evaluated integration scales of on-chip PEs under 3 to 0.3-micron design rules are plotted in Fig. 11.

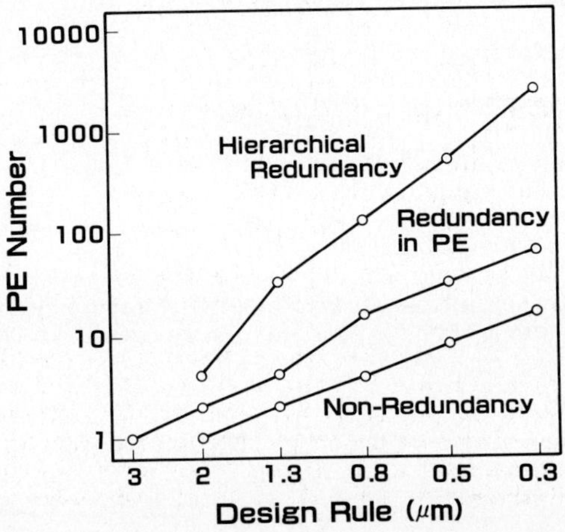

FIGURE 11 Integration Scale Trend Based on Prototype PE Model under Finer Design Rules

TABLE 6 Feasibility Estimation for ULSI/WSI Poly-processor System Based on Effective Active Area Consideration

Design Rule :	3μm	2μm	1.3μm	0.8μm	0.5μm	0.3μm
PE Size[mm²] :	380	171	72	27	11	4
Aeff of PE[mm²] :	54	21.5	8.7	3.2	1.2	0.4
2nd lev. Conf. :	x	4–6	4–5	8–10	16–18	32–34
3rd lev. Conf. :	x	x	8–10	16–18	32–36	64–68
On Chip Net PEs :	1PE	4PE	32PE	128PE	512PE	2048PE
Total Aeff[mm²] :		53.7	54.7	30.4	24.7	34.9
Chip Size[cm²] :		12	43	59	82	105
Spare Amount (Aadd) :		56%	58%	54%	48%	42%

The possible PE numbers are calculated under conditions wherein an Aeff of about 50-square-millimeters or less can be obtained by, at most, a doubling of the area by spares (50 % spares), as detailed in Table 6. This evaluation indicates that integration 30 to 100 times as great will be possible by employment of hierarchical redundancy, in comparison to the non-redundancy cases, in the sub-micron region.

A future ULSI/WSI poly-processor model employing the 0.3-micron design rule is illustrated in Fig. 12. The chip with at least 10 GOPS(giga-operations per second) of processing capability by net 2048 PEs will be possible in a 100 square-centimeter area.

It is estimated that this area can be reduced to 35-square-millimeters Aeff via 3-level hierarchical redundancy. Almost all of this area is occupied by interfaces and switches which are final core areas.

Here, the applied 3-level hierarchical redundancy takes on the following area reduction effects at each level.

* First level (in PE): Reduction of each PE area.
* Second level (in element group): Reduction of core areas in PEs.
* Third level (in set of element groups): Reduction of bus areas. Groups with power-line defects can be replaced at this level.

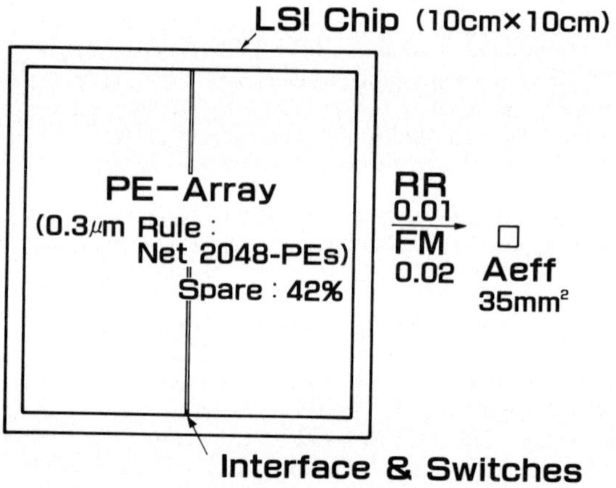

FIGURE 12 A Future ULSI/WSI Poly-processor Chip Model Employing 0.3-micron Design Rule

8. CONCLUSION

The development of the prototype 16-bit processing element chip and yield estimation based on newly proposed effective active area consideration make it clear that defect tolerance technology based on hierarchical redundancy will have a great advantage in integration scale expansion.

Estimates indicate that the 380-square-millimeter chip designed which employs 3-micron design rules has an effective active area equivalent to that of general small-size non-redundancy chip.

This effect still has an advantage in the sub-micron region, in that a ULSI/WSI poly-processor system on a scale 30 to 100 times larger than that of a non-redundancy chip may be attained by employment of 3-level hierarchical redundancy.

The expansion of integration scales in this way allows for a 50 to 70 % reduction in hardware costs and an 80 to 90 % reduction in hardware size. The developed technology can be applied to current or future ULSI/WSI poly-processor systems of various configuration.

ACKNOWLEDGMENTS

This study was conducted at the NTT Musashino Electrical Communications Laboratories in Tokyo. The authors would like to thank Tadamichi Kawada, Yukio Takahashi and Noboru Hagiwara for prototype chip design. They are also grateful to Tsutomu Wada, Nobuhiko Kakuda and Noboru Naitoh for their assistance with fabrication techniques.

REFERENCES

[1] Egawa,Y., Tsuda,N., Masuda,K.,"A 1-Mbit Full Wafer MOS RAM," 1979 IEEE ISSCC Digest of Technical Papers, Feb.1979, pp.18-19.

[2] Egawa,Y.,Wada,T.,Ohmori,N.,Tsuda,N., Masuda K.,"A- 1Mbit Full-Wafer MOS RAM",IEEE Transactions on Electron Devices,Aug.1980, pp.1612-1620

[3] Satoh,T.and Tsuda,N."A New Sorter Configuration using Linear Array Structure",29th National Convention of IPC of JAPAN, 4F-1, 1984, pp.769-770

[4] Satoh,T.and Tsuda,N.,"Two-level Self-testing and Redundancy for a Sort Array",National Convention of IECE of JAPAN, No.408, Part 2, 1985, p.126.

[5] Kawada,T., Takahashi,Y., Tsuda,N., Waki,M., Hagiwara,N.,"A Pattern Matching Processor Array with Defect Tolerance," 1986 IEEE ISSCC Digest of Technical Papers, Feb.1986, pp.90-91.

[6] Tsuda,N., Satoh,T., Kawada,T.,"A Pipeline Sorting Chip" 1987 IEEE ISSCC Digest of Technical Papers, Feb.1987, pp.270-271.

[7] Tsuda,N. and Satoh,T., "Hierarchical Redundancy for a Pipeline Sorting Chip," Proceedings of the Second IFIP WG 10.5 Workshop on Wafer Scale Integration, Egham, England, Sep.1987. Wafer Scale Integration II (R. M. Lea, Eds), North-Holland, 1988, pp.63-74.

Wafer Scale Integration, III
M. Sami and F. Distante (Editors)
Elsevier Science Publishers B.V. (North-Holland)

A Methodology for the Design of Fault-Tolerant, Testable, Wafer-Scale Processor Arrays Using Boundary-Scan Architecture

Najmi Jarwala and Yervant Zorian

AT&T Bell Laboratories
Princeton, New Jersey
USA

Wafer-Scale Technology is the ultimate goal of large area VLSI. It offers higher functionality, performance, reliability and lower cost. However the two main problems that need to be solved before this technology becomes commercially feasible are Defect/Fault-Tolerance and Testability.

This paper proposes a unified solution to these two problems. The IEEE P1149.1 Boundary-Scan Architecture is used as the basis for designing wafer-scale processor arrays that are fault-tolerant and testable. Built-in Self-Test and Boundary-Scan are used to solve the test problems. The Test Access Port defined in this architecture is used to provide 'soft' restructuring using electronic switching as opposed to physical restructuring. This makes it possible to configure the wafer both during manufacture and in the field.

1.0 Introduction

Wafer-Scale technology has long been pursued to achieve lower cost, higher performance, and higher reliability. Until recently, its high cost, lack of demand for high levels of integration, and finally advances in technology made it uneconomical and unnecessary. Today, both the slowing in the development of fabrication technology as well as demands for higher levels of integration have renewed interest in WSI. For some systems, like memories and processor arrays, new manufacturing technologies, new designs, and new interconnect technologies have reduced the cost per function below that of less integrated circuits, making them economically viable.

Several problems remain to be solved before WSI systems can become commercially feasible, mainly yield, testability and power dissipation. Advances in packaging technology have contained the power dissipation problem. This paper will address two other problems: Defect/Fault-Tolerance and Testability.

1.1 Defect/Fault-Tolerance

With current fabrication technology, it is almost impossible to manufacture a defect free wafer. To permit WSI products to have acceptable yields, considerable research has been directed towards redundant processor switch architectures and algorithms that permit the faulty processors to be switched out of the circuit and maximal number of fault free processors to be configured into a useful computational structure like a linear array, mesh, tree, etc. There are several restructuring techniques proposed:

- **Discretionary Wiring** where the individual Processing Elements (PE) and some bus structures are initially fabricated and tested. The final interconnect is by one or more metallization layers where only the fault free PEs are wired together.
- **Electrical Fuses** where polysilicon links are blown by controlled high current pulses so that certain portions of the circuit are disabled.
- **Laser Cutting and Welding** where short laser pulses are used to disconnect or connect circuit elements.
- **Ion Beam Assisted Reactions** where a laser is used to 'write' or etch a conducting line on a substrate.
- **Ion Beam Milling** where a ion-beam is used to cut a via between two metal layers and then metal is sputtered at the step to connect the two layers.
- **EPROM-like Switches** where an E-beam is used to inject a charge in a floating gate and UV light used to erase it. Thus a pass transistor can be used as a switch (although a switch with a high resistance - of the order of several kilo ohms) to control a signal path.

An excellent survey of these and other techniques is presented in [Te89]. All these techniques have some disadvantages. Discretionary wiring has an added impact on the yield and the additional processing steps may introduce defects in previously fault free PEs. Electrical fuses are complex and there is the problem of getting a high current pulse to the right location. Laser programming has a problem (shared by electrical fuses) that it creates an intense thermal disturbance in the vicinity of the fuse. This can impact adversely on the reliability of the circuit. Another potential disadvantage is that physical restructuring may not be effective in accommodating infant mortality failures [Te89]. Finally a significant disadvantage of all these techniques is that they represent hard programming. The circuit can be configured once, during manufacture. If a failure occurs in real time, the residual redundancy cannot be used to tolerate this failure and consequently enhance reliability. This is especially important for WSI because a wafer is likely to represent a large fraction of a systems logic and it is unlikely that its failure would be tolerated.

1.2 Testability

Even if a successful restructuring technique is chosen, there is still a problem of identifying the faulty PEs and interconnect. This problem also exists during operational test. One approach for manufacture test is to fabricate pads on the I/O of each PE so that they can be accessed by wafer probes. One of the primary motivation for WSI is high performance and these pads would load the I/O lines and might result in unacceptable performance penalties. The complexity, maintainability, etc. of the wafer probe is highly questionable. This problem has not been addressed by most of the processor/switch architecture proposals. This paper addresses these two problems. Us-

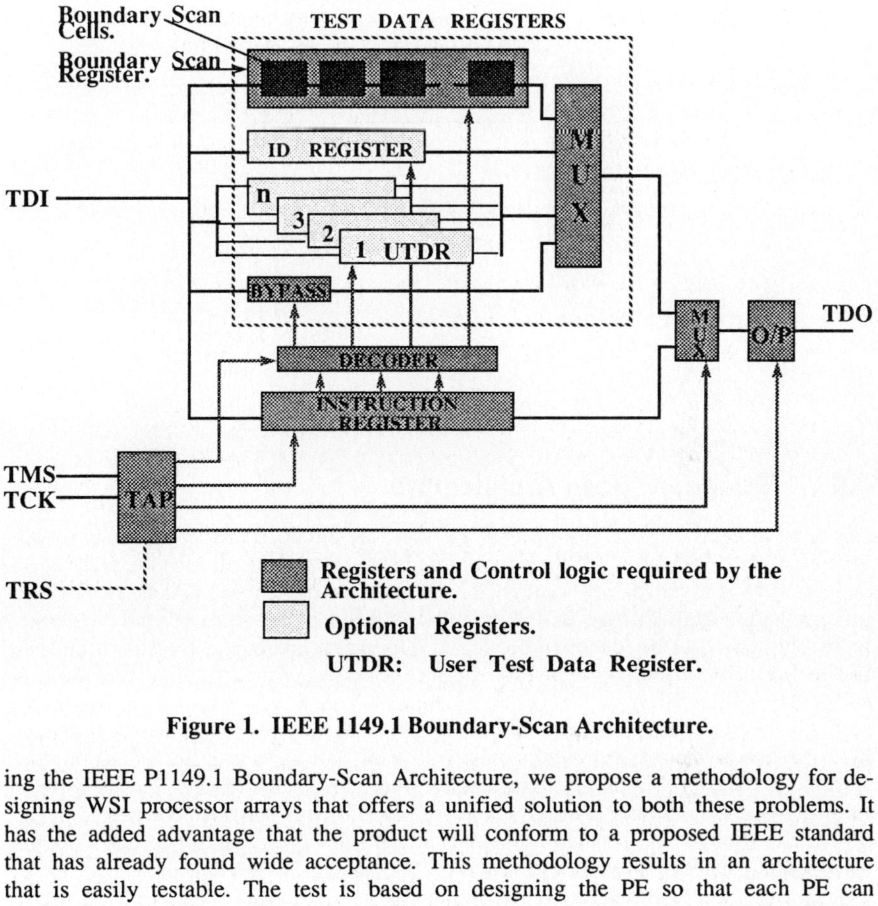

Figure 1. IEEE 1149.1 Boundary-Scan Architecture.

ing the IEEE P1149.1 Boundary-Scan Architecture, we propose a methodology for designing WSI processor arrays that offers a unified solution to both these problems. It has the added advantage that the product will conform to a proposed IEEE standard that has already found wide acceptance. This methodology results in an architecture that is easily testable. The test is based on designing the PE so that each PE can test itself: the Built-in Self-Test (BIST) philosophy. Boundary-Scan (B-S) makes it possible to access and control each PE. It is also used to verify the interconnect between PEs. Testing can be accomplished with ease, both during manufacture and in the field. Additionally, the Test Access Port (TAP) of the B-S is used for 'soft restructuring', that is the PEs can be restructured during manufacture and in the field. This can be done safely, without degrading reliability as the proposed technique has no thermal dissipation.

Section 2 introduces the B-S Architecture. Section 3 outlines the architectural design methodology that is proposed. This methodology is demonstrated in the next two sections. Section 4 describes in detail the design of a PE with BIST and Section 5 discusses restructuring and demonstrates the concepts by applying it to two processor switch array examples. Section 6 discusses the test and reconfiguration process. Finally Section 7 discusses future research areas and offers conclusions.

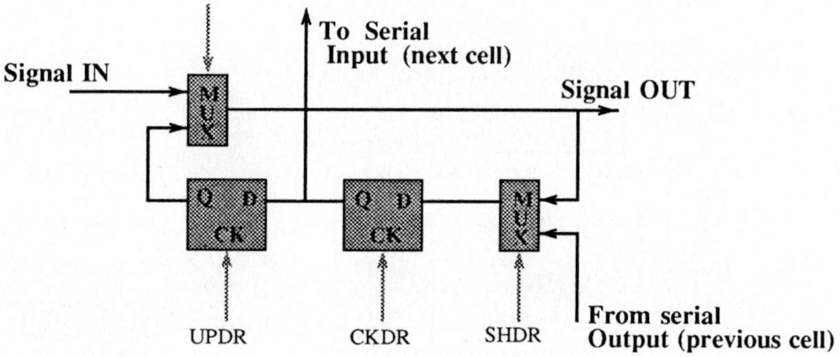

Figure 2. B-S cell permitting controllability and observability.

2.0 Boundary-Scan Architecture

The test problems that face future WSI are faced by circuit boards today. The growing density of circuit boards, surface-mount technology and double-sided circuit boards make it very difficult, if not impossible, to test them by the traditional in-circuit-test, using a bed-of-nails fixture. Just as faulty PEs need to be accessed and identified, faulty chips need to be identified for repair. The board interconnect needs to be tested just as the interconnect between PEs need to be tested. To solve these problems, the Joint Test Action Group (JTAG) has proposed a B-S Architecture. This architecture basically consists of a Test Access Port (TAP) which consists of a four signal interface, a controller, an instruction register, and two or more test data registers (Figure 1). One of these test data registers is the B-S register. This register is formed by serially linking latches (which are part of a B-S cell) that are placed at each device I/O (Figures 2) so that the signals at the I/O can be controlled and observed. This serial path into each device allows us to test the interconnect, apply tests to each device, access BIST and other testability features within devices, and permits us to take ''snapshots'' of the system state in real time [JTAG88].

The four signal interface consists of serial test data path formed by Test Data In (TDI) and Test Data Out (TDO), a control signal - Test Mode Select (TMS) and a Test Clock (TCK). The TAP is a sixteen state finite state machine that controls the operation of the B-S architecture. Figure 3 shows the state transition diagram. The value next to each arc represents the value applied to the TMS line. The state is determined by the sequence applied to the TMS control line. By applying a proper sequence, the data at TDI can either be scanned in to the Instruction Register (IR), or a Test Data Register (TDR) defined by the instruction in the IR. At the same time the selected registers contents are shifted out at TDO. These activities are synchronized to TCK. The TDR can be the required B-S or bypass register, or one of the optional User Test Data Registers (UTDR). The TAP also has a state (Run-Test/Idle) in which BIST can run.

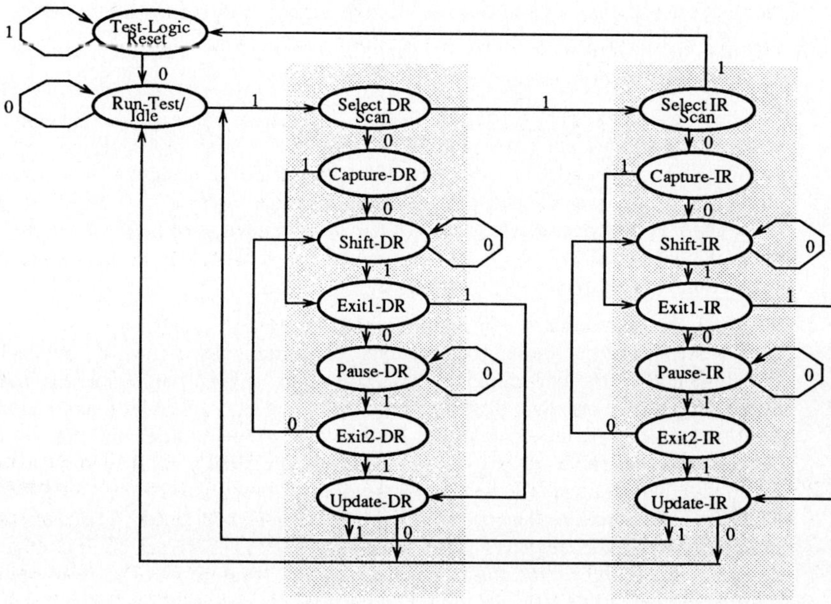

Figure 3. TAP State Transitions

The instruction loaded into the IR not only determines the TDR selected but also the mode. The standard has several predefined instructions. EXTEST selects the B-S register in its external test configuration, which permits interconnect test. The INTEST instruction selects the B-S register in the internal test configuration which allows tests to be applied to the device. The SAMPLE instruction is the normal operating mode of the system logic, and permits the system state to be captured in real time. The BYPASS instruction selects the one-bit bypass register, which allows the user to 'short-circuit' a long B-S chain and access a particular device on a chain, quickly. If BIST is provided then the RUNBIST instruction executes BIST and the response can be scanned out. Besides these instructions, others can be defined.

Consequently, the TAP not only controls the B-S and other BIST logic but can be designed so as to provide the ability to serially read and write into *any* register within the device. This provides a very powerful capability which can be used for applications other than test. We propose to use this capability to provide soft restructuring for wafer-scale architectures.

3.0 Design Methodology

The design methodology at the architectural level is simple. It consists of two main parts: designing the PE with BIST/B-S and designing the restructuring logic. The PE is designed based on the functionality of the processor array. Each PE is then made self-testing. The recommended procedure is to use the divide and conquer methodology described in [ZoJa89], and can be summarized by the following steps:

1. Identify the different modules in the PE and classify into various regular structures and random logic.
2. Develop or select and implement a BIST algorithm for each of the regular structures and the control logic. This stage is crucial as there are many cost/performance trade-off to be made. The BIST algorithms for regular structures must provide the target fault-coverage (typically 99+%) at an acceptable area/performance penalty. One of several approaches may have to be evaluated. Evaluate the fault-coverage of the control logic BIST (by fault-simulation if random pattern test approach).
3. Analyze the requirements for control and access of BIST in each of the subsystems. Reduce these requirements to initializing a set of registers, initiating BIST, and collecting the responses into a single signature that must be no larger than the B-S Register (refer Section 4 and [JTAG88] for details). At this stage a detailed analysis is required to determine the degree of concurrence desirable during BIST. This is principally limited by power dissipation constraints. The current definition of the IEEE P1149.1 defines a RUNBIST instruction that is required to be self contained. Therefore all initialization of the registers used during RUNBIST and scheduling of the different BIST algorithms, must be done by the local logic. The final result must be reduced to a single signature that must reside in the B-S Register.

Implement the IEEE P1149.1 Boundary-Scan architecture. Design the IR so that the required number of instructions to control BIST (determined in step 3) and the restructuring functions (Section 5) are available. Place the register set identified above as the User Test Data Registers (UTDR's).

The second part is the design of the restructuring logic. This part is strongly dependent on the host architecture, the target architecture(s) that are required in real time, the data flow, restructuring algorithms, and the degree of redundancy. Section 5 describes the general approach with some examples. The next section describes in detail the design of a PE and the BIST techniques that are applied to make it self-testing.

4.0 Processor Element Design with BIST

The Built-In Self-Test concept, described above, is used to test each PE. This implies adopting a general testing strategy for PEs, developing specific BIST techniques, and adding extra circuitry in each PE for the purpose of self-test. These issues will be discussed in sections 4.1 and 4.2.

4.1 The Test Strategy for PE's:

The complexity of testing PE's has increased with increasing functionality. [Ro83].

The test strategy proposed reduces this complexity by a divide and conquer approach as the basis for enhancing the controllability and observability of each PE, and then adopting specific self-test techniques to provide autonomous testing of each module. The proposed test strategy is general enough to be adopted for any PE.

The divide and conquer approach partitions a PE into structural modules, and identifying the type of each module (Figure 4). For instance, a module would be an embedded memory of type RAM, ROM, Register File, etc., or a data path module of type ALU, shifter, multiplier, multiplexer, etc., or it would be an internal bus or control logic. Once such modules are identified, a specific BIST technique, which provides high fault coverage, is selected or developed for each type of module. Techniques based on deterministic tests used to test the modules that have regular structures. Test-specific compression techniques [AgZo87] are used to eliminate or drastically reduce the information loss due to output data compression. The implementation of such techniques is described in the following section.

4.2 BIST Techniques for PE Testing:

A common BIST technique cannot be used for the entire PE because it consists of different types of modules with different fault models, and densities. For example, embedded memory modules like RAMs, ROMs, or PLAs have extended fault models and are much denser than other types of modules. Therefore, in order to provide high fault coverage, specific BIST techniques are developed to test each one of them.

Functional and structural regularity is exploited in developing BIST techniques for memory like and other data path modules. Deterministic test algorithms are known for RAMs [DeBe89] [JaSt87] [Ma82], PLAs [LiMc87] [TrFu85], Multipliers [Ho88], Multiplexers [MaMc88], ALUs [CeAb88] and various other modules with regular and semi-regular structures. Similar deterministic algorithms can be developed for regular modules such as CAMs, FIFOs, ROMs, Counters and Shifters.

Due to the regularity of the cells, the test vectors and their responses are very simple and regular. For instance, the set of data vectors needed to test a RAM or a Register File, in general, consist of all zeroes, all ones, walking zeroes and walking ones, and the set of address vectors that need to be applied consist of the incrementing and the decrementing binary vectors. The hardware needed to generate the specific set of vectors can be realized by reconfiguring the existing registers, so that under the BIST mode the external inputs will be blocked and the input registers become test vector generators. Using the same RAM example, it can be shown that the input data register must be switched to a rotating shift register, to generate the first set of vectors, whereas the second set of vectors will be generated by reconfiguring the address register to an up/down counter, during BIST mode. The set of response vectors are as regular as the input data vectors. Therefore, the output response verification blocks also consist of simple hardware structures. The advantage of using output data specific compression is significant in eliminating or at least drastically reducing the aliasing problem. Aliasing [ZoAg86] is caused by information loss due to output data compression. However, since the response vectors of the regular structure modules are regular, it is possible to design output data compression blocks that have either no-loss or minimal loss of information after compression [ZoJa89]. In fact, the reconfigurable registers and some combinational gates can provide that capability [ZoJa89]. Thus, in general the reconfigurable registers (R. Registers), Figure (4), are used as building blocks to implement the specific BIST techniques selected for regular structure modules.

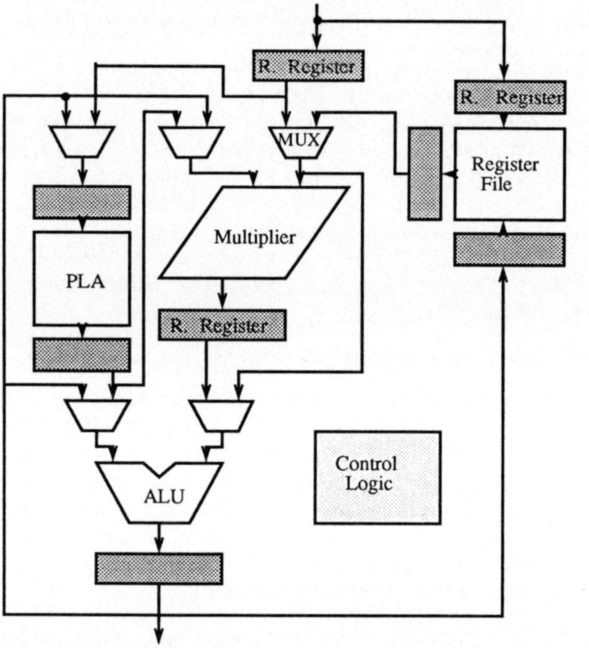

Figure 4. Implementing BIST in a PE

Control logic lacks regularity of function and design. Hence, a generic BIST technique for sequential random logic needs to be implemented. The circular self-test path [PiKr88] with partial scan [PrOb88] can be considered for self-testing the limited amount of random logic that exits in processors. This technique, which is based on pseudo-random test approach, has the capability of providing a trade-off between the hardware overhead and the fault coverage desired.

Finally, the problem of testing internal buses in a PE becomes easy, once all other modules are BISTed. The registers on the input/output of such buses will already consist of reconfigurable registers that may be used to apply tests to the busses and evaluate their response.

5.0 Restructure Logic

5.1 Tradeoff in Physical/Electronic Restructuring

The advantages and disadvantages of physical restructuring versus electronic switching have been discussed in [Te89]. The principal disadvantages of electronic switching are: greater area overhead in implementing the switch and the control logic, speed degradation, and volatility. Wafer level integration is less sensitive to small increases in area than individual chips. Using the IEEE P1149.1 architecture, the additional area required by the control logic is now distributed over another critical function - testability. Performance may not necessarily degrade, especially if multiplexor based switching is used. In that case switching may be combined with signal amplification at low cost. Finally the volatility problem of electronic switching may be solved by using EEPROM technology [Te89].

The advantages of electronic switching include elimination of a sequential step in a batch manufacturing process, safe restructuring during manufacture, utilization of the residual redundancy to provide fault-tolerance during in-service failures, compatibility with aggressive VLSI technologies [Te89] and possibility of reconfiguring the target architecture efficiently during run-time.

It should be noted that a combination of physical restructuring and electronic switching may be the best approach for some systems. For example, power and ground networks cannot be switched electronically.

5.2 Electronic Restructuring using IEEE P1149.1

The best approach for designing the restructuring logic is a function of the PE architecture, the host array architecture, the target architecture(s), the degree of redundancy, the control/data flow, and the reconfiguration algorithm. However using the IEEE P1149.1 architecture to access and control the restructuring mechanism imposes some constraints. First, this methodology is most suited to multiplexor based reconfiguration. Secondly, since there is logic restructuring, global replacement or replacement over a large neighborhood may be difficult. Therefore this technique is most suitable for reconfiguration over a constrained neighborhood

There are two principal elements in reconfiguring an array: eliminating the faulty PE and sometimes some fault-free PEs, and establishing a communication network between selected fault-free PEs.

5.2.1 Eliminating Processor Elements

The most common technique to eliminate faulty or fault-free PEs is to transform them into communication links. Consider a processor array designed for matrix-vector multiplication. There are many important problems, like convolution, filter, and discrete fourier transform, that can formulated as a matrix-vector multiplication problem. The host architecture is a linear array and the target architecture is a linear array [MeCo80]. This makes reconfiguration very simple; after the PE's have been tested, the faulty PE's are configured into communication links. Other methodologies [Ko81], [NeSa85], [GeSa84] also require this function.

This function can be implemented easily with multiplexor based reconfiguration, using the TAP to control the multiplexor. Figure 5 shows a simple structure for achieving this function. The multiplexors route the signals into the computation logic or bypass it. The multiplexors are controlled using the UTDR which is accessed via the TAP. The output of the UTDR could set EEPROM switches to eliminate volatility. The approach shown in Figure 5 is suited for systolic applications, but the concept can be easily extended. If the data path is *b* bits wide, then as many as 4*b* laser cutting/welding operations would be required. Also the multiplexors can include register stages so that, in a pipelined operation, there is no performance penalty beyond the stage delay seen by the first result. Alternately the multiplexors can include buffers to minimize signal propagation delay across the PE..

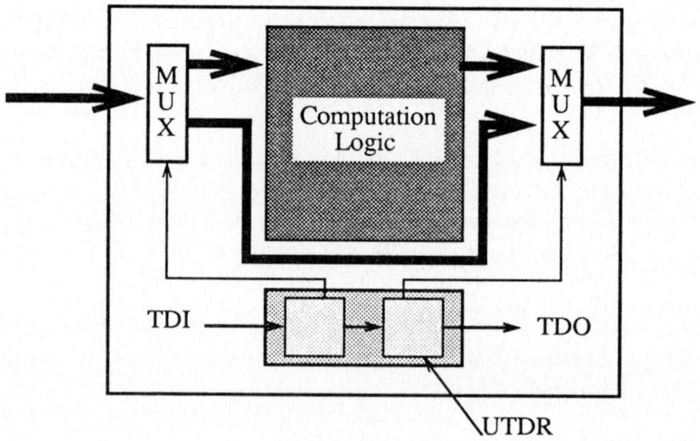

Figure 5. A PE designed to transform to a communication link.

5.2.2 Establishing the Communication Network

Here too the principle remains the same. The TAP is used to access a UTDR which in turn controls multiplexors that route the signals appropriately. As discussed before, this methodology is applicable for constrained neighborhood replacement. There are several algorithms that satisfy this constraint, because in practice, even with physical restructuring, the interconnect cost can become excessive with large neighborhoods. Two classes of algorithms especially suited for this approach are the Diogenes [Ro83] and the Self-Reconfiguration Algorithms reported from the Politecnico di Milano [NeSt86], [NeSa85], [GeSa84]. (For an excellent summary of these algorithms, refer to [Te89].) The second algorithm will be considered in some detail. This algorithm is especially suitable because each PE has a clearly defined **horizontal neighborhood** and **vertical neighborhood** that is constrained. Data input is from the top and left of each PE and data output is from the right and bottom, representing a typical systolic processor structure [NeSt86]. The original algorithm is dynamic and *assumes that all cells have been identified as faulty or functional and that they know their state*. Reconfiguration begins from the input edges of the array, by requests to prolong rows and columns. A functional cell requests from each member of its horizontal neighborhood set, a response indicating whether the contacted cell can accept the requesting cells horizontal connection. The contacted cell responds whether or not it can accept the connection request by executing an algorithm that involves propagating the request further into the array [NeSt86]. A highly dynamic sequence of events is initiated, with row and column prolong request propagating along the array, till the process relaxes to a stable configuration. Figure 5 shows the basic PE configuration for executing the above algorithm. The reconfiguration is multiplexor based and has a 3×3 neighborhood [NeSt86]. If the B-S architecture is used then each PE has an additional six bit UTDR to control the four multiplexors that it interacts with. During manufacture, after BIST has identified the faulty PEs, the algorithm can be run on an *external* host and a single reconfiguration vector can be computed that will achieve exactly the same reconfigured array, as the algorithm running dynamically. This can save a considerable amount of time and provides the option of using dynamic reconfiguration in-service. Also, for a given defect pattern, an external host would be able to run more computationally intensive algorithms, to maximize the harvest. .

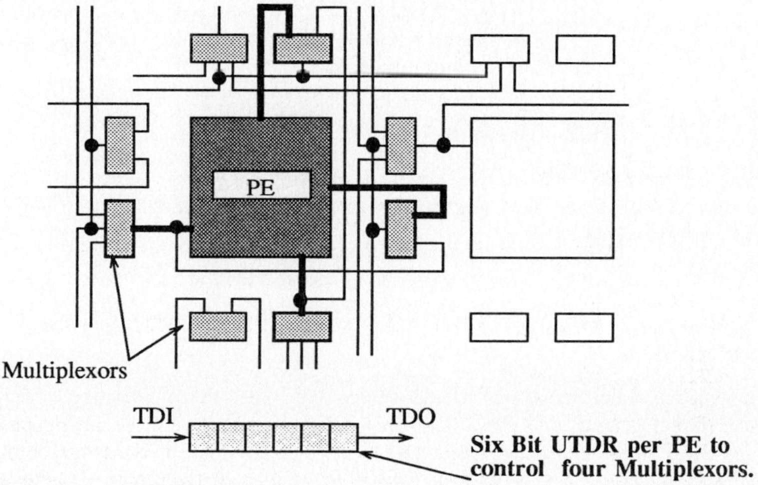

Figure 6. PE multiplexor organization for a 3×3 neighborhood

6.0 Test and Configuration

6.1 Testing the Processor Array

Testing the processor array involves three steps:

1. Verifying the Integrity of the B-S Chain.
2. Testing the Interconnect.
3. Activating BIST and scanning out the resultant signatures.

Testing the integrity of the B-S architecture is supported by the standard [TuYa89], and is summarized here for completeness. The most two significant bits of the IR are initialized to a constant binary '10' pattern during the Capture-IR state. Then by moving to the Shift-IR state and remaining there, the constant patterns loaded into the most two significant bits of the IRs in each PE will be seen at the wafer's serial output. For example, for a wafer containing four PE's each having a four bit instruction register, the pattern at the serial output for a fault-free wafer would be:

10XX10XX10XX10XX

⟵

Shift direction

In the event of a fault, the correct pattern from several components will be observed, followed by faulty data. The location of the faulty data in the pattern identifies the first failing PE. For example, an open in the serial path between PE 2 and 3 would result in the following pattern:

10XX10XX11111111

←

Shift direction

The pull-up on TDI of PE 2 would cause a constant high to be shifted in.

Testing the interconnect requires a sequence of tests that will detect and locate stuck-ats, shorts and opens. Test Algorithms to efficiently accomplish this are presented in [YaJa89] and [JaYa89].

Testing the PE's is accomplished by scanning in the appropriate instructions that will execute BIST and scanning out the responses.

6.2 Configuration

The test procedure described above would detect and locate faulty PEs and interconnect. If the defect pattern, and the redundancy permit, the array can be configured to exclude the faulty subsystems and generate a functional product, as described in Section 5. The test procedure can be executed again to verify that the repair was successful

7.0 Conclusion

This paper proposes the use of BIST and the IEEE P1149.1 Boundary-Scan architecture as the methodology for unifying testability and defect/fault-tolerance in future wafer-scale products. In the long term, the advantages of using electronic switching as opposed to physical restructuring are significant and using a common approach for both testability and fault/tolerance lowers the control cost of both these critical functions. The other advantage is that the product will conform to a proposed IEEE standard which has already gained wide acceptance.

There are many open research issues. The B-S path is hard-core. If a single scan chain is used and it fails, then the entire wafer would have to be discarded. One approach is to have separate scan paths for the rows or columns. This would require synchronization of these chains during interconnect test. A defect on a scan path would then affect a single row or column and the remaining Defect avoidance techniques can also be used in its implementation. Building fault-tolerance in the scan path remains an interesting problem.

The other issue is to develop new restructuring algorithms or adapt existing ones, to account for the neighborhood constraints of using electronic switching. Many existing algorithms already have this constraint implicitly and with minor modifications, could be used.

REFERENCES

[AgZo87] V.K. Agarwal and Y. Zorian, "An Introduction to Output Data Modification", in Perspectives in Computing: Developments in Integrated Circuit Testing, ed. D.M. Miller, *Academic Press*, 1987.

[CeAb88] E. Cerny, M. Aboulhamid, G. Bois and J. Cloutier, "Built-In Self-Test of a CMOS ALU", *IEEE Design & Test of Computers*, pp. 38-48, August 1988.

[Da88] S. Davidson, "Merging BIST and Boundary-Scan at the IC Level", *Proc. National Communications Forum*, pp. 13/3 1-6, Chicago, October 1988.

[DeBe88] R. Dekker and F. Beenker, "Realistic Built-In Self-Test for Static RAMs, *IEEE Design & Test of Computers*, February 1989.

[GeSa84] G. Gentile, M.G. Sami and R. Stefanelli, "Design of Switches for Self-Reconfiguring VLSI Array Structures, *Microprocessing and Microprogramming*, Vol. 14, pp. 99-108, 1984.

[Ho88] S.J. Hong, "An Easily Testable Parallel Multiplier", *Proc. of IEEE Int'l Conference on Computer Aided Design*, pp. 214-219, November 1988.

[JaSt87] S.K. Jain and C.E. Stroud, "Built-In Self-Testing of Embedded Memory", *IEEE Design and Test of Computers*, pp. 27-37, October 1987.

[JaPr88] N. T. Jarwala and D. K. Pradhan, "TRAM: A Design Methodology for High Performance, Easily Testable, Multi-Megabit RAMs," *IEEE Transactions on Computers*, Vol. C-37, No. 10, pp 1235-1250, Oct. 1988.

[JaYa89] N. Jarwala and C.W. Yau, "A New Framework for Test Generation and Diagnosis Algorithms for Wiring Interconnect," *Proceedings, International Test Conference*, Washington D.C., August 29-31, 1998.

[JTAG88] JTAG Boundary Scan Architecture Standard Proposal, Version 2.0, Published March 1988.

[Ko81] I. Koren, "A Reconfigurable and fault-tolerant VLSI multiprocessor array", *Proc. 8th Annual. Symp. Computer Architecture*, pp. 426-441, May 1981.

[LiMc87] D.L. Liu and E.J. McCluskey, "Design of Large Embedded CMOS PLAs for Built-In Self-Test", *IEEE Transactions on Computer Aided Design*, Vol. 7, No. 1, pp. 50-58, January, 1988.

[MaMc88] S.R. Makar and E.J. McCluskey, "On The Testing of Multiplexers", *Proc. of Int'l Test Conference*, pp. 669-679, September 1988.

[Ma82] M. Marinescu, "Simple and Efficient Algorithms for Functional RAM Testing", *IEEE Test Conference*, pp. 236-239, October, 1982.

[NeSa85] R. Negrini, M.G. Sami and R. Stefanelli, "Fault-Tolerance Approaches for VLSI/WSI Arrays", *Proc. IEEE Int'l Phoenix Conference on Computers and Communications*, pp. 460-468, 1985.

[NeSt86] R. Negrini and R. Stefanelli, "Comparative Evaluation of Space-and time- redundancy approaches for *WSI processing arrays, in Wafer Scale*

Integration, G. Saucier and J. Trihle (eds), Elsevier Science Pubs., pp. 207-222, 1986.

[PiKr88] S. Pilarsky and A. Krasniewski, "Effectiveness of Fault Detection for Low-Overhead Self-Testing VLSI Circuits", *International Symp. on Circuits And Systems*, Helsinki, Finland, 1988.

[PrOB88] M.M. Pradhan, E. O'Brien, S.L. Lam and J. Beausang, "Circular BIST with Partial Scan". *Proc. IEEE International Test Conference,* pp. 719-729, Washington, September 1988.

[Ro83] A.R. Rosenberg, "The Diogenes Approach to Testable Fault-Tolerant Arrays of Processors", *IEEE Transactions on Computers*, Vol. 33, No. 10, pp. 902-910, Oct. 1983.

[Te89] S.K. Tewksbury, Wafer-Level Integrated Systems: Implementation Issues, *Kluwer Academic Publishers*, 1989.

[TrFu85] R. Treuer, Hideo Fujiwara, and V.K. Agarwal "Implementing a Built-In Self-Test PLA Design", *IEEE Design & Test of Computers*, April 1985.

[TuYa89] R.E. Tulloss, C. W. Yau, "BIST and Boundary-Scan for Board Level Test: Test Program Pseudocode," *Proceedings, European Test Conference,* April 12-14, 1989.

[ScTu88] H.N. Scholz, R.E. Tulloss, C.W. Yau and W. Wach, "ASIC Implementations of Boundary-Scan and BIST", *8th Int'l Custom Microelectronics Conference*, pp. 43.0-43.9, London, U.K., November 1988.

[Si88] A.D. Singh, "Interstitial Redundancy: An Area Efficient Fault Tolerance Scheme for Large Area VLSI Processor Arrays", *IEEE Transactions on Computers, Vol. 37,* No. 11, pp. 1398-1410, November 1988.

[YaJa89] C. W. Yau and N. Jarwala, ''A Unified Theory for Optimal Test Generation and Diagnosis Algorithms for Board Interconnects,'' *Proceedings, International Test Conference,* Washington D.C., August 29-31, 1989.

[ZoAg86] Y. Zorian and V. K. Agarwal, ''A General Scheme to Optimize Error Masking in Built-In Self-Testing,'' *Proc. of the 16th Int'l Symp. on Fault-Tolerant Computing Systems,* pp. 410-415, Vienna, Austria, July 1986.

[ZoJa89] Y. Zorian and N. T. Jarwala, ''Designing Defect/Fault-Tolerant, Testable Verifiable, VLSI Processors using the JTAG/IEEE 1149.1 Boundary-Scan Architecture.,'' To appear, *Proceedings International Conference on Computer Design, Boston, October 1989.*

Wafer Scale Integration, III
M. Sami and F. Distante (Editors)
Elsevier Science Publishers B.V. (North-Holland)

EVALUATION AND COMPARISON OF SELECTED WSI RECONFIGURATION ARCHITECTURES IN TERMS OF YIELD AND YIELD PER AREA

MATTHIAS WESSELING *ULRICH RAMACHER*

SIEMENS AG ZFE F 1 ME 22, Otto-Hahn-Ring 6

D-8000 Munich 83, F.R. Germany

KARL GOSER

University of Dortmund, D-4600 Dortmund, F.R. Germany

ABSTRACT

The overhead of defect-tolerant circuits, such as test logic, control logic and additional datalines, increases immensely with the amount of redundancy provided by the configuration methods so far known for 2-dimensional arrays of modules. Thus, only arrays of modules with relatively little redundancy are practical. This paper suggests a method of configuration for which the overhead of each module stays small, even with rising redundancy. Furthermore, the modules and wiring have separate redundancy and they are configurated separately. Because of these two aspects a higher yield per area can be reached, especially for great word width, large modules or high defect density.

INTRODUCTION

To be able to obtain an acceptable yield with large area chips, the circuits have to be designed to be defect-tolerant. Defect tolerance is made possible by implementation of redundant devices, which can substitute the defective ones (spatial redundancy), or by redundant algorithms, where the operating modules take on the additional tasks of the defective ones (temporal redundancy). Owing to the latter the yield can be increased, but the performance depends upon the yield of a module.

In order to be able to replace the defective modules in the spatial redundancy, the spare modules must have access to the input-data of the modules that are replaced. This often leads to a great amount of wiring (Diogenes Approach [3,11]) or to a minimal rate of coverage of the spare modules (Row Exclusion Principle [12]).

Contrary to the hard configuration, for example by cutting and welding wires with a laser beam, in this paper we focus our attention on the comparison of soft-configuration schemes. They are exclusively performed by programmable switches.

We present a new approach for configuration of a redundant array of modules. For comparison we will configurate a matrix matrix multiplier, performed by this new method and two further configurations,proposed by Ramacher [6] and Sami [5]. For the latter two configurations, figure 1 shows to which of the modules the data may be transmitted. This wiring demonstrates that Ramacher can configurate only horizontally, Sami also vertically. The fact that Sami postulates self-testing modules in contrast to Ramacher, who suggests a global test, is a further difference. Thus Sami's redundancy is limited to the modules, the wiring is non-redundant. The configuration of Ramacher interlocks a module with the appropriate data-wire to a redundant unit. In this case defects in wiring can be handled.

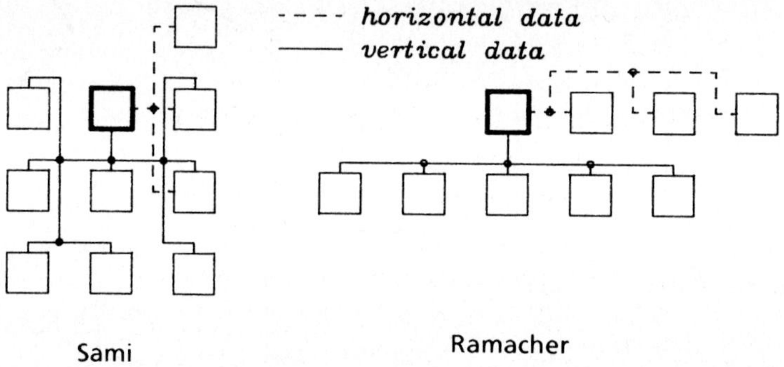

Figure 1: Data overhead for one module

In the next section we describe the architecture of our new configuration method, the flow of data and the wiring. Then we go on to study the significance of parameters, such as module area, data word width, defect density and partitioning of wiring. Finally the results will be compared with the configurations of Ramacher and of Sami, with reference to yield and area.

CONFIGURATION OF MODULES

Figure 2 shows a 4 by 4 functional array of modules each comprising, for example, a multiplier and accumulator. Each row is provided with two spare modules. The data paths are assumed to be functional, their configuration will be considered later. A module of a row is observed to be defective if the C-data path gives a wrong result. In this case, the computation is repeated by the first spare module, the second spare module handles the task of the second defective module, and so on. In order to do this, it is only necessary to choose the right B-data word and to delay the computation so that the result overwrites the C-data word delivered by the defective module. Programmable multiplexers serve to select the B-path

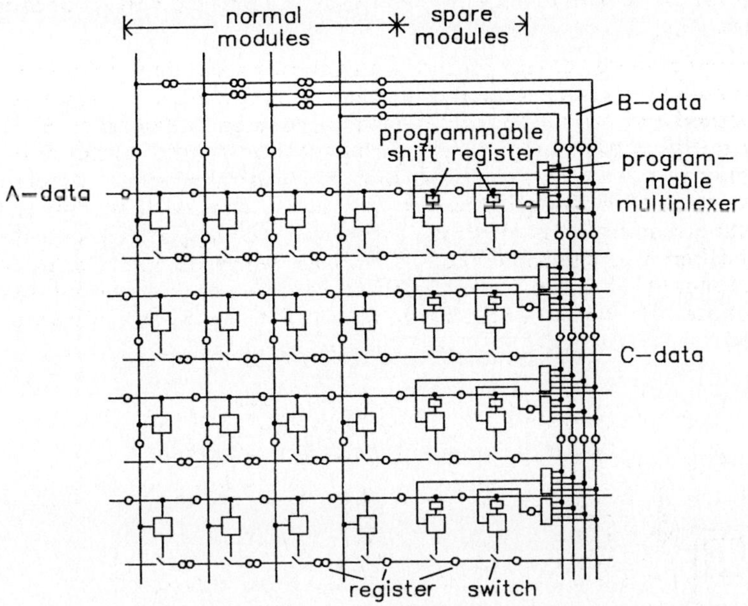

Figure 2: Configuration by selective substitution

required by the spare module. The delay of the repeated computation is performed by means of programmable shift registers for the A-path.

Note that the B-data path to the spare modules has a fixed delay and that only the defective modules are substituted by spare ones. Figure 3 depicts the data flow in the first row of a 3 x 3 array of modules. The elements c_{1j} are computed from the a_{1k} and b_{kj}-elements of A- and B-matrix, respectively, in the following manner. After three cycles, the result c_{11} is computed in the first module. The defective second module´s output XX appears at the fourth cycle and is read into the output register after the fifth cycle. Consequently, after 7 cycles XX has reached the register placed before the spare module. At the same time the spare module has finished the computation of c_{12}. Thus, in the eighth cycle a switch is set for reading the spare module´s output, and XX is overwritten by c_{11}.

The testing of the modules is performed in a manner similar to the one described in [6] . First, the non-redundant part of a row´s modules is tested (note that the data network is assumed to be functional). This is done by feeding a set of test patterns into the array and monitoring the row´s output. This way, defective modules can be located. Next, all possible replacements of the defective modules by spare ones are looked into. Generally, there are R!/(R-D)! such replacements if there are R spare modules and D defective modules. Note that multiplexer, shift register and the spare module share redundancy. As soon as a replacement works the configuration of the row´s modules is stopped and the configuration of the

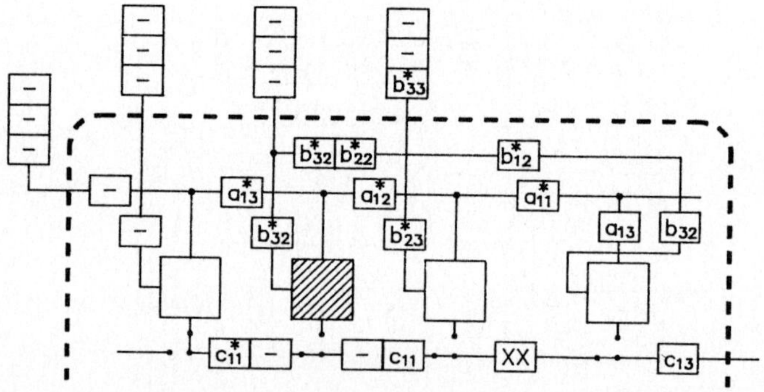

Figure 3a : Data flow in a row with 1 spare module after 7 cycles (XX is the output of the defective module)

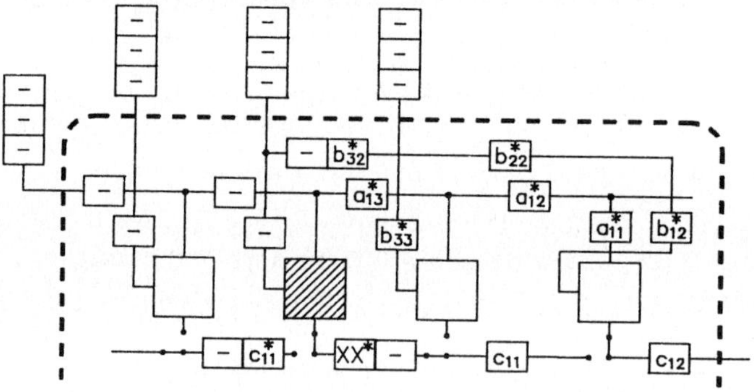

Figure 3b : Data flow after 8 cycles

next row is started. Table 1 shows the time needed to test an 8x8 array if 10^4 patterns are required for a module to be sufficiently tested.

number of redundant modules per row	:	2	4	6
time for configuration	:	24ms	0.2s	5.8s

Table 1 : Time required to test the modules of an 8x8 array (10 MHz clock)

CONFIGURATION OF THE DATA NETWORK

Figure 4 shows how redundancy is applied to the interconnecting lines. Familiar double modular redundancy for segments of the data path suffices to achieve satisfactory yield as will be proven later on. Figure 5 shows a 1 bit switch which is designed for defect-tolerance. By using line drivers with a source impedance much smaller than the one of the input stage of a switch it is guaranteed that a

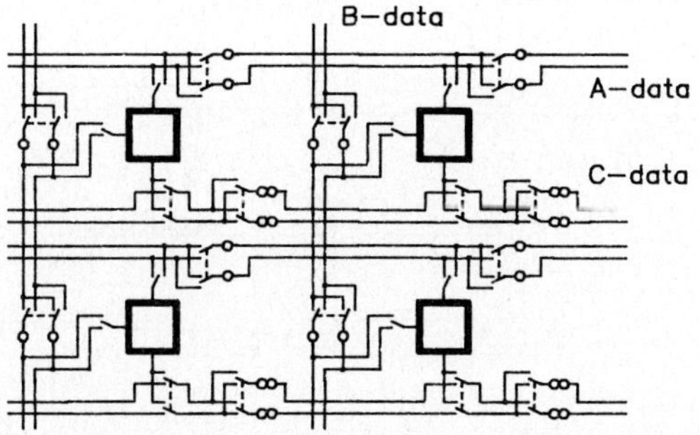

Figure 4 : Principle of redundant implemention of the data network

defect in the switch or/and its control line does not destroy the data words. Whether DMR is to be used at bit level or for several bits must be decided by yield simulation. Although redundancy at bit line increases the line yield most, on the other hand it features the largest switch overhead per path segment. Moreover, it has to be kept in mind that the number of segments in a data path need not correspond to the number of modules in a row. Thus, DMR has to be optimized.

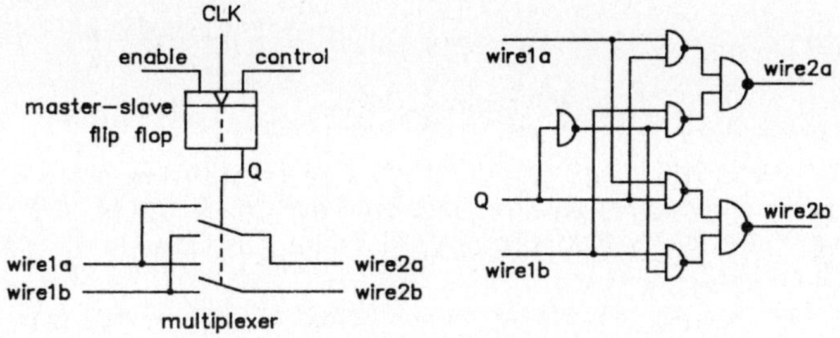

Figure 5 : Defect-tolerant switch design

In order to further reduce the overhead, every switch is supplied with a latch. This allows the array to be configured by a few lines for control and enable, as is shown in figure 6 for the A-data path.

Configuration of A- and C-data is performed in rows, for the B-data in columns. For the row under consideration the enable-line is set, so that the latches can be written; other rows latches are not affected. The control lines of a row are set in response to the evaluation of the test logic which generates the test patterns for the wiring network. These are fed into the data path and compared at its end. Whenever the input does not coincide with the output the bits of the control lines are reset to new values. Otherwise, the latches of the row are disabled and the next row is configured. Note that A-, B- and C-lines should be tested and configured separately. Nevertheless, the time for configuration is quite moderate compared to a system test run in advance of operation (see table 2) .

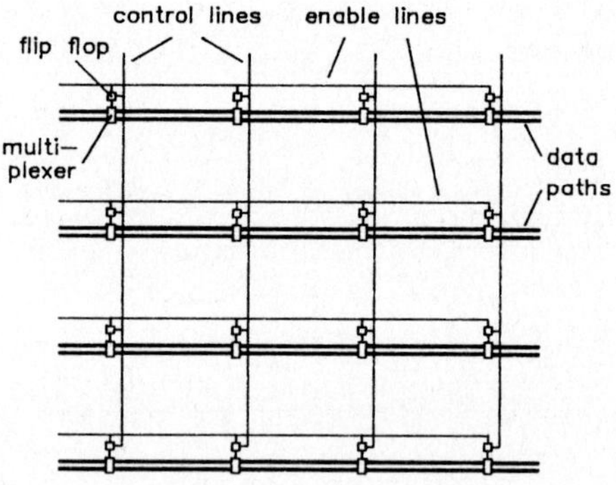

Figure 6 : Global configuration of the A-data paths

number of sections per path	:	4	8	12	16
time for configuration	:	50 µs	0.82 ms	13ms	0.21s

Table 2 : Test time for interwiring of an array of 8x8 modules (10 MHz clock)

YIELD PREDICTION

Several authors started an investigation into WSI defect statistics in order to bring to light the functional dependence of yield on redundancy [7,8,9]. We make use of a model which has been proposed in [10] . Table 3 lists the parameters taken into account for yield calculation.

wafer defect density	V_{DD} & V_{SS} area	transistor area for test logic
cluster coefficient	single module area	wiring area for "
number of rows	vertical word width	mean width of a data line
number of columns	horizontal word width	# of spare modules per row

Table 3 : Parameters of yield calculation

For the calculations we make use of the equation

$$Y=(1+\frac{A\cdot\ d}{c})^{-c}$$

where A stands for area, d for defect density, c for cluster coefficient.

The formula derives from the negative binomial distribution. For c we choose a value of 2.5 for transistor area and 1 for wiring.

We also assume that provisions are made to secure power distribution over the array. Such provisions could consist in locating p-wells (in n-substrate) beneath V_{SS}-lines and including local power transistors for each module. These measures result in a remarkably high yield for the power distribution.

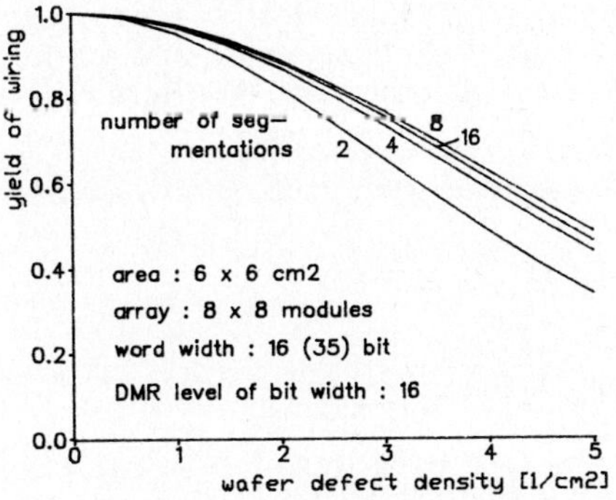

Fig. 7 : Influence of segmentation

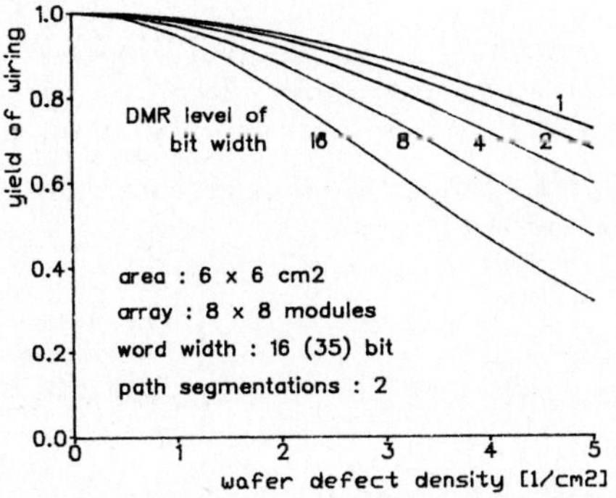

Fig. 8 : DMR at various levels of bit width

Figure 7 shows how segmentation of the data path affects the yield of the wiring net. Note that segmentation will produce poorer results if the segments get smaller than a certain length.

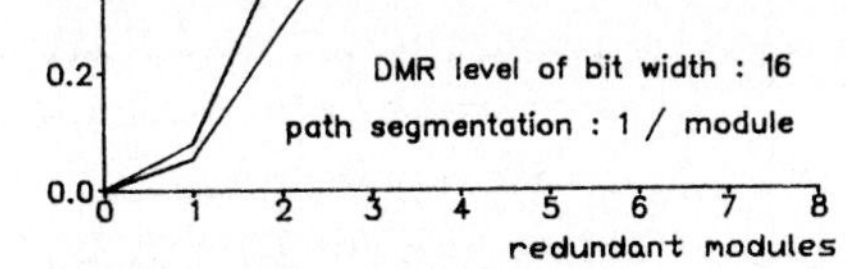

Figure 9 :Separate redundancy for modules and data paths

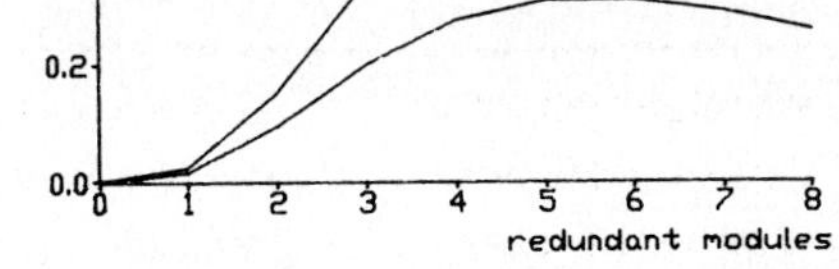

Figure 10 : Joint redundancy for modules and attached data lines

(Array size: 8x8 modules + redundancy, module area: 12mm^2, wafer defect density: 1cm^{-2}, global test logic area: 30mm^2)

Figure 8 demonstrates the effect of establishing DMR at various levels of bit width. The reader should be wary of the assumption that 1bit DMR is always best. It is not; the analysis of larger arrays with different word widths shows that generally the wiring yield has to be optimized for segmentation and DMR level simultaneously.

Figure 9 shows the yield of the system if the data net and the modules are provided with redundancy independently, according to our new approach. The overhead (i.e. wiring, control logic) depends only to a minor extend on the redundancy of the modules. Figure 10 shows the yield if data path and modules share redundancy [6] . It can be seen that the method first mentionned has particular benefits compared with the second one in terms of yield.

Figure 11 shows that an increase in module area is accompanied by an increase of redundancy, for constant yield.

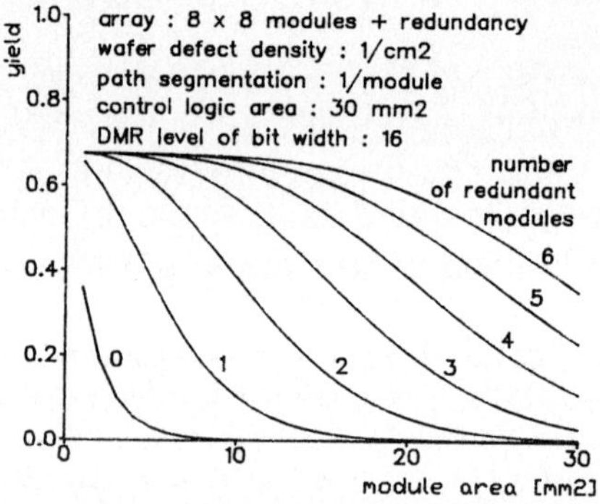

Figure 11 : Dependence of yield on module area

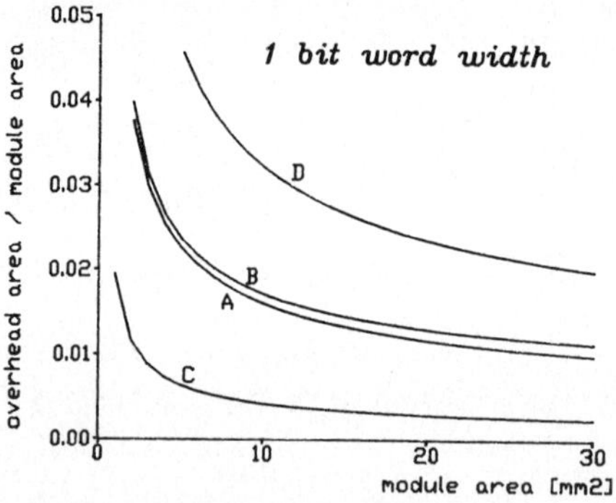

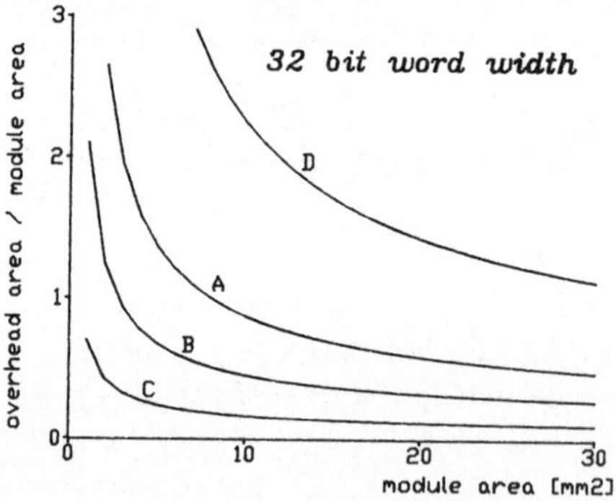

Figure 12 *relative overhead area* *Figure 13*

Finally, we compare this approach with the methods of Ramacher and Sami in respect of area and yield. The examined array consists of 8x8 modules with two redundant columns of modules (Ramacher , Weßeling) or one column and one row (Sami). For all methods the same assumptions apply, concerning word width, defect density, module size and line width. The yield contribution of global test logic is not considered in the following calculations, since Sami postulates self-testing modules. Indeed, local test logic would require considerably more overhead than needed for the other approaches. In figures 12 - 20, curves A, B and C refer to the configurations of Sami, Ramacher and Weßeling. The curve D assumes the configuration of modules according to Sami, and redundant wiring as described above.

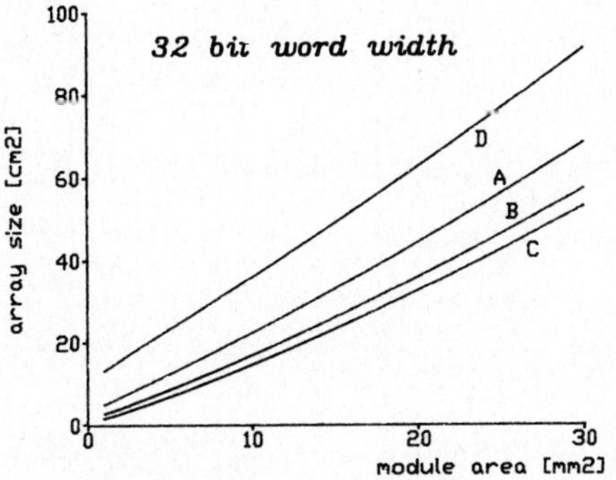

Figure 14 Chip size for various configurations

Figures 12 and 13 demonstrate the ratio of the overhead size to the module size. Here only the area of wiring and control needed for configuration is considered as part of the overhead. This shows that the Sami configuration essentially needs a larger area for the wiring than the others (at 32 bit word width up to 5 times larger than curve C). Naturally, this affects the chip size, the overhead and the module area (figure 14). The configuration D in particular, needs a larger chip size.

Figures 15 - 18 demonstrate the yields calculated for the various configurations. For small word widths, the new configuration offers no advantage over the other ones. But the yield of Weßeling hardly depends at all on the word width, because of the separate redundancy of the wiring, while the yield of the other configurations decreases with increasing word width. Up to a certain word width (< 16 bit) the modified Sami configuration obtains the best yield, while the Weßeling configuration achieves a distinctively greater yield for greater word widths.

Furthermore the yield per area is of interest for our study. Therefore we choose a figure of merit, which relates these yields to the results of the new configuration:

$$\text{figure of merit} := \frac{(\text{yield / area})_i}{(\text{yield / area})_C}, \qquad i = A,B,C,D .$$

In this respect, the advantages of the new approach are most obvious, especially for great word width (figure 19, 20).

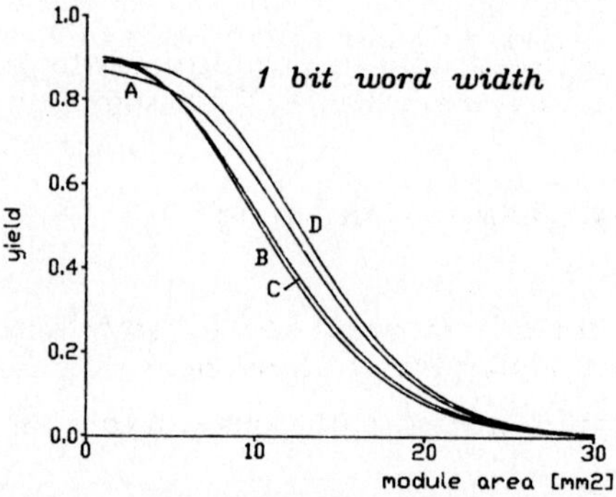

Figure 15

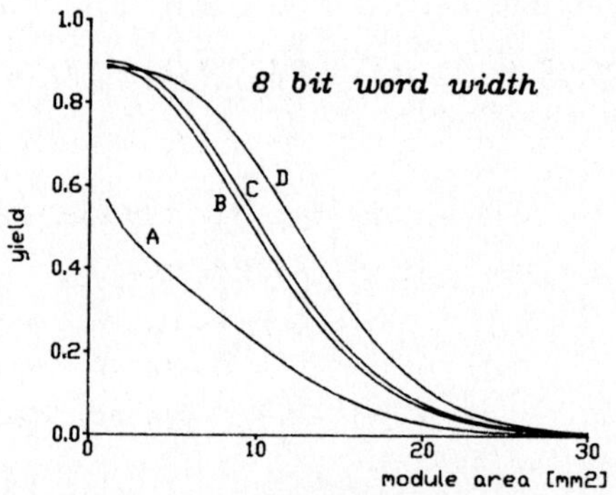

Figure 16

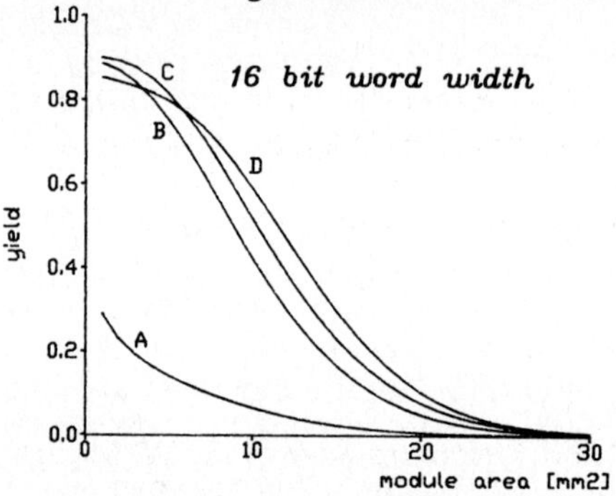

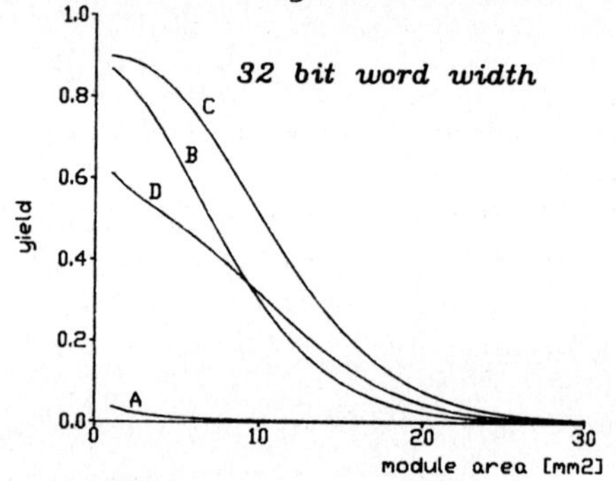

Figure 17 Comparison of various configuration approaches Figure 18
(Curve A = SAMI; B = RAMACHER; C = WEẞELING; D = MODIFIED SAMI: Array size: 8x8 + 16 modules, wafer defect density: 1cm^{-2})

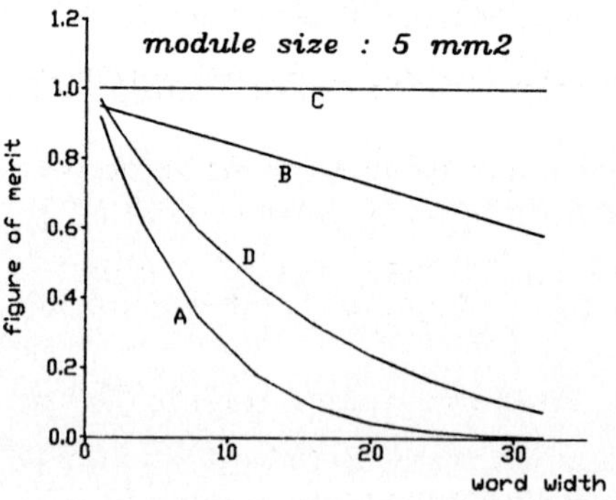

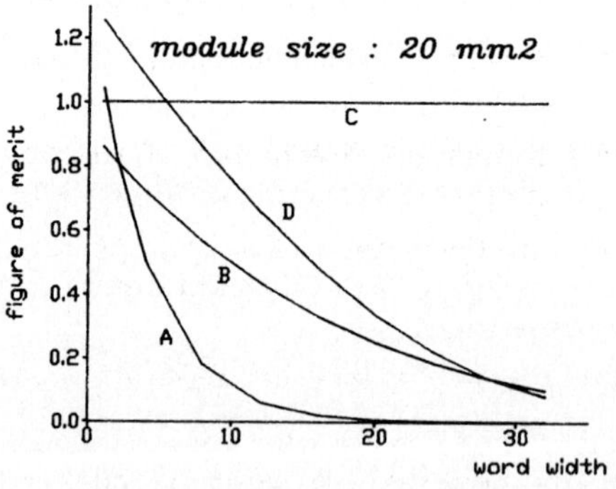

Figure 19 Yield / area in relation to Weßeling Figure 20

CONCLUSIONS

The new configuration proposed in this paper is restricted to architectures where the output data of a module are not requested by other modules. This assumption in mind, the features of the new approach are:

-- separate provision of redundancy for modules and interwiring

-- minimal bypass wiring area

-- increasing redundancy barely affects the overhead required (therefore in most cases the yield can be improved by adding more redundancy)

-- increasing the word width has little effect on the yield (this is in contrast to [5,6])

-- best in terms of yield per area, i.e. costs

-- easyly tested and configured.

Considering systolic architectures which have a module´s output distributed to other neighbouring modules, the results obtained recommend strongly that modules and wiring should be provided with redundancy independently. Double modular redundancy at bit level for the interwiring is predicted to reach very high yield.

REFERENCES

[1] YIELD MODELLING AND DEFECT TOLERANCE IN VLSI, edited by W. Moore, W. Maly and A. Strojwas, Adam Hilger, Bristol and Philadelphia,1988

[2] Proc. of the Int. Conf. on Wafer Scale Integration, San Francisco , Jan. 3-5, 1989

[3] Varman,Ramakrishnan, "A fault-tolerant VLSI matrix multiplier", Proc. of the 1986 Int. Conf. on Parallel Processing, p.351-357, 1986

[4] Proc. of the IFIP workshop on Wafer Scale Integration, Brunel University , Sept. 23-25, 1987

[5] M. Sami, R. Steffanelli, "Reconfigurable architectures for VLSI processing arrays" , Proc. of the IEEE, May 1986

[6] U. Ramacher, J. Beichter, "A self-testing WSI matrix-matrix multiplier", Proc. of I.F.I.P. workshop on Wafer Scale Integration, Brunel University, Sept. 1987

[7] Stapper,Armstrong and Saji, "Integrated circuit yield statistics", Proc. IEEE,Vol.71 No.4, p.453-470, 1983

[8] I. Koren, M.A. Breuer, "On area and yield considerations for fault-tolerant VLSI processor arrays" , IEEE Trans. Comput. , vol. C-33, pp. 21-27, Jan. 1984

[9] T.E. Mangir, "Sources of failures and yield improvement for VLSI and restrucuturable interconnects for RVLSI and WSI" , Proc. IEEE, vol. 72, pp. 690-706, I, pp. 1687-1694, II, 1985

[10] U. Ramacher, "A cost oriented redundancy model for defect-tolerant VLSI/WSI systems", in YIELD MODELLING AND DEFECT TOLERANCE IN VLSI, edited by W. Moore, W. Maly and A. Strojwas, Adam Hilger, Bristol and Philadelphia, 1988

[11] A. Rosenberg, "The Diogenes Approach to testable fault-tolerant networks of processors", IEEE Trans. Comput. , vol. C-32, pp. 902-910, Oct. 1983

[12] C. Jesshope, L. Bentley, "Techniques for implementing two dimensional wafer-scale processor arrays", IEE proceedings, Vol 134, No. 2, March 1987

Wafer Scale Integration, III
M. Sami and F. Distante (Editors)
Elsevier Science Publishers B.V. (North-Holland)

A MODEL FOR THE ANALYSIS, DESIGN AND COMPARISON OF FAULT-TOLERANT WSI ARCHITECTURES

V. S. S. NAIR

Computer Systems Group
Coordinated Science Laboratory
University of Illinois, Urbana, IL 61801

J. A. ABRAHAM

Department of Electrical & Computer Engineering
University of Texas at Austin
Austin, TX 78712

A new model has been proposed for fault-tolerant multiprocessor systems, especially those using algorithm-based fault tolerance techniques (ABFT). Design of complicated ABFT systems was considered to be intractable by previous researchers. In this paper we propose a systematic procedure to design such systems. Complexity of the proposed techniques are less than the previously derived bounds.

1. INTRODUCTION

Fault-tolerant designs have been suggested for WSI architectures for both yield enhancement and long term reliability [1]. Fault tolerance can be achieved either by providing some means of switching a faulty processor out of the processor array, or by providing redundancy in the processor design itself. In either case, a method to detect the faulty module is required. Fault detection can be accomplished by designing self checking processor modules. Various techniques for fault tolerance within non-self-checking modules have been suggested for highly reliable computing structures. The techniques include triplication and voting [2], duplication and comparison, and standby replacement. All these techniques take much more chip area compared with a non-fault-tolerant design [3].

More efficient fault-tolerant designs have been suggested for highly parallel architectures. Algorithm-based fault tolerance (ABFT) [4] is one such technique; it is a system level method of achieving fault tolerance at a low cost. Original algorithms are modified to operate on encoded data so that they produce encoded outputs, from which the useful information can be recovered easily. The task distribution among the processing elements may be done in such a way that any malfunction in a processing element will affect only a small portion of the data, so that errors can be detected and corrected using the properties of the

This research was supported by the National Aeronautics and Space Administration (NASA) under Contract NAG 1-613.

code. ABFT techniques have been proposed for many signal processing computations such as matrix operations [4], Fast Fourier Transforms [5], QR Factorization, and Singular value decomposition [6].

Once an error due to a fault is detected, in order to isolate or repair the fault, the fault tolerance scheme should be able to locate the fault. This makes fault diagnosis an integral part of any kind of fault-tolerant computation. Fault diagnosis in WSI architectures can be very involved due to the large number of processors present on the same wafer itself, and due to the complicated data dependencies among the processors. Various models have been suggested for WSI structures having large number of processors [7] in order to systematize diagnosis. These models are similar to previous models that have been proposed for fault-tolerant multiprocessor systems [8]. In all these models the emphasis was placed on the system interconnection rather than on identifying the faulty module. Therefore, these models are suitable only for the reconfiguration of the processor system, once the fault is detected and located. Fault detection is much more complicated than reconfiguration in the sense that an error due to a fault may propagate to a processor which is not directly connected to the faulty processor. This is the case with most of the repetitive WSI architectures. In order to handle the error propagation, one must take into account the data dependencies among the processors also. The first attempt towards this was done by Banerjee and Abraham [9] who proposed a graph-theoretic model for representing ABFT systems. This model was based on existing models for multiprocessor systems in which processors mutually checked each other [10, 11]. In the graph-theoretic model the data dependency among processors as well as the check operations is modeled as a tripartite graph. Based on this model the authors could analyze a fault-tolerant system for its fault-detecting and correcting capability. The algorithms for analysis are potentially exponential since all possible error combinations at the output of a faulty processor have to be considered. Also, the design of a system using this model proved impractical.

In order to circumvent these problems we developed a matrix-based model [12], for the analysis as well as design of fault-tolerant multiple processor systems. In the conventional analysis of such systems, one assumes the existence of complete tests which always fail when there is a faulty processor [10]. A processor can test another, and if the tested unit is faulty and the tester is fault-free, then the test is guaranteed to fail. However, in systems using ABFT, a particular fault pattern can produce a number of different error patterns. The checking operations detect the errors directly and the faults indirectly. Therefore, fault analysis in such systems is much more complex than conventional fault analysis. The new model is specifically developed for analyzing faults in systems using ABFT; however, application of the model for fault analysis of conventional redundancy techniques (such as duplication or triplication) is a trivial special case.

In this paper we give an overview of the model and a brief description of the analytical algorithms developed in [12], based on this model. Application of the model and the algorithms for the analysis of conventional as well as modern fault-tolerant architectures are illustrated with examples. The design of a fault-tolerant multiprocessor system in which each processor produces more than one data element was considered to be an intractable problem [13]. In that work, for designing a system with required fault tolerance capabilities, first the cardinality of the largest error pattern generated by all possible faults is determined and the system is designed to detect and locate that number of errors. In this paper we propose a direct scheme to design such systems which eventually results in smaller number of checks than the previous methods. The design procedure is illustrated with an example. A comparison between existing schemes and the new scheme is done with respect to the number of checks required for each scheme.

2. OVERVIEW OF THE NEW MODEL

In the multiprocessor system under consideration, the computation is done on m processors $P_1, P_2, \ldots, P_m$. These processors generate n data elements, $d_1, \ldots, d_n$, as results. A fault is defined to be a condition that may cause a malfunction of a particular processing unit. The fault may be transient or permanent. In our analysis, a processor is said to be faulty if and only if there exists at least one error in the data computed by that processor.

In the new model for multiple processor systems, the relationship between processors, data, and checks are represented by three fundamental matrices, the PD (Processor-Data) matrix, the DC (Data-Check) matrix, and the PC (Processor-Check) matrix. $PD_{ij} = 1$ if data d_j is affected by processor P_i, otherwise $PD_{ij} = 0$. Similarly $DC_{ij} = 1$ if data d_i is checked by check c_j. The PC matrix is the product of the PD and DC matrices. The physical significance of the matrices for the fault analysis of the system is discussed in detail in [12]. Two algorithms for analyzing the fault detectability and locatability of fault-tolerant systems have been developed using the model. Various multiprocessor architectures have been analyzed using this model. In the following, we give a brief description of the analysis algorithms.

3. ALGORITHMS FOR FAULT DIAGNOSIS

Based on the new model, efficient algorithms have been developed [12] for evaluating the fault tolerance capabilities of multiple processor systems. These algorithms are much less complex than the previously proposed ones. Much of the complexity alleviation accrues from the application of a special technique called *error collapsing* which avoids exhaustive enumeration of error patterns at the output of a faulty processor, and due to the newly derived simplified sufficiency conditions. For completeness of the paper we give a high level description of the algorithms. A detailed description of these algorithms is given in [12].

Detectability Algorithm

(1) Find whether a fault is detectable if all the data elements produced by the processors involved in that fault are erroneous.

(2) Collapse errors and check the detectability of the fault until no error is left.

(3) Check the detectability of all faults whose cardinality is $\leq t$, in order to check whether the system is t*-fault detectable.*

Locatability algorithm

(1) Check whether a fault is distinguishable from another fault if all the data elements produced by the processors involved in both the faults are erroneous.

(2) Collapse errors and determine whether the faults are distinguishable until no error is left with at least one of the faults.

(3) Check the distinguishability of all faults whose cardinality is $\leq l$, in order to check whether the system is l*-fault locatable.*

Both the algorithms have been implemented as a program. The PD matrix, the DC matrix and the error detectability of the checking operation are the input parameters of the program. The program analyzes the system for its fault detectability first, and then uses that result to evaluate the locatability.

The tool was used to determine the fault detectability and locatability of some realistic systems. Before going into the applications of the model, we need to define the concept of (g, h) *check* as introduced in [13].

DEFINITION 1: A (g, h) *check* is a check that is defined on at most g data elements, such that the check passes if none of the data elements are in error, fails if between 1 and h data elements are in error, and becomes invalid if more than h elements are in error.

4. ANALYTICAL APPLICATION OF THE MODEL

In this section we apply the new model to the analysis of some fault-tolerant WSI processor arrays.

EXAMPLE 1: Consider matrix multiplication using checksums on a mesh connected processor array. The fault tolerance scheme has been proposed by Huang and Abraham [4]. We will briefly describe the system below.

Multiplication of two 3×3 matrices X and Y is done on a mesh connected processor array as shown in Figure 1. We assume that the input data elements are broadcasted on the buses; the processors input the data from the buses. Under a processor failure, we assume that only the corresponding data element of the output matrix Z becomes erroneous. After the computation, the result Z resides in the local memories of the processors and the checking operations are performed as given below.
We have
$DATA\ (P_i) = d_i$ for $i = 1, 2, \ldots 9$,
$CHECK(d_1, d_2, d_3) = C_1$,
$CHECK(d_4, d_5, d_6) = C_2$,
$CHECK(d_7, d_8, d_9) = C_3$,
$CHECK(d_1, d_4, d_7) = C_4$,
$CHECK(d_2, d_5, d_8) = C_5$,
$CHECK(d_3, d_6, d_9) = C_6$.

Now the fundamental matrices of the system will be:

$PD = I_9$, where I_9 is the identity matrix of order 9.

$$DC = \begin{bmatrix} 1&0&0&1&0&0 \\ 1&0&0&0&1&0 \\ 1&0&0&0&0&1 \\ 0&1&0&1&0&0 \\ 0&1&0&0&1&0 \\ 0&1&0&0&0&1 \\ 0&0&1&1&0&0 \\ 0&0&1&0&1&0 \\ 0&0&1&0&0&1 \end{bmatrix}$$

and $PC = PD \times DC = DC$, since $PD = I_9$.

Using the algorithms we found that the system is 3–*fault detectable* and 1–*fault locatable*.

EXAMPLE 2: As a second example we consider the analysis of a TMR system. Here three processors produce three independent output data which are compared by a voter. Each comparison is equivalent to a check operation in the model. Thus, there are $\binom{3}{2} = 3$

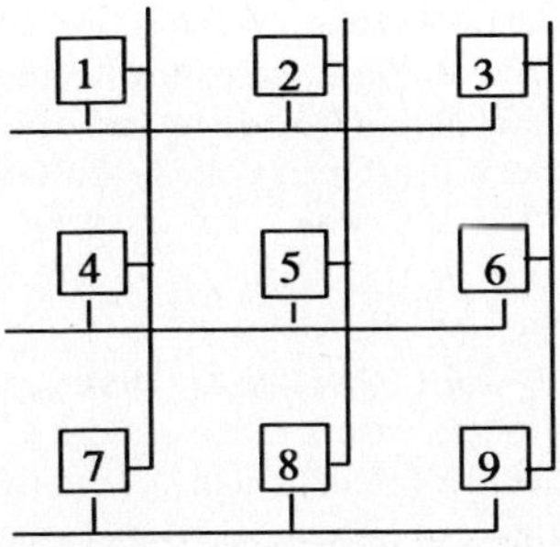

FIGURE 1. Mesh connected processor arrays.

check operations as shown in Figure 2. The fundamental matrices are

$$PD = \begin{bmatrix} 1 & 0 & 0 \\ 0 & 1 & 0 \\ 0 & 0 & 1 \end{bmatrix} \quad DC = PC = \begin{bmatrix} 1 & 1 & 0 \\ 1 & 0 & 1 \\ 0 & 1 & 1 \end{bmatrix}$$

The analysis shows that the system is 2–*fault detectable* and 1–*fault locatable*.

5. DESIGN OF FAULT-TOLERANT SYSTEMS

The task is to design checks for the output data of a non-fault-tolerant system so that the new system has the required fault tolerance capabilities. Design of fault-tolerant systems where each processor producing more than one data element was considered to be an intractable problem [13]. In order to design a system for t–*fault detectability*, first the size of the largest error pattern (let it be equal to s) for all fault patterns was determined and the system was designed for an error detectability of s. Using this approach upper and lower bounds on the number of (g, h) checks were determined in terms of g, h and s [13, 14]. The basic disadvantage of this approach is that the method is efficient only when most of the fault patterns are producing error patterns of cardinality s. On the contrary, if only a few number of fault patterns produce error patterns of cardinality s, the design procedure will be using more number of checks unnecessarily. In this paper we propose a more straightforward design procedure which needs a smaller number of checks than the previous techniques.

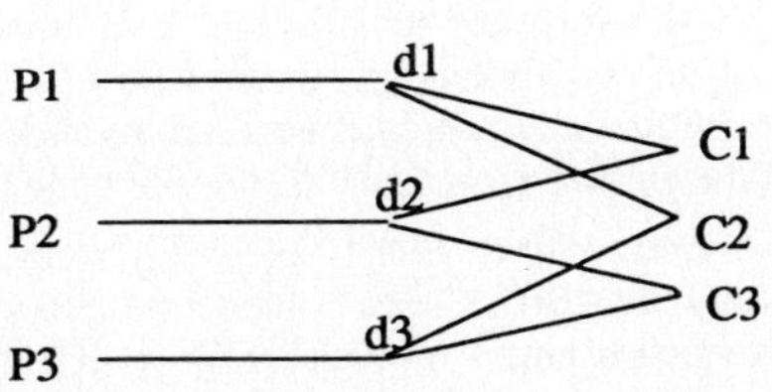

FIGURE 2. Graphical representation of a TMR system.

Using the matrix-based model parameters we define the design problem as follows: Given the PD matrix, find a DC matrix so that the corresponding PC matrix has the required fault diagnosing capabilities.

Since PC = PD*DC, the design involves finding two variables (actually two matrices) from a single equation. Therefore, the solution is not unique. The selection of a particular solution should optimize the number of checks required, and the number of errors detectable and correctable by each check (assuming that the cost of each check increases proportionally to the number of errors it can detect and correct). Our approach to the problem consists of the following steps.

(1) Design a DC matrix for PD = I_m and $h = 1$, such that the system has *t–fault detectability* and *l–fault locatability*. Here I_m is the identity matrix of order m, where m is the number of processors in the system to be designed. We call this system the *unit system* of the actual fault-tolerant system to be designed.

(2) Modify the DC matrix of the unit system according to the given PD matrix in order to get the DC matrix of the actual system.

In the unit system, since the PD matrix is the unit matrix, every processor is producing only one data element. Therefore, the cardinality of the fault patterns will be the same as the cardinality of the resulting error patterns. As mentioned before, in such a situation, the techniques proposed in [13, 14] are efficient and can be used for the design of the unit system. Designs are already available for various values of fault detectabilities and locatabilities. These designs of the unit systems can be used as a template. Now the actual design consists of modifying these template designs to obtain the actual system.

DEFINITION 2: The *Product System* of a given non-fault-tolerant system and the corresponding fault-tolerant unit system is defined as the system obtained by connecting every data element affected by processor P_i in the non-fault-tolerant system to every check element in the unit system which checks the output of processor P_i.

The construction of the product system is illustrated in the following example.

EXAMPLE 3: Let us consider a multiprocessor system with four processors. The first processor produces 4 data elements, the second one 2, the third and the fourth processor produce only one data element each. A unit system is designed for 3–*fault detectability* and 1–*fault locatability*. Construction of the product system is given in Figure 3.

THEOREM 1: If the unit system is *t–fault detectable* and *l–fault locatable*, for t and $l > 0$, then the product system is *t–fault detectable* and *l–fault locatable* if and only if

$$h \geq \max\,[num\,(P_i)].$$

Here h is the error detectability of the checks in the product system and $num\,(P_i)$ is the number of data elements affected by processor P_i.

PROOF:

Proof for the necessary condition, (By contradiction): Let $h < \max\,[num\,(P_i)]$, and the product system be *t–fault detectable* and *l–fault locatable*. Let processor P_j be such that $num\,(P_j) = \max\,[num\,(P_i)]$, where the maximum is computed for $i = 1, 2, 3, \ldots m$. If P_j fails in such a way that all the data elements produced by P_j become erroneous, then all the checks done on P_j in the product system will become invalid. Therefore, such a fault in P_j will not be detected and the system is 0–*fault detectable* which is a contradiction.

Proof for the sufficiency condition: Since the unit system is designed for $h = 1$, for any fault pattern of cardinality $\leq t$, there exists at least one check in the unit system which checks only one processor in that group of processors which are faulty. (In [13] this is referred to as 1-neighbor intersection property). In the product system also, since $h \geq \max\,[num\,(P_i)\,]$, this check will fail and hence the system has the same detectability

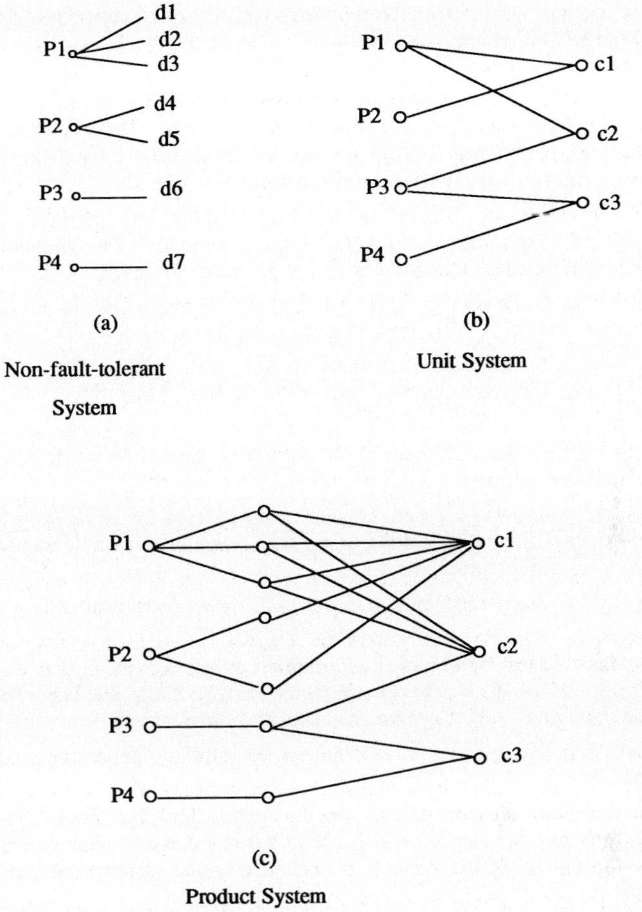

FIGURE 3. Construction of a product system

and locatability as the unit system. Locatability of the product system can be argued in a similar fashion.

5.1. Construction of the actual system

Construction of the actual fault-tolerant system with given (g, h) checks is done by splitting each check in the product system into one or more checks such that every check in the resulting system has at most h data elements from each processor checked by those checks. Now the new system has the same detectability and locatability as the product system. This is because, whenever a check in the product system fails, at least one of the checks formed by splitting that check node will fail in the actual system. Actually, the detectability and locatability of the new system may be higher than the product system. But, we are interested only in the fact that the detectability and locatability of the final system are at least equal to that of the product system.

In the following, we present the whole procedure as an algorithm, in terms of the matrix-based model parameters.

DESIGN ALGORITHM

(1) Construct a DC matrix for the unit system (we call it unit DC matrix), so that the unit system is *t−fault detectable* and *l−fault locatable.*

(2) The DC matrix for the product system (called product DC matrix) is constructed by expanding the columns of the unit DC matrix vertically. The row corresponding to processor P_i in the unit DC matrix is replicated *num* (P_i) times.

(3) The DC matrix of the product system is partitioned into blocks of rows, such that the ith block contains data elements produced by processor P_i.

(4) Each column in the product DC matrix is split into minimum number of columns so that every column has at most h number of 1's in every block.

The resulting DC matrix (and the corresponding PC matrix) has the required fault diagnosing capabilities. The DC matrix can be further simplified by deleting some of the redundant columns as follows.

DEFINITION 3: Column C_i (ie., check C_i) is said to be covered by one or more columns if and only if C_i can be written as a linear combination (with coefficient of multiplication equal to 1) of those columns.

LEMMA 1: If C_i is covered by $C_{j1}, C_{j2}, \ldots C_{jk}$, then C_i is a redundant check and can be deleted.

PROOF: Since C_i can be obtained as a linear combination of $C_{j1}, C_{j2}, \ldots C_{jk}$, whenever C_i fails, at least one of the checks among $C_{j1}, C_{j2}, \ldots C_{jk}$ will fail. Therefore, if the system has checks $C_{j1}, C_{j2}, \ldots C_{jk}$, then check C_i is redundant and hence can be deleted.

COROLLARY 1: Every column is covered by another identical column, if such a column exists.

Once the DC matrix is determined using the design algorithm, the DC matrix is further simplified by deleting all columns which are covered by some other columns. The procedure for modifying the product system to obtain the actual system is illustrated in the following example.

EXAMPLE 4: We consider the same multiprocessor system that we introduced in the previous example. We are supposed to design a fault tolerant system consisting of all these processors such that the system will be 3−*fault detectable* and 1−*fault locatable*. In example 3, we have already seen how to construct the product system. The rest of the algorithmic procedure is shown in Figure 2. The graphical representations of the systems are shown along with their matrix representations. It can be observed in Figure 4(b) that the DC matrix has two identical columns corresponding to checks C_{12} and C_{22}. Either one of those checks can be deleted from the system so that the final system has only 4 checks.

5.2. Comparison with previous schemes

In order to make comparison between the newly proposed scheme and the existing ones, we consider the following example. Comparison is done with respect to the number of checks required in the design.

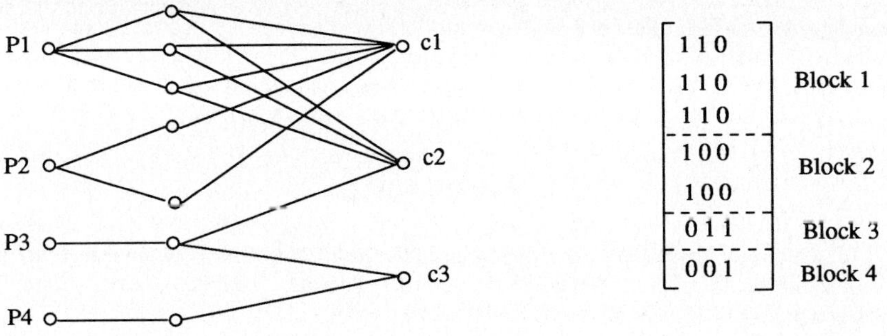

(a)

Product system and the corresponding DC matrix

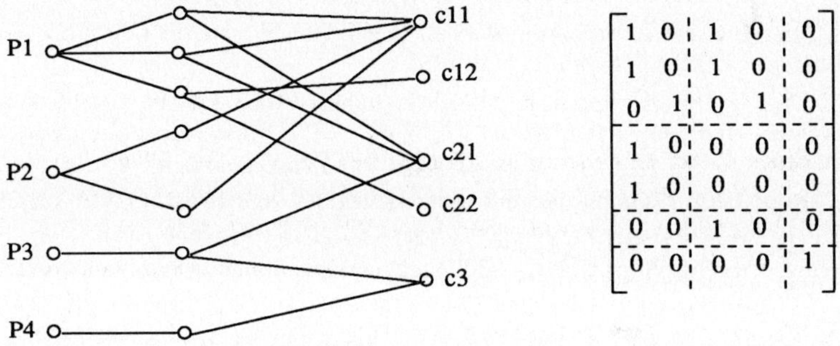

(b)

Fault-tolerant System for h = 2 and its DC matrix

FIGURE 4. Design of the final system from the product system

EXAMPLE 5: Suppose a system involving 500 processors has to be designed for 3–*fault detectability* and 1–*fault locatability*. The checks available are of type (5,1) (ie., g = 5, and h = 1). It is also known that 10 of those 500 processors produce 2 data elements each.

Number of checks required as per the bounds given in [13] : 408
Number of checks required as per the bounds given in [14] : 363
Number of checks required for our scheme : 200

6. CONCLUSIONS

In this paper we have given an overview of a matrix-based model developed for fault-tolerant multiprocessor systems (FTMS). A brief description of the analysis algorithms based on this model is also given. A systematic and straightforward design methodology is proposed for the design of FTMS where individual processors may produce more than one data element. The new scheme is more efficient than the previous schemes with respect to the number of checks used in the overall system. Examples are provided for the illustration of the design methodology and for the comparison of various schemes.

REFERENCES

[1] W. Chen, J. Mavor, P. B. Denyer, and D. Renshaw, ''A WSI approach towards defect/fault-tolerant reconfigurable serial systems,'' *IEEE Journ. Solid-State Circuits*, vol. 23, No. 3, pp. 639-645 , June. 1988.

[2] S. E. Butner, ''Triple time redundancy, fault-masking in byte-sliced systems,'' in *Tech. Rep. CSL TR 211* , Comp. Sys. Lab., Dept. of Elect. Eng., Stanford Univ. California, Aug. 1981.

[3] W. R. Moore, ''A review of fault-tolerant techniques for the enhancement of integrated circuit yield,'' *Proc. IEEE*, vol. 74, No. 5, pp. 684-698, May 1988.

[4] K. H. Huang and J. A. Abraham, ''Algorithm-based fault tolerance for matrix operations,'' *IEEE Trans. Computers*, vol. C-33, pp. 518-528, June 1984.

[5] J. Y. Jou and J. A. Abraham, ''Fault-tolerant FFT networks,'' *IEEE Trans. Computers*, vol. 37, pp. 548-561, May 1988.

[6] C. Y. Chen and J. A. Abraham, ''Fault-tolerant systems for the computation of eigenvalues and singular values ,'' *Proc. SPIE,Advanced Algorithms and Architectures for Signal Processing*, vol. 696, pp. 228-237, Aug. 1986.

[7] A. L. Rosenberg, ''Graph-theoretic approaches to fault-tolerant WSI processor arrays,'' in *Wafer Scale Integration*, Bristol, U.K, pp. 10-22, 1986.

[8] J. P. Hayes and R. M. Yanney, ''Fault recovery in multiprocessor networks,'' in *Proc. 8 th Int. Symp. on Fault-Tolerant Computing*, Toulons, France, pp. 123-128, Jun. 1978.

[9] P. Banerjee and J. A. Abraham, ''Concurrent fault diagnosis in multiple processor systems,'' in *16th Int. Symp. on Fault-Tolerant Computing*, Vienna, Austria, pp. 298-303, 1986.

[10] J. D. Russel and C. R. Kime, ''System fault diagnosis: Closure and diagnosability with repair,'' *IEEE Trans. Computers*, vol. C-24, pp. 1078-1088, 1973.

[11] J. D. Russel and C. R. Kime, ''System fault diagnosis: Masking, exposure, and diagnosability without repair,'' *IEEE Trans. Computers*, vol. c-24, pp. 1155-1161, 1975.

[12] V. S. S. Nair and J. A. Abraham, ''A model for the analysis of fault-tolerant signal processing architectures,'' in *Proc. 32 nd Int. Tech. Symp. of SPIE*, San Diego, pp. 246-257, Aug. 1988.

[13] P. Banerjee and J. A. Abraham, ''Bounds on algorithm-based fault tolerance in multiple processor systems,'' *IEEE Trans. Computers*, vol. C-35, pp. 296-306, 1986.

[14] D. J. Rosenkrantz and S. S. Ravi, ''Improved bounds on algorithm-based fault tolerance,'' in *Proc. 26 th Annual Allerton Conf. on Communication, Control, and Computing*, Monticello, Illinois, pp. 388-397, Sept. 1988.

Wafer Scale Integration, III
M. Sami and F. Distante (Editors)
Elsevier Science Publishers B.V. (North-Holland)

TESTABILITY AND DESIGN FOR TESTABILITY FOR WSI SYSTOLIC HEXAGONAL ARRAYS

Donatella Sciuto

Dept. Automazione Industriale, Universitá di Brescia, Via Valotti 9, 25060 Brescia, ITALY

The problem of the definition of sufficient conditions for testability of wafer-scale hexagonal arrays is afforded. Two sets of conditions are derived that make two-dimensional hexagonal arrays of combinational cells testable for single faulty cell.
Then, design for constant testability is afforded, based on the addition of states to the transition table of the cell. These states achieve full controllability and observability of the test results by regenerating the test input and propagating to the primary outputs the faulty states.
Aim of both classes of testing techniques introduced, is to extend the definition of testability conditions to array constituted by cells of higher complexity than the basic additive inner-product cell.

1. Introduction

Wafer scale integration will make available large arrays for computational intensive operations such as digital signal processing. The regularity, nearest neighbour communication and parallel computation have made systolic and semi-systolic arrays very suitable for wafer scale integration. Interest in such structures has fostered a great deal of research [11] and realization of arrays with up to 5000 cells on a single wafer seems a reality [12]. For some digital signal processing operations, such as matrix multiplication and some classes of filtering operations the hexagonal interconnection network has been proposed in literature as the most efficient [1,2,8].

One of the main drawbacks of wafer scale integration derives from the low yields achieved during the manufacturing process [10]. To overcome such problem fault tolerance techniques have been introduced in the array design and in particular techniques such as restructuring and reconfiguration have been identified as the most suitable [10]. However, before reconfiguration, the array should be tested and the faulty cells localized.

The present paper will analyze testing with respect to hexagonally interconnected arrays. Testing hexagonal cells is made difficult by the asymmetric nature of the interconnection network: at every time step three different signal flows interact in each cell. The test results presented in this paper derive from the definition of a technique that allows to linearize a non-linear problem determined by the hexagonal topology of the interconnection network of the array. The testing technique defined is based on the type of interaction among the three signal flows, i.e. on the function performed by the cells. If such interaction satisfies the conditions introduced in section 4, a test procedure with complexity proportional to the number of cells in the array can be devised. The conditions identified here are less restrictive than those previously published [4].

Finally in section 5 a design for testability methodology independent of the function performed by the cell is proposed. It is based on the addition of states to the transition table of the basic cell that allows to achieve controllability and observability of the array. This procedure allows testing in time independent of the number of cells but related only to the number of states in the transition table.

The purpose of both classes of approaches presented in this paper is to extend testability conditions to cells whose operation complexity is greater than the basic additive inner-product cell. For this particular class of cells it can be proved that C-testability is an intrinsic characteristic [4].

2. Definitions and Notation

In an hexagonally connected array, each cell produces (at each time step t) a right-moving, (denoted by $x^o(t)$), a vertical-moving (denoted by $y^o(t)$) and a diagonal-moving (say $z^o(t)$) output depending on its left (x), vertical (y) and diagonal (z) inputs.

Let X be the set of right-moving signals, Y the set of vertical-moving signals and Z the set of diagonal-moving signals. Let $x^o : X \times Y \times Z \rightarrow X$ be the right-moving mapping, $y^o : X \times Y \times Z \rightarrow Y$ be the vertical-moving mapping, and $z^o : X \times Y \times Z \rightarrow Z$ be the diagonal-moving mapping. Thus, the x, y and z outputs of a basic cell which receives the right input x_i, the vertical input y_j and the diagonal input z_k are denoted by $x^o(x_i, y_j, z_k), y^o(x_i, y_j, z_k), z^o(x_i, y_j, z_k)$ respectively. Let the state pair x_c/x_e represent correct right-moving signal x_c and erroneous right-moving signal x_e.Analogously when the vertical or diagonal moving signals are considered. The transition from a state (say x_i) to a second state (say x_c) is denoted as $x_i \rightarrow x_a$. The transition from a state (say x_i) to either a state (say x_c) if the basic cell operates correctly, or to another state (say x_e) if the cell is faulty, is denoted by $x_i \rightarrow x_c/x_e$. The input-output relationships can also be defined through the transition table [3,8]. The transition table of a basic cell is a table in which columns represent the vertical states (referred as y-states) and the rows represent both horizontal and diagonal states (referred as x-states and z-states respectively). An entry in row i, k and column j represents the right horizontal output, the vertical output and the diagonal output of the basic cell when it receives as input x_i, y_j, z_k. Let's assume that there are m different x-values, n y-values and l z-values.

In the presence of a fault, the cell remains combinational, i.e. no sequential characteristic appears in a faulty cell. It is assumed that to completely test a cell, all input combinations $X \times Y \times Z$ must be applied, i.e. exhaustive testing is required. This assumption makes testing of the cells independent of their actual implementation and of the fault model for permanent faults. It is also assumed that to completely test the array, every cell must be completely tested. The single cell permanent combinational fault model is assumed.

The array is defined to be testable if any input combination can be applied to any cell of the array and any fault can be propagated to an observable output of the array [5].

The horizontal, vertical and diagonal inputs of cell (i,j) at time t are denoted by $x_{i,j}$ (t), $y_{i,j}$ (t) and $z_{i,j}$ (t) respectively. Hence, cell (i,j) at time $t+1$ will produce horizontal, vertical and diagonal inputs $x_{i,j+1}$ $(t+1)$, $y_{i+1,j}$ $(t+1)$ and $z_{i+1,j+1}$ $(t+1)$.

An input or output labelled x/x' where $x' \neq x$ means normal input or output x, and faulty input or output x'. If instead of only one element x', the entire set X, without considering the correct element x, is considered, then it will be written as $x/*$. These definitions are also valid if the diagonal and vertical moving signal are replaced to the horizontal moving signal. Finally, let's define x^o $(x_1/x_2, y, z) = x'_1/x'_2$ as x^o (x_1, y, z) $=$ x'_1 and x^o (x_2, y, z) $=$ x'_2 and $x_1 \neq x_2$ and $x'_1 \neq x'_2$. Analogously for y_1/y_2 or z_1/z_2 and if y^o and z^o are defined.

3. General testability conditions

The analysis of general testability of hexagonal arrays proposed in this section, aims at the construction of a test procedure through the definition of controllability and observability test sequences in the most general case in which the three signal flows interact completely.

The set of testability conditions is composed of three controllability conditions and three observability conditions, i.e. one controllability condition and one observability conditions for each signal flow direction.

Considering the horizontally moving signal flow the controllability condition C1 requires that if a particular horizontal output x has to be obtained, it is possible to apply a diagonal input z and a horizontal x', no matter what the vertical input is, but additionally, if some horizontal input different from x' is applied, a different horizontal output, different from the expected one x, should be obtained. This condition can be expressed mathematically as follows:

C1: for every $x \in X$, there exist $x' \in X$ and $z \in Z$, such that $x^o(x'/*, Y, z) = x/*$.

Differently, condition C2, that considers the diagonally moving signal, requires that there is a diagonal output for every input combination, but if a different horizontal input from x' is applied, a right output different from the expected one x must be obtained. This condition can also be written synthetically as follows:

C2: for every $z \in Z$, there exist $z' \in Z$, $y' \in Y$, $x' \in X$ such that $z^o(x', y', z') = z$ and $x^o(x'/*, y, z') = x/*$.

When the vertically moving signal is considered, two symmetrical conditions, similar to the condition stated for the horizontally moving signal, can be introduced. The first condition introduces the independence from the diagonally moving signal, while the second condition introduces the independence from the horizontally moving signal. These conditions can be stated mathematically as follows:

C3: for every $y \in Y$, there exist $y' \in Y$ and $x \in X$, such that $y^o(x, y'/*, Z) = y/*$.

C'3: for every $y \in Y$, there exist $y' \in Y$ and $z \in Z$, such that $y^o(X, y'/*, z) = y/*$.

As it will be shown in the definition of the test procedure, only one of the two conditions C3 and C'3 is necessary and sufficient for controllability.

Let's now consider the conditions that allow to observe the test results by propagating them, without masking, to the primary outputs of the array. One condition for each signal flow has been identified. For the horizontally moving signal propagation of x/* for some diagonally moving signal z has to be guaranteed, no matter what the vertical input is. Thus:

O1: for every $x \in X$, there exist $z \in Z$ such that $x^o(x/*, Y, z) = x'/*$.

For the diagonal and vertical signal flows, two analogous conditions have been derived, that can be respectively stated mathematically as follows:

O2: for every $z_1, z_2 \in Z$, with $z_1 \neq z_2$, there exist $x \in X$, $y \in Y$ such that

$$z^o(x, y, z_1) \neq z^o(x, y, z_2)$$

O3: for every y_1, $y_2 \in Y$, with $y_1 \neq y_2$, there exist $x \in X$, $z \in Z$ such that

$$y^o(x, y_1, z) \neq y^o(x, y_2, z)$$

Condition O2 (O3) means that for any two distinct diagonal (vertical) inputs z_1, z_2 (y_1, y_2), there exist inputs x, y (x, z) that generate two distinct diagonal (vertical) outputs. Thus any two distinguishable values of the diagonal (vertical) inputs can be propagated as distinct values of the diagonal (vertical) output.

The sufficiency of the conditions stated above can be proved by showing how to drive a set of test inputs to any cell (i,j) and how to take the test outputs from cell (i,j) to the observable outputs.

Therefore the test procedure can be subdivided into two distinct parts:

a. first show how the controllability conditions have to be applied to generate the appropriate sequence of values of the external inputs to the array such that it is possible to get the correct test inputs to the cell under test (i,j).
b. Then, show how to apply the observability conditions to find the appropriate values of the external inputs of the array so that the faulty test outputs can be propagated to the observable cells at the boundaries of the array.

To test cell (i,j) it is necessary to setup its test inputs, let's say (x_0, y_0, z_0). If condition C3 is applied to get y_0 then the setup procedure is different at the first step, but after that it is the same for all three test inputs.

Let's consider only the diagonal input z_0. Condition C2 guarantees the existence of $x_{i-1,j-1}(t-1)$, $y_{i-1,j-1}(t-1)$ and $z_{i-1,j-1}(t-1)$ such that $z_0 = z_{i,j} = z^o(x_{i-1,j-1}(t-1), y_{i-1,j-1}(t-1), z_{i-1,j-1}(t-1))$. Hence it suffices to apply input $x_{i-1,j-1}(t-1)$, $y_{i-1,j-1}(t-1)$ and $z_{i-1,j-1}(t-1)$ at cell $(i-1, j-1)$ at time $t-1$. Condition C3 (or C'3) assures that it is possible to find the necessary inputs in order to generate $z_{i-1,j-1}(t-1)$. Condition C1 assures that it is possible to define the necessary inputs in order to generate $x_{i-1,j-1}(t-1)$. In order to find the diagonal input $z_{i-1,j-1}(t-1)$ condition C2 has to be applied again. Therefore the problem can be solved similarly. This means that it is required to apply to cell $(i-2, j-2)$ the inputs $x_{i-2,j-2}(t-2)$, $y_{i-2,j-2}(t-2)$ and $z_{i-2,j-2}(t-2)$ that will allow to generate $z_{i-1,j-1}(t-1)$ through condition C2. Condition C1 assures again the possibility to determine the horizontal input $x_{i-2,j-2}(t-2)$. Condition C3 guarantees the existence of a set of inputs such that $y_{i-2,j-2}(t-2)$ will be generated. Then by applying iteratively conditions C1, C2 and C3 the directly controllable inputs can be reached and the values of the sequences of external inputs to be applied such that generation of the correct signals in all the cells involved in the setup process of the test inputs can be determined.

This solves the controllability problem.

To perform the propagation of results to the observable outputs, let's assume that the normal horizontal, vertical and diagonal outputs of cell (i, j) under the test inputs (x_0, y_0, z_0) are respectively x, y, z.

Assume that the test for the fault z/z' in the diagonal input is executed. Propagation of the diagonal output is performed by using condition O2 an then using the controllability conditions necessary to setup the correct inputs for the application of O2. A problem arises if condition C3 is used for the vertical-moving signal. In fact in that case in order to generate y such that propagation of the diagonal-moving signal z can be performed in a distinguishable way, it is first necessary to generate a right moving signal from cell (i, j), the cell under test. But this could cause error masking. In fact if cell (i, j) is faulty, the right moving signal generated will be different from the expected one and the vertically moving signal will also be different from the expected one. Therefore a different diagonal-moving signal will be propagated at the external outputs, showing that a fault is present in cell (i, j). In the other case, if condition C'3 holds, no problem of this type arises.

Condition O2 assures the possibility to propagate the diagonally moving signal $z(t)$ such that $z^o = (x(t), y(t), z(t)/z'(t)) = z(t+1)/z'(t+1)$. This can be obtained by applying recursively the controllability conditions for the horizontal and vertical moving signals in order to obtain the $x(t)$ and $y(t)$ at cell (i,j). Then using inductively condition O2 and the controllability conditions necessary to drive the correct inputs for condition O2 to be applied at the cells of the propagation path of the diagonally moving signal, it can be guaranteed that the diagonally moving output can be propagated to an observable output.

Finally condition O3 guarantees the existence (analogously to the cases analyzed above) of a right moving input $x(t)$ and a diagonally moving input $z(t)$ such that distinguishability of vertical outputs can be always maintained. The test procedure is analogous to the one for the diagonal moving signal.

This solves the observability problem. Hence the following theorem is proved.

Theorem 1. Any hexagonal array of combinational cells that satisfies the set of unrestricted testability conditions, is testable for any single faulty cell.

The testing time has to be evaluated. The procedure given above for the solution of the observability problem, must be repeated for any possible diagonal signal $z' \neq z_0$, then it is possible to state that cell (i, j) has been completely tested for input z_0. A similar procedure has to be performed for x_0 and y_0 and then it is possible to declare cell (i,j) tested for input (x_0, y_0, z_0). This is then repeated for each possible input combination in $X \times Y \times Z$. Cell (i,j) is then tested completely. Therefore the total testing time is given by $(|Z|^2 \times |X| \times |Y|)t$, where t is the time required to control and propagate a set of values for cell (i,j). Time t can be evaluated in the order of $O(N\sqrt{N})$, where N is the total number of cells in the array.

4. Linear Time Testability

Linear testing of hexagonal arrays has been presented in [4]. In this case independence of one signal flow from the other two is required. The conditions given in the following allow to achieve linear time testing without requiring such a strict dependence.

The conditions given below for linear time testability can be proved necessary if the following three general assumptions hold [5].

Assumption 1. There must exist a way to initialize one of the signals of the array.

For example let's assume that the horizontally moving signal can be initialized: thus, the horizontal input of a cell (i, j) can be controlled by controlling only the horizontal input of the leftmost cell and the diagonal inputs. Vertically moving signals do not matter for this purpose. If this assumption is dropped, there is no direct way to drive the fault-free array to a state where the inputs of all cells are known. Hence, it will be assumed that the following condition hold (but anyone of the controllability conditions given for the unrestricted case in the previous section would satisfy this assumption):

C1: for every $x \in X$ there exist $x' \in X$, $z \in Z$ such that $x^o(x', Y, z) = x$.

Assumption 2. When testing for faults initiated at some cell (i, j), these faults are not propagated to an observable output through that cell.

This condition is necessary because if a fault is allowed to be propagated to an observable output through the same cell that caused it, there is no guarantee that the fault will not be masked.

Assumption 3. Any test inputs, in any order, can be applied to any cell of the array.

This assumption is justified by the following reason. If the array operates at the word level, the cells may be sequential at the bit level (even if they operate combinationally at the word level): hence the order of test inputs may be important.

To obtain linear time testability it is necessary to introduce three conditions, one for each signal flow direction, which allow simultaneous controllability and observability of signals. This is motivated by the following reasoning. To test a particular cell for inputs (x, y, z) two procedures must be performed: first, apply controllable inputs to the array so that the inputs of the cell under test become indeed (x, y, z) and then propagate its test results to the observable outputs. Obviously this takes a time in the order of the number of cells in the array. Therefore this cannot be done independently for each cell, otherwise the time required is proportional to N^2, where $N = n \times n$ is the total number of cells in the array. Hence, at the same time that a fault is propagated to an observable output through some cell (i, j) propagation of the required input to some other cell through the same cell (i, j) should be provided.

If for example propagation of a fault x_1/x_2 through some cell (i, j) is considered, at the same time from cell (i, j) generation of an output z such that an input z' can be generated for another cell in a successive time step has to be provided. The same must hold for signal y so that an output y' is generated by the same cell (i, j) in order to produce another input y for some other cell. Therefore it is necessary to introduce the following condition:

OC1: for every $x_1, x_2 \in X$ with $x_1 \neq x_2$, for every $z \in Z$ there exist $z' \in Z$ and for every $y \in Y$ there exist $y' \in Y$ such that $x^o\ (x_1, y', z') \neq x^o(x_2, y', z')$, $z^o\ (x_1, y', z') = z$ and $y^o\ (x_1, y', z') = y$.

OC1 states that for every fault x_1 / x_2 in a left input and for every desired output y and z there exist a vertical input y' and a diagonal input z' such that they propagate the fault to the right output; they produce z as a diagonal output and they produce y as vertical output. If condition OC1 does not hold for these x_1, x_2, z, the operation described above cannot be done.

OC2 and OC3 are the symmetrical versions of OC1 for the other two signal flows, and the same arguments hold for their validity.

OC2: for every $z_1, z_2 \in Z$ with $z_1 \neq z_2$, for every $x \in X$ there exist $x' \in X$ and for every $y \in Y$ there exist $y' \in Y$ such that $z^o\ (x', y', z_1) \neq z^o\ (x', y', z_2)$, $x^o\ (x', y', z_1) = x$ and $y^o\ (x', y', z_1) = y$.

OC3: for every $y_1, y_2 \in Y$ with $y_1 \neq y_2$, for every $x \in X$ there exist $x' \in X$ and for every $z \in Z$ there exist $z' \in Z$ such that $y^o\ (x', y_1, z') \neq y^o\ (x', y_2, z')$, $x^o\ (x', y_1, z') = x$ and $z^o\ (x', y_1, z') = z$.

Given these conditions it can be proved, as described in [9], that the following theorem holds.

Theorem 2. The set of conditions C1, OC1, OC2 and OC3 is a sufficient set of conditions for testability of the entire array in a time proportional to the number of cells of the array.

Let's consider the propagation of the vertically moving signals in an hexagonal array. Since the diagonal signals are always ready at the cell one time unit before necessary (because their path from the directly controllable inputs to any given cell is shorter of one with respect to the horizontally or vertically moving signals) then, in order to avoid any conflict in the production of diagonally moving signals necessary to the propagation of the vertically moving signal to the observable outputs, one time unit delay is introduced between one input and its following input at the controllable boundaries of the array.

Test of one cell is performed in $2n$ time units. By pipelining, test of one row (or column) of cells is performed in $T+2n$ time units, if T is the number of tests required. Therefore $O(N)$ time is necessary to test an array of $N = n \times n$ cells.

5. Design for C-Testability of hexagonal cells

The proposed approach to C-testability of generalized hexagonal arrays is based on adding a constant number of additional states in each direction. This technique has already been applied to the case of two-dimensional orthogonal arrays [3,7].

In the following it will be assumed that the number of states in the horizontal and vertical directions respectively $m, n >> 4$, while in the diagonal direction it will be assumed that $l >> 2$. The total number of states m' along a direction (i.e. the sum of the original states and additional states) is equal to the number of binary combinations of the patterns in the lines along that direction, i.e. $m' = 2^{x_L}$, where x_L is the number of lines in the x-direction. This assumption is not restrictive, because it has been proved [3,7] that redundant states can be easily taken care of. It has also been proved that the number of redundant states does not affect the complexity of the testing process. The redundant states are characterized by a property which allows them to pass through unmodified, independent of the other two inputs.

The general approach to C-testability can be characterized by introducing four additional horizontal states (x_{m+1}, $x_m + 2$, $x_m + 3$, $x_m + 4$), four additional vertical states (y_{n+1}, y_{n+2}, y_{n+3}, y_{n+4}) and two additional diagonal states (z_{l+1}, z_{l+2}) to the original *mnl* transition table of the basic cell, such that the additional x-, y- and z-states satisfy 6 properties that can fully determine their behavior under all possible input combinations. These 6 properties are shown here only graphically (figures 1a and 1b), but their mathematical formulation can be found in ([6], [9]).

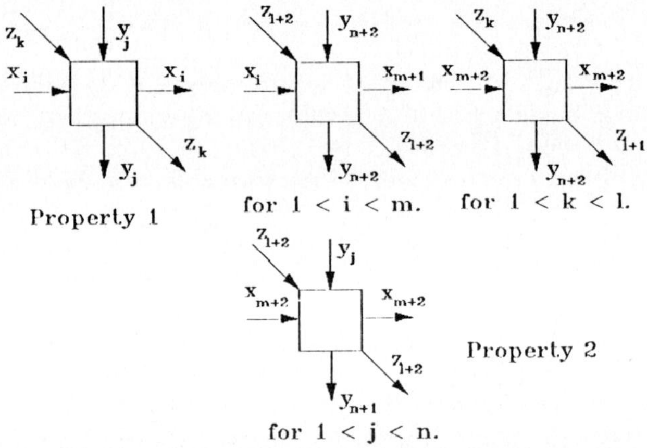

Figure 1a: Properties 1 and 2 that characterize the design for C-testability

Property 1 states that in all cases not considered by the following five properties, given a horizontal input, a vertical input and a diagonal input, the outputs of the basic cell are unchanged. Property 2 implies that it changes the original x-state (or y-state or z-state) into x_{m+1}, (or respectively into y_{n+1} or z_{l+1}) while the other two additional states along their directions do not change.

Property 3 reverses the effect of property 2 by regenerating an original state and keeping unchanged the additional states. Property 4 states that if all inputs are in the second additional state, then their outputs remain unchanged. Property 5 implies that the third additional state is unchanged (for two directions) under the conditions of property 5 while modifying the other original state. The effects of property 6 are similar to those of property 5 while generating a different original state and preserving the fourth original state intact. Properties 5 and 6 allow to create two different cycles, for each original state input, in the cellular automata that represents the basic cell function.

Note that all properties are symmetric for the horizontal and vertical signals, while different properties apply for the diagonal signal. The fact that the properties are symmetric is an important feature that is exploited in the exhaustive proof of the sufficiency of the six properties, since the same cases analyzed for the horizontally moving signals can be applied similarly to the vertically moving signals.

In the proof [6] property 5 is used to drive a x-state x_i to x_{m+1}, that is a distinguishable state. Then property 3 is used to drive x_{m+1} to x_r; x_r is driven to x_i by property 6,

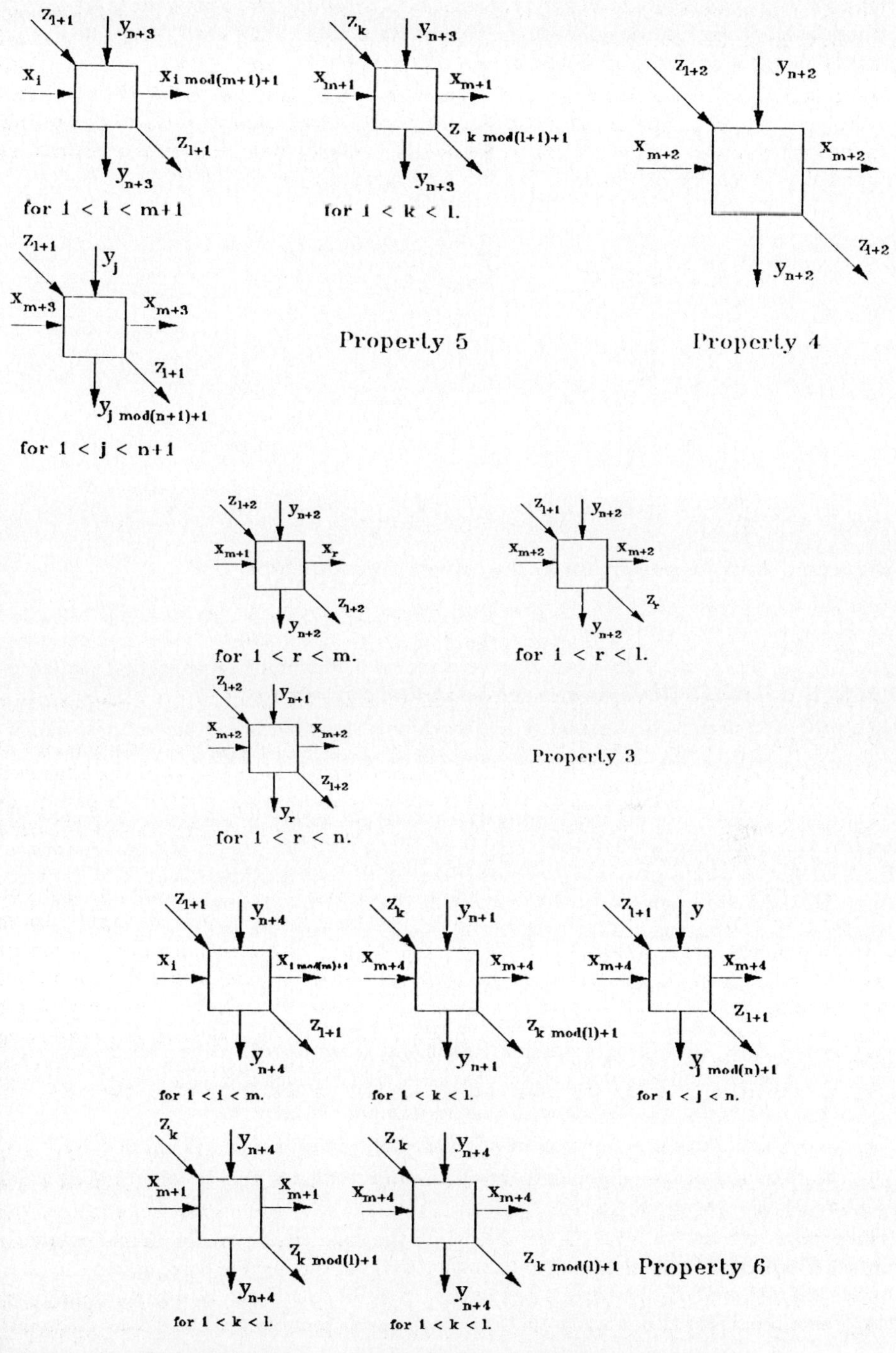

Figure 1b: Properties 3, 4, 5, 6 of the C-testability approach

therefore x_i is regenerated. Property 2 is used to drive an erroneous x-state to x_{m+1}. Property 1 is used to propagate the erroneous state after it has been driven to x_{m+1}. This is shown graphically in figure 2.

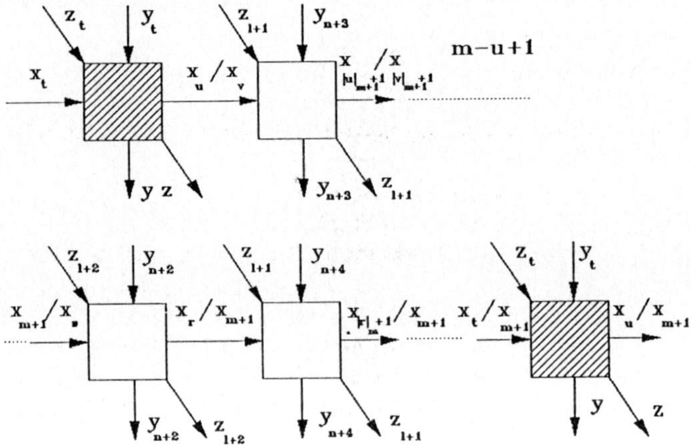

Figure 2: Test regeneration and propagation sequence

Another advantage given by the previous properties is that y^o $(x_i, y_j, z_k) = y_j$, z^o $(x_i, y_j, z_k) = z_k$, for $1 \leq i \leq m+4$, $n+2 \leq j \leq n+4$ and $k \geq l+1$. Therefore $y_{n+2}, y_{n+3}, y_{n+4}, z_{l+1}, z_{l+2}$ can be used to form a test input regeneration sequence due to this feature. This sequence can be applied to every row.

The proposed design for testability approach to C-testable hexagonal arrays, which is characterized by the six properties requires at most $(2m+1)(2n+1)(2l+1)(m+4)(n+4)(l+2)$ test vectors.

This can be proved [6] by considering the maximum length of each test sequence in each of the three directions of the signal flow. The maximum length of a test sequence for the right moving signal x is $(2m+1)$ (case 1 [6]). The maximum length of the test sequence for testing any vertical transition is $(2n+1)$ by symmetry. The maximum length of the test sequence for testing the diagonal transition is $(2l+1)$. Hence the maximum length of test vectors to test the array is $(2m+1)(2n+1)(2l+1)(m+4)(n+4)(l+2)$ because the modified transition table has $(m+4)(n+4)(l+2)$ states.

Finally, let's consider an application of this methodology. In [8] a semi-systolic transversal filter with programmable coefficient is presented. The cell design can be considered as an hexagonal cell by characterizing the signal flow follows:

- the horizontally moving signal y travels unmodified: $x^o = x$;
- the vertical output s depends from all the cell's inputs: $s = (x \cdot y) \oplus a \oplus b$;
- the diagonal output c depends on all the input signals and is computed as $c = (x \cdot y) \cdot a + (x \cdot y) \cdot b + a \cdot b$.

This original cell design is not C-testable, since the conditions given in [4] are violated: in fact there is more than one signal that is modified during the computation in the basic cell. By applying the given design for testability approach and thus modifying the transition table and adding the four horizontal, four vertical and two diagonal additional states, the array becomes testable with a number of test vectors independent of the number of cells in the array, but depending only on the number of states in the original transition table. Since these cells work on binary signals, then $m = 2$, $n = 2$ and $l = 4$. Thus the worst case number of test vectors is equal to $5^2 \times 9 \times 6^3 = 6300$.

6. Concluding Remarks

The paper has presented two different approaches to the problem of testing WSI arrays with complex interconnections such as hexagonal arrays.
The first approach is based on the generation of test procedures of different complexity depending on the characteristics of the signal flow of the basic cell. Conditions and test procedure to achieve a test complexity proportional to the number of cells in the array has been proposed.
The second approach is based on a technique of design for testability. The basic cell is described only by its state transition table without any knowledge of the function performed and the state table is augmented in order to allow state controllability and observability. In this case constant testability is achieved.

References

[1] Kung S.Y., "VLSI Array Processors", Prentice Hall, Englewood Cliffs, NJ, 1988.

[2] Rote G., "A Systolic Array Algorithm for the Algebraic Path Problems (shortest paths, matrix inversion), Computing, (34), pp.192-219, 1985.

[3] Elhuni H., Vergis A., Kinney L., "C-testability of Two-Dimensional Iterative Arrays", IEEE trans. on CAD of ICAS, vol.CAD-5, N.4, pp.573-581, 1986.

[4] Elhuni H., Kinney L., "Techniques for Testing Hexagonally Connected Systolic Arrays", Proc. ICCAD 1986, 1986.

[5] Vergis A., Steiglitz K., "Testability conditions for bilateral arrays of combinational cells", IEEE Trans. Computers, vol.C-35, n.1, January 1986, pp.13-22.

[6] Sciuto D., "Testing and Reconfiguration Techniques of VLSI Processor Arrays", PhD Dissertation, University of Colorado, USA, December 1988.

[7] Huang W-K, Lombardi F., Sciuto D., "Design and Analysis of C-Testable Arrays", Wafer Scale Integration, M. Lea ed., December 1988.

[8] Noll T., "Semi-Systolic Maximum Rate Transversal Filters with Programmable Coefficients", Systolic Arrays, 1986, pp.103-112.

[9] Sciuto D., "Linear and Constant Testability of Hexagonally Connected Arrays", Proc. ISCAS 89, Portland, OR, May 1989.

[10] Negrini R., Sami M.G., Stefanelli R., "Fault Tolerance Through Reconfiguration in VLSI and WSI Arrays", The MIT Press, 1989.

[11] "Wafer Scale Integration II", (R.M. Lea ed.), IFIP, North-Holland, 1988.

[12] Petrolly J.P., "A Systolic Device for Image Transformation", in Wafer-Scale Integration (G. Saucier and J. Trilhe eds.), IFIP, North-Holland, 1986, pp.59-76.

Wafer Scale Integration, III
M. Sami and F. Distante (Editors)
Elsevier Science Publishers B.V. (North-Holland)

ON TESTABLE REDUNDANT PLAS FOR WSI IMPLEMENTATION AND MANUFACTURING*

Yi-Nan Shen and Fabrizio Lombardi

TEXAS A&M UNIVERSITY
Department of Computer Science
College Station Tx 77843-3112

*This research supported in part by grants from AT&T and NATO.

ABSTRACT.

The yield enhancement of programmable logic arrays requires full diagnosis i.e. not only fault detection, but fault location and identification for single and multiple faults must be accomplished. This paper presents the basic structure of a testable and repairable PLA (TRPLA) and the design modifications which are required for a full diagnosis and yield enhancement. The testing process is fully analyzed and the conditions for diagnosis are presented. It is proved that identification in the presence of multiple faults (crosspoint, stuck-at, bridge) is possible with high coverage. The criteria which permit diagnosis, are based on a hierarchical organization of the testing process; significant improvements over previous redundant structures can be achieved.

1. INTRODUCTION.

Today's computer systems requires complex, yet modular devices for an efficient imbedding of large logic structures. One of the devices which achieves this objective at modest overhead, is the programmable logic array (PLA) [1,2]. Examples of the application of PLAs for the design of general purpose processors (such as microprocessors) as well as dedicated systems, can be found in [6]. PLAs can be manufactured using complex integration technologies such as VLSI and WSI [2]. Demand for high density PLAs is expected to significantly grow over the next few years as technology makes remarkable improvements for speed and yield enhancement [6]. A major design problem for providing cost effectiveness using PLAs is the ability of attaining high yield at manufacturing time, while still preserving a compact density [11]. Yield is commonly defined as the ratio of the number of working (deliverable) devices over the number of manufactured devices [19]. A high yield can not always be guaranteed [15]. As device geometry shrinks, the probability of having defective chips increases. This can result in a significant reduction of yield to a point of unfeasability [9]. Redundancy is commonly employed to provide spare resources in the presence of faults. Spares are used to circumvent and replace faulty components, thus enhancing the probability of producing working devices out of the faulty devices. This process is commonly referred to as repair [5]. Redundancy has been extensively used for yield improvement of chips such as memory devices by row/column deletion [5].

The addition of redundancy however, is not free from possible negative effects. Switching and routing of signals inherently increase the time delay, thus reducing the operational speed of the chip. A large amount of redundancy can also cause a decrease in yield, as the added complexity can outstrip the ability of repairing a chip by increasing the number of faults [19]. Usually, a fault intolerant area is required. This must be kept at a minimum level to reduce any hard core [11]. A further requirement for yield enhancement is to provide good testing capabilities. An extensive literature can be found for PLA testing [3,4]; this encompasses Built-In Testing (BIT) and Self-Testing (BIST) under both single and multiple fault models [7,8,10].

However, different criteria must be met at production-time. They radically differ from those associated with run-time [20]. These new criteria can be formulated as follows.

1. The multiple fault model must be applied. At run-time, the single fault model is a more reasonable assumption [20].
2. The time required for testing at production-time is not as stringent as at run-time. This permits a more comprehensive testing phase of the chips.
3. Fault identification/location as well as detection are required for production-time repair of PLAs. This suggests that while fault detection can be accomplished using the commonly-used LSSD technique of [21], more stringent conditions must be met to identify and locate faults.
4. Overhead due to added fault intolerant circuitry as well as redundancy in the components should not be excessive, or at least not to significantly affect the overall reliability of the whole chip [9]. The homogeneous layout of a PLA should be preserved [1].
5. Effective solutions for CAD implementation should be provided.

The objective of this paper is to analyze and present the basic principles of yield enhancement of WSI PLAs, with particular reference to testable redundant PLAs (TRPLA).

2. REVIEW AND PRELIMINARIES.

Repair of PLAs has been explicitly proposed in [9]. The basic PLA structure is argumented by spares for all lines (input, output, product). Repair rules under the common PLA fault model [10,12] as well as routing and switching circuits, are presented. These rules specify the spare requirements once faults are located. Faults that are analyzed, include stuck-at, crosspoint and bridge in both planes [9]. Faulty lines are replaced by spares and switching the appropriate signals through selectors at the input/output boundaries. It has been proved that using this method an efficient floor plan of a RPLA (Repairable PLA) can be achieved [9]. This also results in a moderate overhead for providing sparing capabilities; the largest single area overhead is accounted for by a spare input line. Successive simulation results have proved [11] which the majority of faults that can occur in a RPLA, can be repaired using a spare product line. A limited amount of input/output spare lines must be however, provided to guarantee a high level of repair capability. The main disadvantage of this configuration [9] is that fault location and diagnostic issues are not addressed for actually implementing the repair of the RPLA.

This aspect is analyzed in [15]. A layout which is basically similar to the one in [9] (Figure (1)), is proposed: a different utilization of the programmable switches, however, is used. The switches are used to either connect, or cut interconnections in the augumented PLA. Rerouting of signals is performed by cutting links to switch out faulty lines and connecting links to route signals to the spare lines. Three augumented lines are provided to enhance the diagnosis of the PLA structure [12,14]. The first augumented ouput line Z^* is provided for observability. Z^* is the logical OR of all product lines. The other two lines , I_1^* and I_2^*, are added between the input decoders and the AND plane to enhance the ability of identifying and locating faults in the AND plane. A further characteristic of this structure [15] is that spare lines are not utilized in the diagnostic capabilities of the augumented RPLA. This is a direct result of the diagnostic algorithm. The diagnostic algorithm of [15] is based on the following steps.

1. Test of design of diagnosability circuitry. This step tests the fault intolerant part of the RPLA as consisting of the augumented Z^* line and the shift register. This consists of a sequential shift of appropriate values through the register and the activation of each product line through the observability point provided by Z^*.

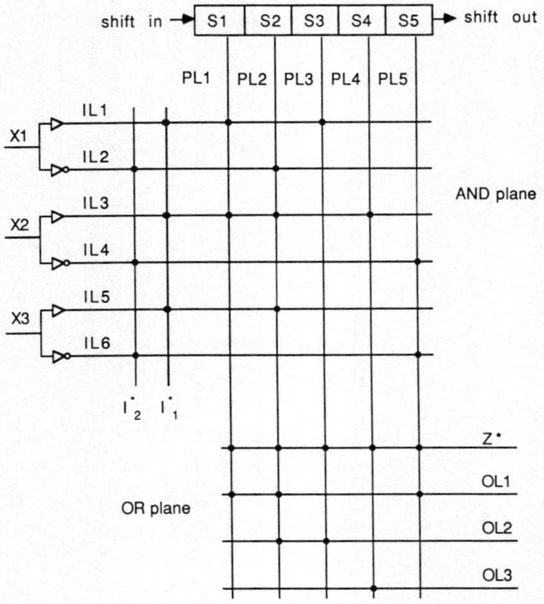

Fig.1 Example of augmented PLA by [15]

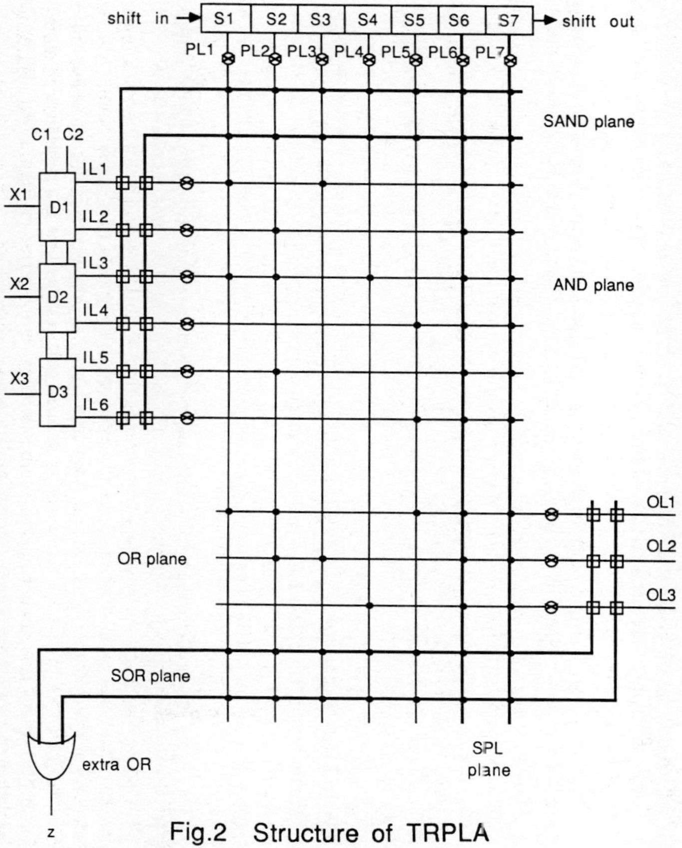

Fig.2 Structure of TRPLA

2. AND plane diagnosis. This step is based on the sequential activation of each product line. The results are observed at Z^* using I_1^* and I_2^*. The four possible cases for observability are analyzed. For each additional case, further tests are applied to reduce the ambiguity in the fault location.
3. OR plane diagnosis. This step is similar to the step for AND plane diagnosis. Appropiate settings of the product lines are used with the results of the previous step to locate the faults.

The structure of [15] is elegant; however, it suffers from many shortcomings in the repair process. Consider the following scenario (using the notation of [15]). If the true bit of X_1 is s-a-1 and a device in the crosspoint between this input line and I_1^* is also faulty, then the IS1[1] fault is not detectable. Thus, Algorithm test_AND(i,v) of [15] incorrectly diagnoses R_3 and R_5 to have a PS0 fault. This occurs because v=0. This diagnosis makes a potentially repairable device to be discharded.

3. THE STRUCTURE OF THE TRPLA.

Examples of Testable PLA structures can be found in [13,14,16]. These devices stem from the basic configuration proposed in [3]. This PLA is provided with a shift register *SR* in the upper part of the AND plane and with a decoder between the primary input lines and the generation of the inputs to the AND plane.

The proposed configuration of a TRPLA is shown in Figure (2). This device complements the testable capability of the PLA of [3] with the RPLA configuration of [9]. The TRPLA shown in Figure (2) also requires the addition of an OR gate. The output of this gate is directly observable, while its inputs are given by the spare product and ouput lines. It is this characteristic that set apart this configuration from the one proposed in [15]. In a field PLA, this is connected to ground via either a fuse or a suitable arrangement (as shown in Figure (3)) using nMOS transistors. The structure of the decoder D is shown in Figure (4). When the control line C_1 is low, the output IL_{2i-1} is pulled to ground.

The physical configuration of PLAs is commonly of a NOR-NOR type [3]. The upper plane is referred to as the AND plane, while the lower plane is referred to as the OR plane. In this paper, for ease of description, the logical AND-OR description of the planes is utilized in place of its physical description. This results in an analysis which is independent of the physical implementation of the PLA [12].

The layout of a TRPLA can be divided into the following planes:

1. OR-plane, whose rows consist of output lines.
2. AND plane, whose rows consist of input lines.
3. Spare OR (SOR) plane, whose rows consist of output spare lines.
4. Spare AND (SAND) plane, whose rows consists of input spare lines.

The columns of the above planes are given by the product and the spare product lines. Note that in this paper the word input line is used in place of bit line. The input lines of [15] are referred as primary input lines.

A TRPLA has n input lines, p product lines and m output lines. The number of primary input lines is therefore $n/2$. The number of available spare input lines is denoted by n_a, while the number of available spare product and output lines are given by p_a and m_a respectively. The ith row of the AND plane (spare AND plane) is denoted as IL_i(SIL_i). The jth column of the TRPLA (i.e. a product line) is denoted as PL_j. The column corresponding to the rth spare product line is denoted as SPL_r. SPL_r can also be denoted as PL_{p+r}. The kth row of the OR plane (spare OR plane) is denoted as OL_k (SOL_k). The TRPLA shown in Figure (2) has n=6, n_a=2, p=5, p_a=2, m=3 and m_a=2.

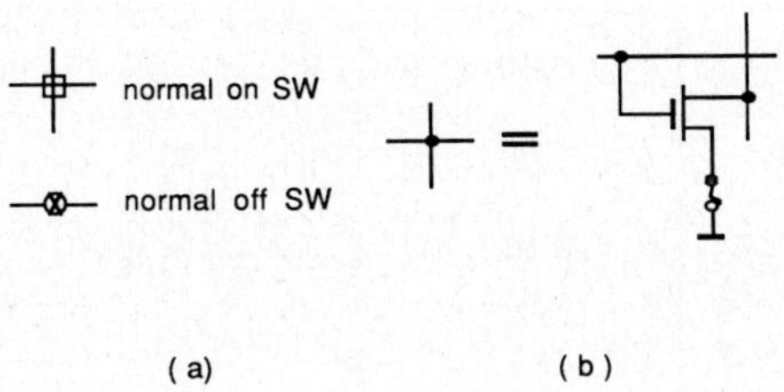

Fig. 3 Switches in a Testable Redundant PLA (TRPLA)

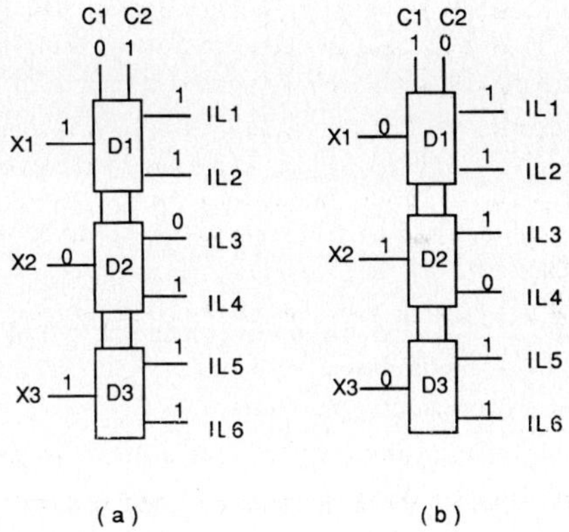

Fig. 4 The outputs of the augmented decoder

TABLE 1 Test Set

Test	input line	shift register		z	output line
	$IL_1 \cdots IL_i \cdots IL_n$	PL_1 PL_j PL_p	PL_{p+1} PL_{p+r} PL_{p+pa}	z	$OL_1 \cdots$ $OL_k \ldots OL_m$
I^1	1 . . . 111 . . . 1	0 . . . 000 . . . 0	0 . . . . . 000 0	0	0 . . . . 0 . . . 0
I^2	1 . . . 111 . . . 1	0 . . . 000 . . . 0	1 . . . . . 111 1	1	1 . . . . 1 . . .
I^{3a}_j	0101. 01	0 . . . 010	 0	$z_a \wedge z_b = 0$	
I^{3b}_j	1010. 10	0 . . . 010	 0		
$j = 1,2, \ldots, p + pa$					
I^4_j	1 . . . 111 . . . 1	0 . . . 010	 0	1	$e_{1j} \ldots e_{kj} \ldots e_{mj}$
$j = 1,2, \ldots, p + pa$					
I^5_{ir}	1 . . . 101 . . . 1	0 . . 000 . . . 0	0 . . . . . 010 0	0	0 . . . 0 . . .0
$i = 1,2, \ldots, n$ $r = 1,2, \ldots, pa$					
I^6_{ij}	1 . . . 101 . . . 1	0 . . 010	 0	$\bar{d}_{ij}$	
$i = 1,2, \ldots, n$ $j = 1,2, \ldots, p + pa$					
I^{7a}_{ij}	1 . . . 101 . . . 1	0 . . . 010	 0	$z_a \oplus z_b = d_{ij}$	
I^{7b}_{ij}	1 . . . 111 . . . 1	0 . . 010	 0		
$i = 1,2, \ldots, n$ $j = 1,2, \ldots, p + pa$					
I^8_j	1 . . . 111 . . . 1	0. . . 0110	 0		
$j = 1,2, \ldots, p + pa - 1$					

* d_{ij} (e_{ij}) = 1 -- a device in the crosspoint (i, j) of AND (OR) plane
= 0 -- no device . . . .

** I^5_{ir} is a subset of I^6_{ij} , therefore during this test, only I^6_{ij} need to be . applied. I^7_{ij} does not to be applied during testing.

*** $I^3_j = I^{3a}_1 I^{3b}_1 I^{3a}_2 I^{3b}_2 \; \ldots\ldots \; I^{3a}_{p+pa} I^{3b}_{p+pa}$

And so is I^7_{ij}

The z_a (z_b) is the output , when I^a(I^b) is applied .

**** I^7 and I^8 are auxilary test sequences and I^8 is only used for diagnosing multiple faults.

The assumed fault model includes s-a-0 (1), bridge and crosspoint (missing and extra devices in the AND and OR planes), as in previous papers [8,12,15]. The ANd-OR logical description is assumed in the analysis. If the TRPLA is implemented in nMOS technology, the 0-strength is stronger than the 1-strength when the bridge fault in an AND gate is considered [17]. This implies that the strength of the *SR* in the TRPLA should be enought to pull down (i.e. to ground) the voltage level of the product lines, even if there exists a s-a-1 in a product line. This assumption is required for the *SR* to inactivate a product line. Note that this was inherently assumed also in [15]. This requirement can be acccomplished by enlarging the area of the pull-down MOS transistor in Figure (3). Faults in the *SR* can easily diagnosed [15] and are not dealt in this paper; the reader should refer to [13] for a detailed discussion of this aspect.

Definition 1. A test vector to a TRPLA is a $(n+p+p_a)$-tuple $\{IL_1, \dots, IL_i, \dots, IL_n, PL_1, \dots, PL_j, \dots PL_p, PL_{p+1}, \dots, PL_{p+r}, \dots, PL_{p+p_a}\}$.

The state of PL_j $(j=1, \dots, p+p_a)$ can be set to any desired value by scanning suitable data through the shift register. The state of IL_i can be obtained by providing the input test vector at the primary input lines (i.e. the X's) and selectively adjusting the control lines C_1 and C_2. The augumented decoder D works as follows. If $C_1=C_2=0$, the PLA works in normal mode. In the test mode, if $C_1=1$ and $C_2=0$, $IL_{2i-1}=IL_{2i}=1$ for $X_i=0$ and $IL_{2i-1}=1$ $IL_{2i}=0$ for $X_i=1$. The same holds for $C_1=0$ and $C_2=1$.

4. THE SINGLE FAULT CASE.

The test set shown in Table (1) is used under single fault assumption (excluding I^8). This set is made of various sequences which are basically the same as the ones used in [3,12]. However, these sequences are used not for fault detection only, but also for fault location and identification. This process is referred to as full diagnosis. The basic principles of the proposed diagnostic procedure follows a hierarchical scheme. This is as follows. The stuck-at fault can be detected relatively easy using the test set shown in Table (1), i.e. the test that can detect some crosspoint faults, can also be used to detect some stuck-at faults, but the reverse is not true. Therefore, the stuck-at fault must be diagnosed first. Having diagnosed the stuck-at fault and if it is found non existent, the remaining faults can be considered in the diagnosis. In the Lemmas that follow, it is assumed that the test sequence I^j is applied only if all the test sequences I^i $(i < j)$ have been successfully passed. It is also assumed that the output z has already been diagnosed as fault free.

The process for full diagnosis under single fault assumption [22] is based on the following testing conditions.

1. I^1 can diagnose a s-a-1 in SOL_t $(t=1, \dots, m_a)$ or a s-a-1 in OL_k $(k=1, \dots, m)$.
2. I^2 can diagnose a s-a-0 in IL_i $(i=1, \dots, n)$ or a s-a-0 in OL_k $(k=1, \dots, m)$ or a s-a-0 in z.
3. The pair of test sequences $\prod_{j=1}^{p}(I^{7a}{}_{ij}, I^{7b}{}_{ij})$ can diagnose that most of the IL_i are s-a-0 fault free. The diagnosis of a s-a-0 in IL_i can be achieved by modifying at most one SPL_j.
4. The test pair $(I^{3a}{}_j, I^{3b}{}_j)$ can diagnose a s-a-1 in PL_j.
5. The test $I^4{}_j$ can diagnose a s-a-0 in PL_j, a bridge fault between PL_j and $PL_{k'}$ (where $k'=j\pm1$), a crosspoint fault at (k,j) in the OR plane; the test sequence $I^4{}_j$ $(j=1, \dots, p)$ can also diagnose a bridge fault between OL_k and OL_{k+1}.
6. The test sequence $I^5{}_{ir}$ $(r=1, \dots, p_a)$ can diagnose a s-a-1 in IL_i.
7. The test $I^6{}_{ij}$ can diagnose a crosspoint fault at (i,j) in the AND plane. The test sequence $I^6{}_{ij}$ $(j=1, \dots, p+p_a)$ can diagnose a bridge fault between IL_i and IL_{i+1}.

Conditions (1) through (7) characterize input test sequences which are mostly function independent [3,12]. Only for the test sequence $\prod_{j=1}^{P} I^4{}_j$ and $\prod_{i=1}^{n}\prod_{j=1}^{P} I^6{}_{ij}$, the output responses are function dependent. The following theorem is easily proved.

Theorem 1. The test sequences I^1, I^2, (I^7 if I^2 detects faults), I^3, I^4, I^5, I^6 can diagnose all single stuck-at, bridge, and crosspoint faults in a TRPLA.

5. THE MULTIPLE FAULT CASE.

Fault location in the presence of multiple faults is a very challenging, yet important task [4]. This is imperative due to large number of multiple faults, e.g. if n is the number of lines, there are only $2n$ single s-a-0(1) faults, but there are 3^n-1 multiple s-a-0(1) faults [4]. Diagnosis is made more difficult by such negative coupling effects between faults as masking, equivalence and dominance [23]. The basic objective in a TRPLA is the improvement in yield enhancement and the requirement of fault location and fault diagnosis; this is reflected in the proposed approach by considering only a limited number of s-a-0(1) and bridge faults as meaningfull to the repair process. Faults with specific characteristics can be located at once after certain faults have been diagnosed. This effectively characterizes a hierarchical organization of the testing and diagnostic processes. The test sequences of Table (1) are applicable inclusive of I^8.

Prior to describing the proposed method, it is important to point out that the definitions and analysis are based on a well defined test set $\{T\}$ [23]. For a large TRPLA an exhaustive test set is unrealistic. Therefore, T is one or all of the test sequences shown in Table (1).

Definition 2. Let the test sets for faults f_1 and f_2 be T_1 and T_2 respectively, $T_1, T_2 \in T$. f_1 is equivalent to f_2 if $T_1 = T_2$. f_1 is dominated by f_2 if $T_1 \supseteq T_2$. f_1 maskes f_2 if T_2 is no longer the test for f_2 under the existance of f_1. All the above occurrences are broadly referred to as the covering of f_2 by f_1.

Definition 3. If f_1 covers f_2 and the diagnosis and repair of f_1 will cause f_2 not to exist, then f_1 is said to strongly cover f_2.

An example of strongly covering occurs in a PLA when a s-a-0(1) line fault covers all crosspoint faults on the same line [15].

Definition 4. Let a multiple fault set $\{F\}$ make all SPL_r to have faulty outputs. The set F is referred to as the complete covering product lines set ($CCPS$). Let a multiple fault set $\{F\}$ make the ith row of the SPL_r to have a faulty output. The set F is referred to as the complete covering product line set for R_i and is denoted as $CCPS_i$ (the subscript i is changed to k if the faulty row is in the OR plane).

As an example, assume that thre are two SPL_r (r=1,2). A $CCPS$ is a faulty set F which contains a s-a-0 of SPL_1 and a s-a-1 of SPL_2. If F contains a s-a-0 of SPL_1 and a missing fault in crosspoint (3,2), this set is a $CCPS_3$.

Definition 5. Let a multiple fault set $\{F\}$ make the jth column of all SOL_k to have a faulty output. This set F is referred to as the complete covering output lines set for C_j ($CCOS_j$).

The proposed approach for diagnosis and repair in the presence of multiple faults aims at distinguishing the covering properties of the faults. This approach is based on the following basic procedural algorithm and associated conditions.

Algorithm: Multiple Faults.

Condition 1: Apply test I^k. If output is α, then no fault f_1 exists.

Condition 2: Apply test I^k. If output is α AND no fault f_1 exists, then fault f_2 exists and no f_3 exists.

Condition 3: If some new faults have been detected or excluded, then repeat from Condition (2).

The above algorithm follows from the basic results of Section (4) as the diagnosis of some faults can follow from the exclusion of others. Some faults can be excluded very easily: for example if the output of OL_k is 0(1), then there is no s-a-0(1) fault in this line, or if the output line OL_k and OL_{k+1} are different, then no bridge fault exists. Consider for example Condition 2. This condition implies that if f_1 dominates or is equivalent to f_2 under test I^k, then the fault f_2 can be diagnosed only after excluding f_1. Equivalently, if f_1 masks f_3 under test I^k, then f_3 can be excluded only if no f_1 exists.

Definition 6. An output is valid only if the output is α ($\alpha \in \{0,1\}$) and the output has no s-a-α fault.

Only valid outputs are usefull in detecting faults other than output faults. Therefore, validity is used as the main criterion for diagnosis and to achieve high fault coverage. Multiple fault diagnosis is based on applying the following test sequences (they are applied in the same order of appearance [22]).

1. I^1 diagnoses s-a-1 of z and s-a-1 of OL_k ($k=1, \ldots, m$).
2. I^2 diagnoses at least one s-a-0 at IL_i and is effective if there is no $CCPS$.
3. $I^{3a}{}_j$ and $I^{3b}{}_j$ (for $j=1, p+p_a$) diagnose s-a-1 of PL_j and are effective if no $CCOS_j$ exists.
4. $I^6{}_{ij}$ (for $i=1,n$ and $j=1, p+p_a$) diagnoses s-a-0 at IL_i; $I^4{}_j$ (for $j=p+1, \ldots, p+p_a$) proves that it exists no s-a-0 at IL_i; $I^4{}_j$ (for $j=1,p+p_a$) diagnoses s-a-0 at PL_j or bridge fault at PL_J. This sequence is effective if no $CCPS_i$, or $CCOS_j$ exists.
5. $I^5{}_{ij}$ (for $i=1,n$ and $j=p+1,p+P_a$) diagnoses s-a-1 fault at IL_i and is effective if no $CCOS_j$ and no $CCPS_i$ exist.
6. $I^6{}_{ij}$ proves that there is no s-a-0 at IL_i and together with $I^4{}_j$ (for $i=1,n$ and $j=1,p+p_a$), diagnoses crosspoint faults in the AND plane. The diagnosis of bridge faults at IL_i is accomplished by using $I^6{}_{ij}$ $I^6{}_{i+1,j}$ (for $i=1,n$). This sequence is effective in diagnosing a crosspoint fault in the AND plane if no $CCOS_j$ exists.
7. S-a-0 faults of IL_i are diagnosed by $I^{7a}{}_{ij}$ and $I^{7b}{}_{ij}$ (for $i=1,n$ and $j=1, \ldots, p+p_a$)
8. The auxiliary test for diagnosing bridge faults at PL_j consists of $I^8{}_j$ (for $j=1, p+p_a$). This is effective if no $CCOS_j$, no $CCOS_{j+1}$ and no bridge fault among three continuous PL lines exist.

To evaluate the proposed approach, two figures of merit are considered, namely detection coverage and area overhead. Detection coverage is given by one minus the probability that at least one fault is not detected using the procedures and algorithms presented in the previous sections. For single fault, it follows from the basic results of [3,12] that the test procedure of Section (4) yields a detection coverage of 1. For multiple faults, a more detailed analysis is required. The only cases that multiple faults cannot be detected, is in the presence of $CCPS_i$ and $CCOS_j$. Let N_t be the total number of crosspoint in the TRPLA, N_f be the number of faulty crosspoints, where $N_t=[(n+n_a)+(m+m_a)](p+p_a)$. The probability of a $CCPS_i$ can be found to be $P_0=\dfrac{N_f!}{N_t^{\,p_a}(N_f-p_a)!}$ for $N_t >> p_a$. The probability of the occurrence of any $CCOS_j$ ($j=1, \ldots, p+p_a$) is $P_2=\dfrac{N_f!(p+p_a)}{N_t^{\,m_a}(N_f-m_a)!}$ The following Theorem can be easily proved.

Theorem 2. The detection coverage under multiple faults is given by $1-(nP_0+P_2)$.

Note that bridge faults and s-a-α faults have been considered in an equal manner, because they require the same repair rules and they can be easily distinguished [9]. The second figure of merit which is considered, is the area overhead for a TRPLA. Compared with the RPLA structure of [9], the TRPLA shown in Figure (2), requires an additional $(p+p_a)$ lines shift register and an n lines augumented input decoder. The area for this first device is

$170 \times 8 \times (p+p_a) = 1360(p+p_a)$ using the lambda based rules of [2]. The argumented input decoder requires $135 \times 16 \times n = 2160n$ area (where n is half the primary input lines) more than a conventional decoder. The area required for the additional OR gate to produce the z signal can be neglected due to its small value compared with the overall TRPLA structure.

The floor plan of the TRPLA is shown in Figure (5). Table (2) shows the values of the original PLA area (A), the area required by the spares (A_s), the area required by the proposed approach for the testing hardware (A_t). All area values are in 1000 lamda [2]. The first two cases shown in this table correspond to the BELLMAC-32A and the AT&T WE32000 microprocessors [6] respectively.

Table 2.

n	p	m	n_a	p_a	m_a	A	A_s	A_t
50	190	67	0	3	1	1563.5	63.824	370.48
50	190	67	1	6	3	1563.5	152.72	374.56
60	200	60	0	3	1	1682.2	64.512	405.68
60	200	60	1	6	3	1682.2	155.328	409.76

It is obvious that the increase in area in a TRPLA is mainly caused by the testing hardaware and in particular the shift register. This overhead however, decreases proportionally as the size of the PLA increases. Note that a reduction in hardware is also possible using the improvement techniques described in [16].

6. SIMULATION RESULTS.

The proposed approach has been exensively evaluated; faults have been injected using a random distribution. The simulated PLAs are the commonly used benchmarks for ESPRESSO [18]. All simulated PLAs are in minimized form.

The following statistical information is provided.

1. Probabilty of correctly diagnosing a fault free line (P_{CD}).
2. Probability of correctly diagnosing a faulty line (P_D).
3. Probability of not-detecting a faulty line (P_U).
4. Probability of incorrectly diagnosing a fault free line as faulty (P_{MISD}).
5. Probability of uncertainty on the status of a fault free line (P_{UNC}).
6. Probability of wrongly diagnosing a fault type of a faulty line (P_{WF}).
7. Probability of a line being covered by a fault (P_{COV}) (this is only possible when multiple faults are considered).
8. Probability of uncertainty on the status of a faulty line (P_{UNCF}).

The above probabilities are identified for two classes of faults (given by a further subscript).

1. Crosspoint faults (c).
2. S-a-0(1) and bridge faults (sb).

Note that $P_{CDi}+P_{Di}+P_{Ui}+P_{MISDi}+P_{UNCi}+P_{WFi}+P_{COVi}+P_{UNCFi}=1$, for $i=c$ or sb. This implies that the fault coverage C is given by $(P_{CDc}+P_{CDsb}+P_{Dc}+P_{Dsb})/2$, while $C_c=C_{CDc}+C_{Dc}$ and $C_{sb}=C_{CDsb}+C_{Dc}$. The first fault file consists of 5 random crosspoint faults in the AND plane, 3 random crosspoint faults in the OR plane and 3 random faults (either stuck-at or bridge faults) in any plane. Table (3) shows the values of C_c, C_{sb} and C for the simulated PLAs.

Table 3.

PLA	C_c	C_{sb}	C
Z5xp1	89.68	96.21	92.94
adr4	93.42	98.96	96.19
alu2	88.12	95.01	91.56
alu1	68.10	87.06	77.58
alu3	87.68	95.67	91.67
co14	63.29	86.84	75.06
dc1	64.42	89.94	77.18
dc2	78.49	92.70	85.59
dk17	71.44	90.18	80.81
dk27	75.23	91.83	83.53
dk48	78.08	94.35	86.21
f51m	90.99	97.39	94.19
in3	82.23	93.95	88.09
rd53	76.30	91.26	83.78
in5	72.37	89.97	81.17
in6	85.99	94.83	90.41
in7	77.96	82.38	85.17
radd	90.31	97.31	93.81
rck1	67.80	84.86	76.33
risc	92.11	98.25	95.18
root	85.06	94.37	89.71
sqn	79.63	93.21	86.42
sqr6	86.40	95.71	91.05
wim	65.18	90.07	77.62
z4	91.57	97.82	94.69

Tables (4), (5), (6) and (7) shows all the statistical information provided by the proposed diagnostic process for a second fault file made of 1 crosspoint fault in the OR plane, 1 crosspoint fault in the AND plane and 1 either stuck-at or bridge fault in any plane.

Table 4.

PLA	P_{CDc}	P_{Dc}	P_{Uc}	P_{MISDc}
alu3	95.41	0.10	0.00	0.00
apla	92.49	0.20	0.00	0.00
co14	67.12	0.27	0.00	0.00
dc1	88.45	0.98	0.00	0.00
dc2	93.90	0.19	0.00	0.00
dk17	90.85	0.30	0.00	0.00
dk27	84.62	0.50	0.00	0.00
dk48	91.28	0.16	0.00	0.00
f51m	94.37	0.09	0.00	0.00
in3	94.88	0.02	0.00	0.00
rd53	86.24	0.37	0.00	0.00
in5	94.47	0.05	0.00	0.00
in6	93.17	0.04	0.00	0.00
in7	90.52	0.05	0.00	0.00
radd	91.38	0.11	0.00	0.00
rck1	89.12	0.06	0.00	0.00
risc	94.68	0.13	0.00	0.00
root	95.60	0.15	0.00	0.00
sqn	95.16	0.25	0.00	0.00
sqr6	95.59	0.15	0.00	0.00
wim	87.55	0.97	0.00	0.00
z4	90.76	0.16	0.00	0.00

Table 5.

PLA	P_{UNCc}	P_{WFc}	P_{COVc}	P_{UNCFc}
alu3	2.51	0.00	1.98	0.00
apla	2.94	0.00	4.36	0.01
co14	15.49	0.00	17.03	0.09
dc1	0.00	0.00	10.58	0.00
dc2	2.24	0.00	3.66	0.00
dk17	2.97	0.00	5.87	0.01
dk27	5.56	0.00	9.31	0.01
dk48	4.53	0.00	4.02	0.01
f51m	3.30	0.00	2.23	0.00
in3	2.34	0.00	2.76	0.00
rd53	8.29	0.00	5.07	0.04
in5	2.07	0.00	3.42	0.00
in6	2.68	0.00	4.11	0.00
in7	3.66	0.00	5.77	0.00
radd	5.69	0.00	2.83	0.00
rck1	3.92	0.00	6.88	0.01
risc	2.18	0.00	3.01	0.00
root	7.78	0.00	7.13	0.03
sqn	1.42	0.00	3.16	0.00
sqr6	1.09	0.00	0.28	0.00
wim	4.43	0.00	7.04	0.00
z4	6.24	0.00	2.84	0.00

Table 6.

PLA	P_{CDsb}	P_{Dsb}	P_{Usb}	P_{MISDsb}
alu3	97.93	0.32	0.00	0.00
apla	97.16	0.52	0.00	0.00
co14	84.17	0.55	0.00	0.00
dc1	97.03	1.45	0.00	0.00
dc2	97.83	0.48	0.00	0.00
dk17	97.11	0.67	0.00	0.00
dk27	94.06	0.79	0.00	0.00
dk48	96.76	0.43	0.00	0.00
f51m	97.53	0.30	0.00	0.00
in3	97.73	0.19	0.00	0.00
rd53	94.15	0.59	0.00	0.00
in5	97.99	0.26	0.00	0.00
in6	96.97	0.20	0.00	0.00
in7	96.17	0.25	0.00	0.00
radd	96.03	0.29	0.00	0.00
rck1	94.63	0.31	0.00	0.00
risc	98.02	0.44	0.00	0.00
root	97.73	0.40	0.00	0.00
sqn	98.09	0.55	0.00	0.00
sqr6	98.24	0.39	0.00	0.00
wim	96.23	1.23	0.00	0.00
z4	95.59	0.36	0.00	0.00

Table 7.

PLA	P_{UNCsb}	P_{WFsb}	P_{COVsb}	P_{UNCFsb}
alu3	1.47	0.00	0.28	0.00
apla	1.67	0.00	0.65	0.00
co14	11.79	0.00	3.49	0.00
dc1	0.29	0.00	1.23	0.00
dc2	1.31	0.00	0.35	0.00
dk17	1.81	0.00	0.41	0.00
dk27	3.81	0.00	1.34	0.00
dk48	1.97	0.00	0.84	0.00
f51m	1.99	0.00	0.18	0.00
in3	1.62	0.00	0.46	0.00
rd53	8.29	0.01	5.07	0.04
in5	2.07	0.00	3.42	0.00
in6	1.96	0.00	0.87	0.00
in7	2.69	0.00	0.89	0.00
radd	3.53	0.00	0.15	0.00
rck1	3.59	0.00	1.47	0.00
risc	1.15	0.00	0.39	0.00
root	1.45	0.00	0.42	0.00
sqn	1.00	0.00	0.36	0.00
sqr6	1.09	0.00	0.28	0.00
wim	1.89	0.00	0.65	0.00
z4	3.76	0.00	0.30	0.00

Table (8) shows C_{sb} for two pathological cases of PLAs versus the number of stuck-at/bridge faults only (denoted by F_{sb}); rd53 has a low fault coverage due to the high density of crosspoints in the identity matrix, while in3 shows a very sparse identity matrix after minimization. Note that the fault coverage for crosspoint faults C_c only, has been found by simulation to be always 100%.

Table 8.

F_{sb}	C_{sb}(rd53)	C_{sb}(in3)
1	99.20	51.04
2	98.68	51.72
3	91.62	44.20
4	86.41	40.42
5	82.90	40.70
6	83.72	36.50
7	77.97	37.34
8	78.42	40.78
9	75.66	35.64
10	69.41	39.67

The following conclusions may be drawn from the simulation results.

1. It has been found that P_{MISD} is always 0, independently of the fault types and their number. This means that the incorrect diagnosis of a fault free line as faulty never occurs in practice. This is obviously a positive feature of the diagnostic process.

2. The same consideration of (1) above also applies to P_U, (i.e. the probability of not-detecting a faulty line) when crosspoint faults are considered. The presence of stuck-at and bridge faults can however, result in $CCOS$ s and $CCPS$ s with obvious negative effects (as proved previously). This consideration also applies to P_{WF} as P_{WFc}=0 always.
3. The PLAs which have been simulated are the ones given by the benchmarks for the ESPRESSO system [18]. Their identity matrices do not necessarily satisfy the conditions given in [4,14] for testability. Hence, a sometime low value for fault coverage is possible.
4. The overall performance of PLAYER is very satisfactory considering that for the majority of the benchmark PLAs $P_{CD}+P_D \geq 90\%$ under all types of faults in the fault model. For all PLAs, simulation has shown that $P_{CDsb}>P_{CDc}$ as the probability of a bridge fault to affect 3 or more lines is very remote.
5. The minimization process by ESPRESSO [18] does not necessarily result in high fault coverage for a TRPLA. This is evident in Table (8), when in3 for example, is considered. Simulation results [22] have proved that fault coverage for the unminimized form of this device is significantly different from the one shown in Table (8). This is caused by the different and sometime contrasting requirements of logic minimization and testability for fault location and identification under multiple faults.

7. REFERENCES.

[1] Fleisher, H. and L.I. Maissel, "An Introduction to Logic Arrays," *IBM J. of Res. & Develop.* Vol. 19, pp.98-109, 1975.

[2] Mead, C.A. and L.A. Conway, "Introduction to VLSI Systems" Addison Wesley, reading, Mass., 1980.

[3] Fujiwara, H. and K. Kinoshita, "A Design of Programmable Logic Arrays with Universal Tests," *IEEE Trans. on Comput.*, Vol. C30, No. 11, pp.823-828, 1981.

[4] Agarwal, V.K., "Multiple Fault Detection in PLAs" *IEEE Trans. on Comput.*, Vol. C29, No. 6, pp. 518-522, 1980.

[5] Schuster, S.E., "Multiple Word/Bit Line Redundancy for Semiconductor Memories," *IEEE J. of Solid State Ccts.*, Vol. SC13, No. 5, pp. 698-703, 1978.

[6] Law, H.-F. S. and M. Shoji, "PLA Design for BELLMAC-32A Microprocessor," *Proc Int. Conf. Cir. Comput.*, pp. 161-164, 1982.

[7] Daehn, W. and J. Mucha, "A Hardware Approach to Self-Testing of Large Programmable Logic Arrays," *IEEE Trans. on Comput.*, Vol. C30, pp. 829-833, 1981.

[8] Ramanathan, K.S. and N.N. Biswas, "A Design for Complete Testability of PLAs," *Proc. Test Conf.*, pp. 67-74, 1982.

[9] Wey, C.L., Vai, M.K. and F. Lombardi, "On the Design of a Redundant Programmable Logic Array (RPLA)," *IEEE J. Solid State Circuits*, Vol. SC22, No. 1, pp. 114-117, 1987.

[10] Somenzi, F. and S. Gai, "Fault Detection in PLAs," *Proc. IEEE*, Vol. 74, No. 5, pp. 655-668, 1986.

[11] Wey, C.L., "On Yield Considerations for the Design of Redundant PLAs," *IEEE Trans. on CAD of IC and Syst.*, Vol. 7, No. 4, pp. 528-537, 1988.

[12] Treuer, R., Fujiwara, H. and V. K. Agarwal, "Implementing a Built-in Self-Test PLA Design," *IEEE Design and Test of Computers*, No. 4, pp. 37-48, 1985.

[13] Hassan, S.Z. and E. J. Mc Cluskey, "Testing PLAs using Multiple Parallel Signature Analyzers," *Proc. FTCS13*, pp. 422-425, 1983.

[14] Fujiwara, H., "A New PLA Design for Universal Testability," *IEEE Trans. on Comput.*, Vol. C33, No. 8, pp. 745-750, 1984.

[15] Kuo, S.Y. and W.K. Fuchs, "Fault Diagnosis and Spare Allocation for Yield Enhancement in Large Reconfigurable PLAs," *Proc. ITC*, pp. 944-951, 1987.

[16] Khakbaz, J., "A Testable PLA Design with Low Overhead and High Fault Coverage," *IEEE Trans. on Comput.*, Vol. C33, No. 8, pp. 583-588, 1984.

[17] Maly, W., "Fault Models for NMOS PLAs" *Proc. IEEE CICC*, pp. 467-470, 1986.

[18] Brayton, R. et all, "Logic Minimization Algorithms for VLSI Synthesis," *Kluwer Academic Publ.*, Boston, Mass, 1984.

[19] Mangir, T.E., "Sources of Failures and Yield Improvement for VLSI and Restructurable Interconnections for RVLSI and WSI," *Proc. IEEE*, Vol. 72, No. 6, pp. 690-709, 1984.

[20] Sami, M.G. and R. Stefanelli, "Reconfigurable Architecture for VLSI Processing Arrays," *Proc. IEEE*, Vol. 74, No. 5, pp. 712-722, 1986.

[21] Williams, T., "VLSI Testing," *North Holland Publ.*, Amsterdam, The Nederlands, 1986.

[22] Lombardi, F., "PLAYABIT: An Environment for PLA Yield Enhancement," Internal Report, Department of Computer Science, Texas A&M University, 1988.

[23] Breuer, M.A. and A.D. Friedman, "Diagnosis and Reliable Design of Digital Systems," *Computer Science Press*, Rockville, Md. 1976.

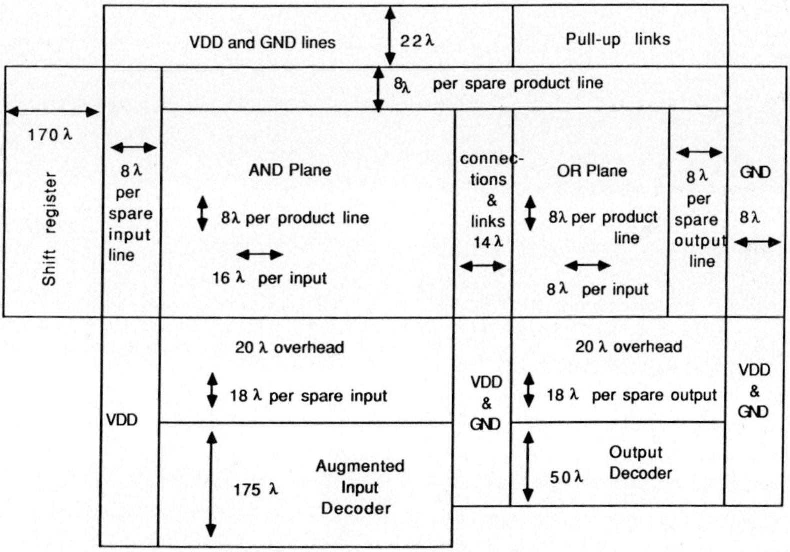

Fig.5 Floor plan of a TRPLA

Wafer Scale Integration, III
M. Sami and F. Distante (Editors)
Elsevier Science Publishers B.V. (North-Holland)

SWITCHES ACTIVATION AND DESACTIVATION USING THE E.BEAM

P. Girard and B. Pistoulet
Laboratoire d'Automatique et de Microélectronique de Montpellier
U.S.T.L., pl. Eugène Bataillon, 34060 Montpellier cédex, France

In this paper experimental data about negative and positive electron beam charging of a floating gate MOS transistor are reported. The main parameters governing both phenomena are established and discussed with theory. A complete switching of a device is shown in both senses.

I. INTRODUCTION

These last years, many efforts have been devoted to the Wafer Scale Integration problem (1,2,3). Various routing possibilities are given using either electrical programming, laser fuses antifuses, electron beam charging of floating gate MOS transistors. In this last case, the test and restructuring is potentially suitable in the same tool, the Scanning Electron Microscope. However, if e-beam writing is now admitted (4,5), the erasure problem is still formulated. Authors generally propose ultra violet irradiation (4,5) but more recently experimental data using e-beam erasure have been reported (6). The aim of this paper is to present experimental results and discussions about both negative and positive e-beam charging conditions of floating gate MOS devices. The surprising electron positive charging results from the fact that in the low energy range, the insulator surface emits more electrons than it receives. The first part of this paper is devoted to the negative charging region and the obtained data are discussed with theory. The second part concerns the positive charging region. Then the parameters governing this phenomenon are established. Finally a switching from the on to the off state of a cell and reversely is shown .

II. EXPERIMENTS AND DISCUSSIONS

Devices have been made in a 22.5 nm gate oxide thickness technology. N channel floating gate MOS transistors operating in the depletion mode devices are generally used. So they are initially sensitive to the negative charge deposition on the top oxide. The oxide thickness above the polysilicon gate is 700 nm while the surrounding area merges around 2000 nm. Generally the charge deposition is obtained with a beam scanning on surfaces much larger than the 25 x 25 μm^2 gate area. The spot mode of operation, when used, is clearly pointed out. The electrical characteristics of the devices are obtained under vacuum, the samples remaining inside the SEM chamber.

II.1 HIGH ENERGY REGION

The switching from the on to the off state of a n channel depletion mode floating gate transistor is now currently admitted (5). It requires primary beam energies larger than the second cross over (7,8).On the one hand, since electrical damages to the circuits need to be avoided, the primary energy then used need to be close to the energy used in e-beam testing (9), i.e. around 1 keV. On the other hand the suitable charging conditions require in principle high beam energies.

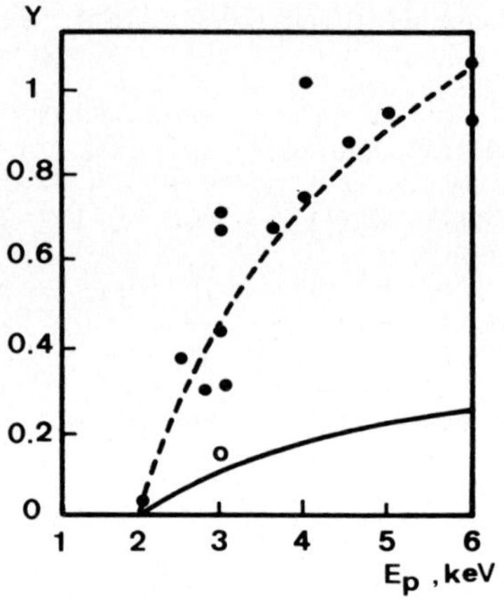

Fig.1 : Experimental charging yields determination in the scanned mode of operation. The (o) represents a spot mode measurement. The theoretical yields correspond to the lowest and the dashed curves, respectively when the surface exchanges are neglected or taken into account.

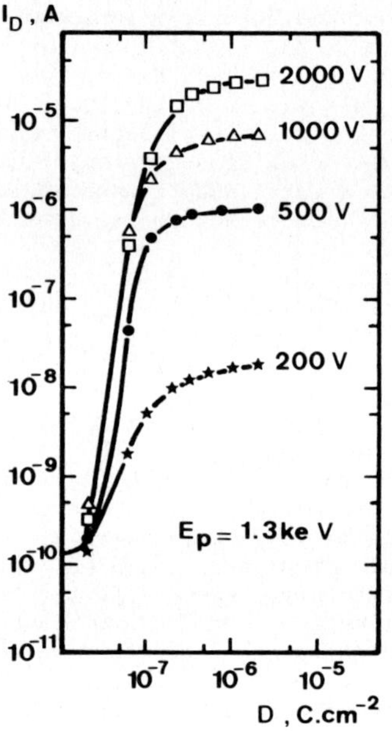

Fig.2 : Switchings of a n channel enhancement transistor at various extraction voltages.

Under these conditions, charging yield determinations are necessary. The methodology has been described elsewhere (10). The fig. 1 shows the experimental yield y, i.e. the ratio between the deposited and incident charges (absolute value) versus the primary beam energy. Due to the structure used, a 1.0 value is obtained at 6 keV but at 3 keV a value of 0.4 still remains .

The theoretical approach (7) gives $y = (1 - \eta - \delta)$ (1) where the backscattered coefficient η is assumed to be a constant in the low energy range considered but the secondary emission yield δ depends strongly on the primary energy E_p.

It can be written :

$$\delta(E_p) = 1.11\ \delta_{max} \left(\frac{E_p}{E_{p_{max}}}\right)^{-.35} \left[1 - e^{-2.3\left(\frac{E_p}{E_{p\,max}}\right)^{1.35}}\right] \tag{2}$$

Using $\delta max = 1.35$, $Ep\ max = 0.35$ keV, $\eta = 0.18$, the theoretical yield is deduced and reported on fig. 1. The experimental value corresponds to yexp = 4y. This "amplification" of the charging yield may be explained in the following way. First, the capacitances between surface and ground (or substrate) are different above the gate and outside. Surface voltages developped by the charge are different and surface charge transfer occurs from the outside into the gate. The apparent or experimental yield is then increased. On the one hand the coefficient 4 is consistent with oxide thickness and primary electrons penetration in the energy range considered and the theoretical parameters used correspond to the magnitudes found into the litterature (11). On the other hand the charging yield in the spot mode of operation fits with the theoritical curve (see fig.1), in this case no surface charge exchange occurs.

II.2 LOW ENERGY REGION

Below the second cross over, since the number of reemitted electrons is larger than the number of incident ones, a net positive charge then results. This effect is used in electron beam testing conditions in order to examine buried line potentials via the capacitive coupling (11).

On fig. 2, the drain source currents evolutions at 5 V polarization are reported versus the incident doses falling here on a n channel enhancement mode floating gate device . The increase of the electron currents is consistent with electrostatic coupling between the channel and a positive surface charge.

The effect of a supplementary positive extraction voltage is clearly seen. On the one hand the capacitor bridge between the extraction grid and the grounded substrate increases the surface voltage. On the other hand an electrostatic field is generated normally to the sample surface, the low energy secondary electrons trajectories are driven towards a zero angle of incidence and then the secondary emission yield is changed. This explains qualitatively the increase of the current with the extraction grid polarization. The charging yield becomes now :

$$y = \eta + \delta - 1 \tag{3}$$

Now basing on experiments achieved with n channel depletion mode transistors previously switched off with a 3 keV beam investigations may go further. It has been shown that the floating gate variations increase first linearly with the incident dose and then saturate (6). Higher is the extraction voltage higher are the gate voltage variations.

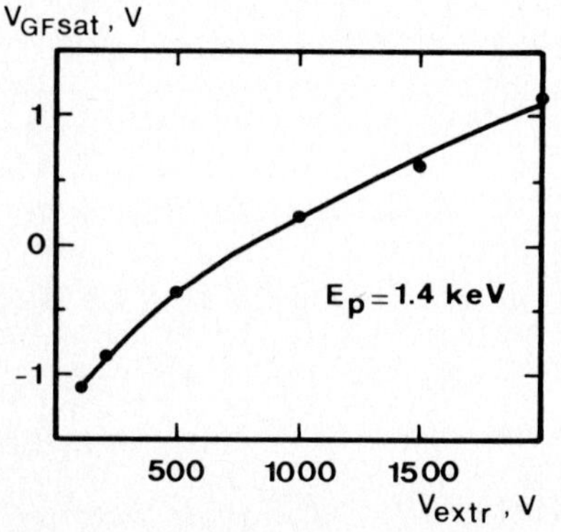

Fig. 3 : Influence of the extraction voltage on the saturation voltage of the floating gate.

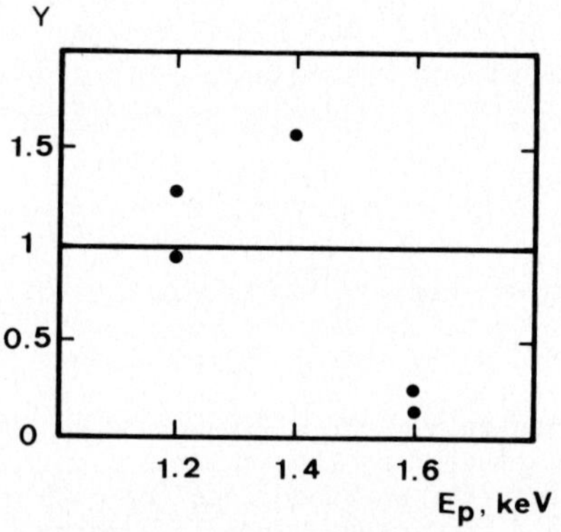

Fig.4 : Yields behaviour at high extraction voltages versus the primary beam energy.

The floating gate voltages at the saturation level are reported on fig. 3 versus the extraction voltage. The 1 volt floating gate voltage obtained is sufficient to completely switch on the device. Then results are consistent with the fig. 2 data.

Since the induced surface voltage versus the series capacitor bridge is at most 0.3 V, this effect may be considered as a second order one. The main effect results from the extraction voltage on secondary electrons (SE) trajectories. When the surface becomes slightly positively charged, the lowest energy SE are drifted back to the surface. The extraction field tends to impose a direction normal to the surface and then increases the escape events. The secondary emission coefficient then increases.

The positive surface plays the role of a high energy pass filter

$$F(Vs) = \int_{qV_s}^{50\ eV} N(E)\ dE \Big/ \int_{0}^{50\ eV} N(E)\ dE \qquad (4)$$

with

$$N(E) = 1{,}5\ E\ \exp(2 - (8/3\ E)^{.5}) \qquad (5)$$

Let us name G(Vextr) the influence of the extraction voltage on the secondary emission yield.

This quantity becomes :

$$\delta\ (Ep,\ Vs,\ Vextr) = \delta_o(Ep)\ \ G(Vextr)\ F(Vs) \qquad (6)$$

The surface saturation voltage Vs is reached when the total number of reemitted electrons is equal to the number of incident ones. Then, combining the relations (3) and (6) :

$$F(Vs) = \frac{1 - \eta}{\delta o(Ep)\ G(Vextr)} \qquad (7)$$

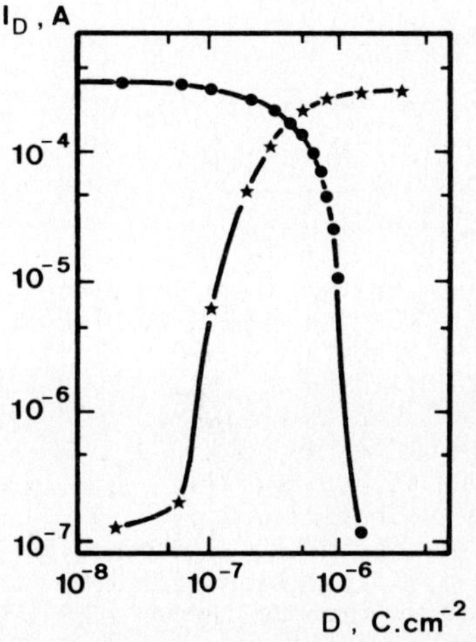

Fig. 5 : Complete and successive switchings of a floating gate device (n channel depletion mode transistor).

Higher the initial secondary emission and extraction voltage are, the higher the surface voltage becomes.

The experimental yields at high extraction voltages are reported on fig. 4. The primary energy dependence is clearly seen and consistent in the upper range with theory (7). It results that the best positive charging compromize results in a 1,4 keV primary energy and on extraction field of around 1 kV/mm.

In order to illustrate the negative and positive charging 5 V drain source currents are shown of fig. 5. Complete switchings are clearly seen.

III. CONCLUSION

In this paper, we have shown that both switchings from the on to the off state of a n depletion floating gate device has been obtained using the same tool, the Scanning Electron Microscope. The main parameters governing the two steps have been established and the effective charging yields determined. The results are in agreement with theory.

So the idea of associating e-beam testing and reconfiguration does not seem unrealistic. Next steps may result in an association of the elementary gate with a logic in order to increase reliability or easiness of operation.

AKNOWLEDGEMENTS

It is a pleasure to aknowledge MM M. VALENZA and R. LORIVAL for their experimental assistance.

REFERENCES

1- Raffel J.I., Anderson A.H., Chapman G.H., Konfle K.H., Mathur B., Soares A.M., Wyatt P.W., IEEE J. of Sol. St. Circuits, SC 20, 1, Feb. 1985, 399.

2- Trilhe J., VLSI 87, Ch. Séguin ed., Elsevier Science Publishers BV (North Holland), IFIP 1988, 293.

3- Chen W., Mavor J., Denyer P.B., Renshaw D., IEEE J. of Sol. St. Circuits, 23-3, june 1988, 639.

4- Shaver D.C., J. Vac. Sc. Techn. 19(4), 1010.

5- Girard P., Pistoulet B., Valenza M., Lorival R., Wafer Scale Integration II, R.M. Lea editor, Elsevier Science Publishers B.V. (North Holland), IFIP 1988, 211.

6- Girard P., Pistoulet B., Valenza M., J. de Physique, Coloque C4, Supplément au n° 9, tome 49, sept. 1988, C4-805.

7- Seiler H., J. Appl. Phys. 54 (11), November 1983, R1.

8- Goldstein J.L., Yakowitz H., Plenum Press, New York (1975).

9- Girard P., Scann. Microsc. 2, 1, 1988, 151.

10- Girard P., Scann. Microsc. Vol. 2 n° 4, 1988, 1995.

11- Goerlich S., Herrmann K. D., Reiners W., Kubalek E., Scann. Microsc. 1986,II, 447.

12- Kollath R., Hambuch der Physik, 21, 241 (1956) Springer Verlag, Germany.

Wafer Scale Integration, III
M. Sami and F. Distante (Editors)
Elsevier Science Publishers B.V. (North-Holland)

WAFER SCALE L-U DECOMPOSITION ARRAY WITH A NEW RECIPROCAL CELL

V. K. Jain **D. L. Landis** **C. E. Alvarez**

Department of Electrical Engineering
University of South Florida
Tampa, Florida 33620, U.S.A.

This paper presents a wafer scale architecture for L-U decomposition using a newly developed reciprocal cell. Since the speed of an L-U array is dictated by the reciprocal cell, the design objective of the new cell is to obtain fast performance. The theory of operation of the new cell and its unique implementation are described in detail. The array employs only two types of cell, MA and R, thus facilitating restucturing through laser cutting and linking. A WSI self-test architecture is also described which supports laser restructuring interconnection tests as well as system level testing.

1. INTRODUCTION

Considerable interest exists in massive parallel processing and its application to signal processing [1]-[6]. Typical applications include image filtering, phased-array radar, speech recognition and the like. A large subset of applications involves one common hurdle, namely the solution of a linear system of equations $\mathbf{A}\ \mathbf{p} = \mathbf{y}$ where $\mathbf{A}$ is a square matrix, $\mathbf{y}$ is a data vector, and $\mathbf{p}$ is the vector of unknowns. An efficient way of solving this set of equations is through forward and back substitution steps, after performing an L-U decomposition on the system matrix: $\mathbf{A} = \mathbf{B}\ \mathbf{C}$, where $\mathbf{B}$ is lower triangular, and $\mathbf{C}$ is upper triangular. Specifically, the solution is obtained as follows:

Solve the following for $\mathbf{q}$ by forward substitution

$$\mathbf{B}\ (\mathbf{q}) = \mathbf{y}$$

Then, solve the following for $\mathbf{p}$ by back substitution

$$\mathbf{C}\ \mathbf{p} = \mathbf{q}$$

The paper presents a systolic architecture using a new reciprocal cell designed by us. The objective of the new cell is to obtain fast performance, since the speed of the array is governed by the reciprocal cell, and simultaneously to achieve good error performance. The need for fast reciprocal arithmetic arises in a broad class of array processing architectures such as L-U decomposition [2]-[3], lattice filtering [4], and

This research was supported in part by DARPA Grant No. MDA 972-88-J-1006.

Kalman filtering [5],[6]. The traditional approaches to finding the reciprocal are a) Newton-Raphson, and b) Implicit Division methods. We present here a new approach to the computation of the reciprocal of a number. Developed specifically for VLSI implementation, it is based on a second order interpolation technique, leading toarchitectural simplicity, high speed and stable answers. An analysis of the accuracy of the cell indicated that the cell incurs a third of an LSB error (in the answer for 1/x) on an average.

A WSI self-test architecture is also described which supports laser restructuring interconnection tests as well as system level testing. This WSI self-test architecture requires each component (MA or Reciprocal cell) to have a) a simple set of boundary scan I/O pads in addition to a standardized self-test interface, and b) BIST circuitry.

2. LU DECOMPOSITION ARRAY

2.1 Array

This section briefly presents a wafer scale architecture for L-U decomposition using the newly developed reciprocal cell. Since the speed of an L-U array is dictated by the reciprocal cell, our systolic design yields fast performance. The array employs only two types of cell thus facilitating restucturing through laser cutting and linking. Our systolic array implementation is after Kung [2]. A single assignment formulation of the L-U algorithm is given below:

For k from 1 to N
For i from k to N
For j from k to N

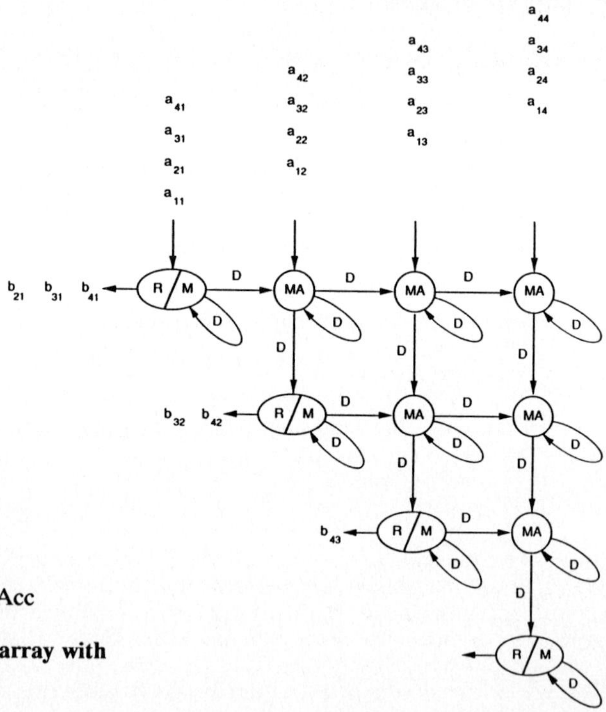

R Reciprocal cell
M Multiply cell
MA Multiply and Acc

Fig. 1 L-U decomposition array with new reciprocal cell

$$c(i,j,k) \quad \leftarrow \quad \begin{cases} a(i,j,k-1) & \text{if } i=k \\ c(i-1,j,k) & \text{otherwise} \end{cases}$$

$$b(i,j,k) \quad \leftarrow \quad \begin{cases} a(i,j,k-1)/c(i,j,k) & \text{if } j=k \\ b(i,j-1,k) & \text{otherwise} \end{cases}$$

$$a(i,j,k) \quad \leftarrow \quad a(i,j,k-1) - b(i,k,j) * c(i,j,k) \tag{1}$$

The array for L-U decompposition algorithm is given in Fig. 1.

2.2 Restructuring for WSI

Only two types of cells are employed, R and MA. The R cell represents the new reciprocal design being presented in this paper. The MA cell provides simple arithmetic multiply and accumulate operation. For the M function needed along the diagonal of the logical array (see Fig. 1), the MA cell is utilized by bypassing the accumulation. Restructuring for defect toleranceby laser linking and cutting [1],[9], [10],[11] will be performed in house. The 'laser linking and cutting' facility is being installed at the University of South Florida as an integral part of DARPA wafer scale integration project.

3. FAST RECIPROCAL CELL

3.1 Overview

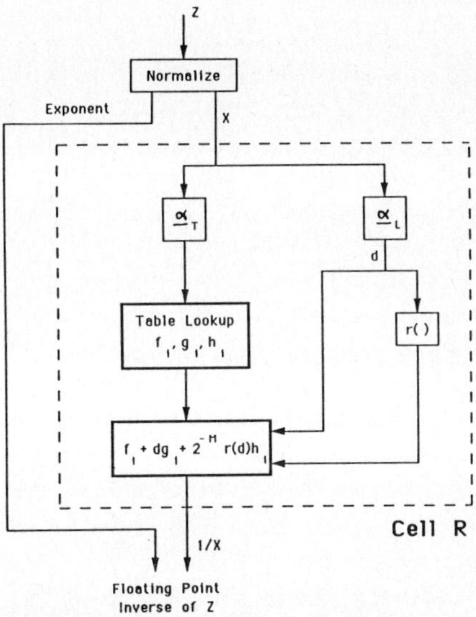

Fig. 2 Block diagram of new reciprocal cell

In contrast to existing iterative algorithms, our proposed algorithm is noniterative, as evidenced in the block diagram of Fig. 2. The magnitude of the input, $Z = |Z_{in}|$, is first normalized so that the result X satisfies $0.5 < X < 1.0$. The new arithmetic is based on a second order polynomial interpolation to the reciprocal function over the interval X_i and X_{i+1}, where i and i+1 are indices of a 2^{M-1} point grid over 0.5 to 1.0. The interpolation formula, derived in [8] is given below:

$$1/X \quad = \quad f_i + \; d\, g_i \; + 2^{-M}\, r(d)\, h_i \qquad (2)$$

The operations involved are one table lookup, one explicit and one implicit multiplication, and two additions. The significance of the various symbols is explained in sub-section 3.2 below.

The architecture of the new cell is given in Fig. 3. Our present design is for a 16 bit input and a 16 bit output word. The value of M is 7 which gives rise to a 64 word lookup table; each lookup word consists of three components: f_i, 16 bits, g_i, 9 bits, and h_i, 3 bits. The transistor count is approximately 7000. Note that the multiplier of this cell requires a 9x9 precision.

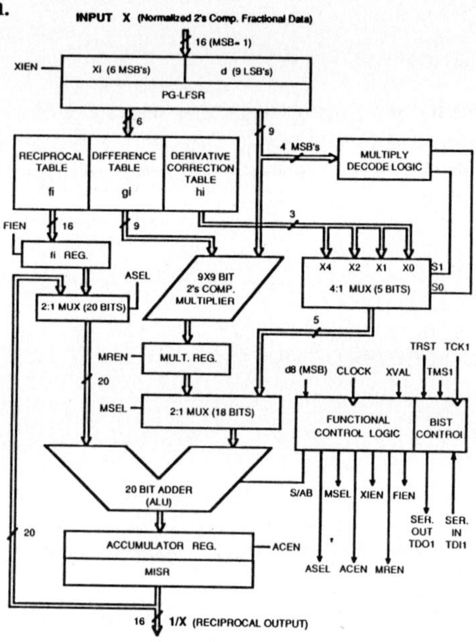

Fig. 3 Reciprocal cell architecture

3.2 Theory of New Reciprocal Cell

The binary representation for the normalized input X is shown below:

$$\begin{aligned} X: \quad & [a_{N-1}.....a_{N-M}\;\; a_{N-M-1}....a_0] \\ = & \;[\;\underline{a}_M \quad \underline{a}_L\;] \\ = & \;[\;1\; \underline{a}_T\; \underline{a}_L\;] \end{aligned} \qquad (3)$$

Note that $\underline{a}_M$ consists of the most significant M bits of X, while $\underline{a}_L$ consists of the least significant L bits. In turn, $\underline{a}_T$ is composed of the highest M-1 bits excluding the MSB (which is of course 1). The bits of $\underline{a}_T$ can be thought to place a 2^{M-1} point grid on 0.5 to $1\text{-}2^{-M}$; thus, the high word $\underline{a}_M$ has an arithmetic value $X_i = 0.5 + i/2^M$ where i is the integer value of $\underline{a}_T$. Clearly, X lies between X_i and X_{i+1}.

Our arithmetic is based on a second order polynomial interpolation to the reciprocal function over the interval X_i and X_{i+1} as shown in Fig. 3. The polynomial is so chosen that it matches the reciprocal function at both X_i and X_{i+1}, and further its slope matches the slope of the reciprocal function at X_i. As shown in [8], this polynomial is

$$P_2(x) = f_i + d(f_{i+1}-f_i) + D\,[f_i' - \tilde{f}_i'] \, d \, d_c \tag{4}$$

where $D = 2^{-M}$, d is the value of the low word x_L with the binary point artificially placed to the left of its MSB, $d_c = 1 - d$, $f_i' = f'(x_i)$, and $\tilde{f}_i' = [f_{i+1} - f_i]/(x_{i+1}-x_i)$. In fact, as discussed below, $(d\,d_c)$ can be replaced by a logically-simple-to-implement function, r(d), so that (4) can be rewritten as in (2), where

$$g_i = f_{i+1} - f_i \tag{5}$$

$$h_i = [\, f_i' - \tilde{f}_i' \,] \tag{6}$$

3.3 Simplification of the term $(d\,d_c)$

As the example above shows, the term $Dh_i(d\,d_c)$ tends to impact the result in the lowest two bits. It is, therefore, reasonable to use lower precision versions for d and d_c. We have used a four bit version as shown below, which we will, hereafter, call $\hat{d}$.

$$d: \quad [\, a_8 \; a_7 \; a_6 \; a_5 \quad a_4 \; a_3 \; a_2 \; a_1 \; a_0 \,]$$

$$\hat{d}: \quad [\, a_8 \; a_7 \; a_6 \; a_5 \,]$$

Recall that the binary point is at the left of the msb to yield the numerical value of d or $\hat{d}$. We will denote the integer value of the four bit pattern as $I_{\hat{d}}$. Similar comments hold for $d_c = 1-d$. Table 1 below gives the integer product $I_{\hat{d}} I_{\hat{d}_c}$ as well as a four bit approximation to it, $I_{\hat{d}\hat{d}_c}$. In interpreting the table it is important to bear in mind that the underlying quantities of interest are real numbers obtained by placing the binary point to the left of the binary string, and that the table folds about row eight, i.e., $I_{\hat{d}} = 9$ can be read from the row for which $I_{\hat{d}_c} = 9$ or $I_{\hat{d}} = 7$.

Inspection of the last column reveals that the quantity $d{*}d_c$ can be approximated as .0000, .0001, .0010, .0011, or .0100. A further simplification will now be made. The two entries 0011 in the table will be changed as follows: the entry 0011 in row 4 will be changed to 0010, and the entry 0011 in row 5 will be changed to 0100. Simulation reveals that the impact on performance is very slight, while theimplementation is simplified as discussed in the next section. Thus the function r(d) is characterized by the listing in Table 2. In a typical example [8] the use of the function r(d) yields $Dh_i r(d)$ = -0.000015, which compares favorably with the true (and much more expensive) value -0.000017.

Table 1

$\hat{d}$	$\hat{d}_c$	$\hat{d}\,\hat{d}_c$	Rounded to 4 bits $\widehat{\widehat{dd_c}}$	Binary
0	16	0	0	0000
1	15	15	1	0001
2	14	28	2	0010
3	13	35	2	0010
4	12	48	3	0011
5	11	55	3	0011
6	10	60	4	0100
7	9	63	4	0100
8	8	64	4	0100

Table 2

$\hat{d}$	$\widehat{\widehat{dd_c}}$	r(d)
0	0	.0000
1, 15	1	.0001
2, 3, 4, 12, 13, 14	2	.0010
5, 6, 7, 8, 9, 10, 11	4	.0100

4. RECIPROCAL CELL ARCHITECTURE

4.1 Architecture

As indicated in the simplified block diagram of Fig. 2, the fast reciprocal cell algorithm requires a table lookup operation followed by a simple arithmetic computation. A uni-processor implementation of the algorithm, as described in equation (1), would require the following sequence of operations upon receipt of a normalized input X:

1. table lookup (f_i, g_i, h_i)
2. multiplication ($d\ g_i$)
3. computation of r(d)
4. multiplication and shift ($2^{-M}r(d)\ h_i$)
5. addition ($f_i + dg_i$)
6. addition $[(f_i + dg_i) + 2^{-M}r(d)\ h_i]$

Concurrency of computation can clearly be used to speed up this calculation. The block diagram of Fig. 3 illustrates the internal details of the overall architecture.

A pipelined approach is used, with input and output data interfaces provided through explicit latches. This is consistent with the systolic architecture requirements of the overall wafer scale design. During the first clock cycle, the input data X is used as an address for the lookup table to produce the f_i, g_i, and h_i terms. On the second clock cycle, the d*g_i term is computed by the explicit 9x9 multiplier. Concurrently with the explicit multiplication, the r(d)*h_i implicit multiplication is performed and this term is added to f using the adder / accumulator. In the final step, the explicit product is added to the accumulator result to form the completed reciprocal. Careful examination of the data flow reveals that the final accumulation can be performed concurrently with the table lookup operation of the next reciprocal computation. Thus in a real time pipelined data application, a new input can be accepted (and a new reciprocal output produced) every two clock cycles. However, the fill and flush times for the pipeline will be three clock cycles; a fact which must be taken into account in the overall system data flow organization.

The test chip implementation of this architecture has been designed using a 2 micron CMOS technology. Estimated propagation delay time for the 9x9 explicit multiplication is 40 ns. This exceeds the ROM table lookup and the combined explicit multiplier/adder delay times, and will set the maximum achievable clock frequency. Thus predicted performance is a 80 ns cycle time for each new reciprocal computation (with a 120 ns fall through delay for the first result).

4.2 Statistical Error Analysis

A detailed statistical analysis has been performed [8] upon the detailed reciprocal implementation to assess its accuracy. In the discussion here, a hat on a symbol will signify its quantized or finite arithmetic value. Then we have

$$\hat{P}_2(x) = \hat{f}_i + d(\hat{f}_{i+1}-\hat{f}_i) + 2^{-M}[\hat{f}_i' - \tilde{f}_i'] \, d \, d_c + z_1 + z_2 + z_3 \tag{7}$$

where z_1, z_2, and z_3 are multiplication arithmetic noise components. We shall ignore this noise inasmuch as an 18 bit result is retained, thus keeping two additional (lower order) bits relative to the final result. Also note that we have retained $d\,d_c$ instead of r(d) in the expression because of the very small size of the third term, so that only a fourth order error (fourth order in (d D)) is caused.

Then the average error can be shown to be [8]

$$\begin{aligned} \text{Average bias} &= -2D \; E[d\, d_c] \; E(-1/X^3) \\ &= -15(D)/192 \\ &= -0.08(\text{lsb}) \end{aligned} \tag{8}$$

The results on the variance are presented in [8]. We give below samples of a few simulation runs (using a computer model of the cell described in Fig. 3).

Table 3

X	Binary Word	Cell Performance 1/x	1/x Binary	Error Measured in LSBs
0.50780	1000 0001 1111 1111	1.96929	01.11111000001001	1
0.68848	1011 0000 0100 0000	1.45248	01.01110011110101	1
0.73578	1011 1100 0101 1100	1.35910	01.01011011111011	0
0.75000	1100 0000 0000 0000	1.33333	01.01010101010101	0
0.86078	1101 1100 0101 1100	1.16174	01.00101001011001	0
0.95323	1111 0100 0000 0111	1.04906	01.00001100100011	0

Statistically speaking, the cell incurs a third of an LSB error (in the answer for 1/x) on an average.

5. WSI SELF-TEST SYSTEM ARCHITECTURE

5.1 Test Architecture

The reduced controllability and observability implicit at the WSI level requires some form of design for testability and/or self test discipline. Our WSI self-test architecture requires each component (MA or Reciprocal cell) to have a) a simple set of boundary scan I/O pads in addition to the standardized self-test interface, and b) BIST circuitry. The basic elements and requirements of this self-test architecture are enunciated below.

All WSI cell designs will include a standardized test interface based upon the proposed IEEE P1149.1 standard [13]. A common probe pattern (physical probe pattern) is defined for all cells, utilizing a standardized placement of probe pads and test interface signals. This allows a single probe card to be used to test an entire wafer irrespective of the number of different cell species.

All cells incorporate boundary scan [14] to facilitate internal and inter-chip testing, as well as Built-In Self-Test circuitry.

5.2 Application of WSI Self-Test Architecture to RVLSI

Fig. 4 illustrates the overall WSI self-test system concept. As indicated, a common seven pin interface is included in each wafer level cell. By placing the test interface in the corner of the chip layout, a common probe card can be used to probe this interface, irrespective of the actual cell dimensions. Thus a complete wafer test can be performed without physically removing the wafer from the tester or having to interchange multiple probe cards.

Naturally, the test process must be organized as a sequence of tests of individual cell types, with delays between these tests to load the unique test program

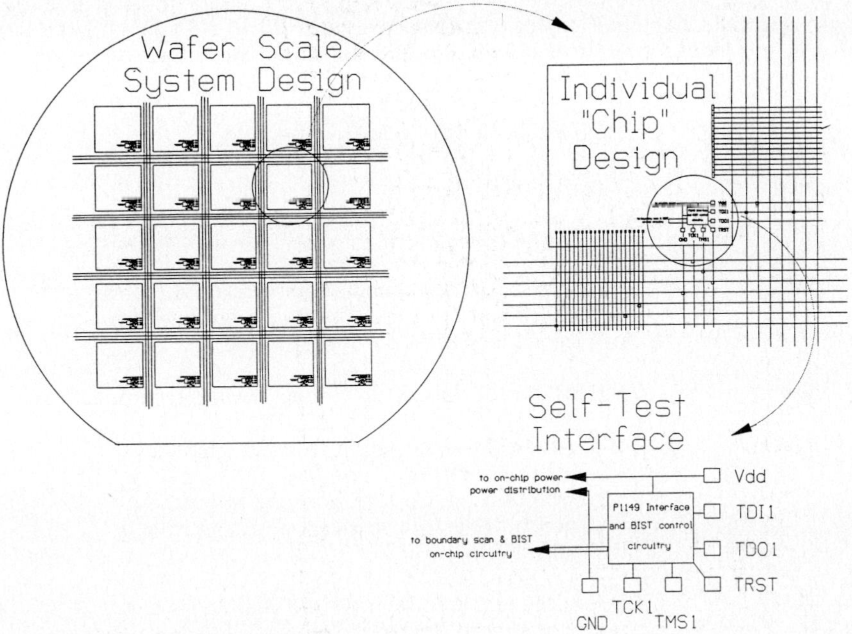

Fig. 4 Wafer level self-test concept

associated with each cell type. Note that the incorporation of BIST within cells can drastically reduce the test time, test program length, test cost, and test data storage requirements. As indicated, the standard interface includes both power and ground pins for the unit under test. This is because the physical power and ground connections are made during the restructuring process, so hat devices with catastrophic Vdd - Gnd shorts do not disable an entire wafer. The signal designations on the self-test interface follow the proposed IEEE P1149 standard for the minimum serial data subset, with signal definitions as given in Table 4 below.

5.2 Application to R-Cell

As stated earlier, we are including both on-chip Built-In Self-Test and standardized self-test interface circuits. As indicated in Fig. 3, the 15-bit input register includes a pseudo-random 15-bit linear feedback shift register pattern generator mode of operation. This will allow exhaustive self-testing by generating all (2^{15} - 1) possible input sequences (the reciprocal of 0 is considered separately as a special case). At the output boundary register, a 20-bit multiple input signature register (MISR) mode of operation is included for test result compression (again using LFSR techniques). The figure also indicates the inclusion of proposed IEEE P1149 standard test interface circuitry for testability bus access.

The Test Bus Core Interface (TBCI), as illustrated in Fig. 5 provides the rudimentary test access port interface required to control on-chip Built-In Self-Test.

It is designed to support generic WSI cell testing, and is being implemented on the reciprocal cell. The TBCI implements the proposed IEEE P1149 minimum serial data set test bus interface which is based upon the JTAG Boundary Scan standard.

Table 4
Self-Test Interface Signal Definitions

Signal	Functional Description
Vdd, GND	These provide the power up capability during the test.
TCK1	TCK1 is the test clock, and under control of the test interface circuitry provides those chip level system clock signals necessary for test/self-test.
TMS1	TMS1 is the test mode select input for the chip under test.
TDO1	TDO1 is the test data output from the chip under test. This test data may come from the boundary scan registers within the chip, or from the test interface circuitry itself (e.g., the signature resulting from a BIST operation).
TDI1	TDI1 is the test data input to the chip under test. This input is used in conjunction with TCK1 to load either functional test patterns into the boundary scan path; or to load test instructions into the self-test interface circuitry (e.g., a BIST initiation command.

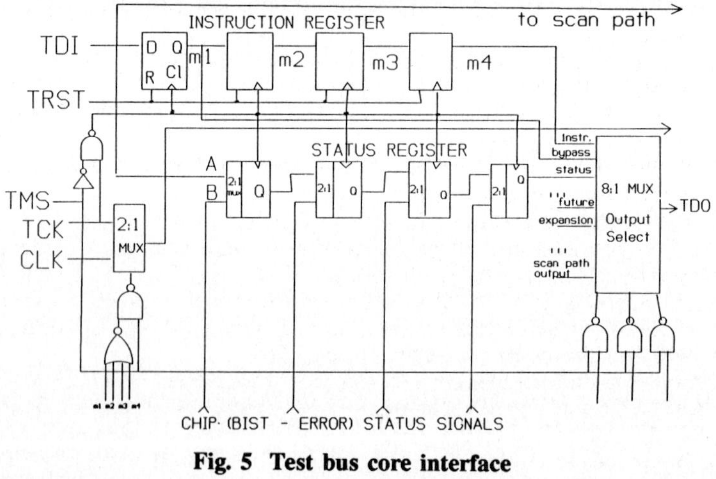

Fig. 5 Test bus core interface

6. CONCLUSIONS

This paper has described a WSI architecture for L-U decomposition and has presented an algorithm for fast computation of the reciprocal of a number. This arithmetic is based on a second order polynomial interpolation. For this arithmetic, the architecture of a reciprocal (VLSI) cell was also given. High speed, combined with inherent stability in inversion, is thereby achieved. As an application of the cell, a systolic array for L-U decomposition of a square matrix was included. A slight enhancement of the cell architecture is apparent. The number X=0.5 can be handled as a special case; this will effectively make available one more bit of precision in the output. Also presented was a WSI self-test architecture which supports laser restructuring interconnection tests as well as system level testing.

ACKNOWLEDGEMENTS

The authors are deeply indebted to Dr. Earl E. Swartzlander, Jr. for his helpful comments on this research.

REFERENCES

[1] Proceedings of the International Conference on Wafer Scale Integration, January 1989 (Published by IEEE Computer Society Press).
[2] S. Y. Kung, VLSI Array Processors (book). Prentice-Hall, 1988.
[3] C. L. Nikias, A. P. Chrysafis, and N. Ventsanopoulos, "The LU decomposition theorem and its implications to the realization of two-dimensional digital filters," IEEE Trans. on Acoustics, Speech and Signal Proc., pp. 694-711, June 1985.
[4] H. Kimura, and T. Ostada, "Cannonical pipelining of lattice filters," IEEE Trans. on Acoustics, Speech and Signal Proc., pp. 878-887, June 1987.
[5] H.-G. Yeh, "Systolic implementation of Kalman filters," IEEE Trans. on Acoustics, Speech and Signal Proc., pp. 1514-1517, Sept. 1988.
[6] S. Y. Kung, and J. N. Hwang, "An efficient tri-array systolic design for real-time Kalman filtering, " Proc. Conf. on Acoustics, Speech and Signal Proc., pp. 2041-2044, 1988.
[7] J. E. Robertson, "A new class of digital divison methods," IRE Trans. of Elec. Comp., Vol. EC-7, pp. 219-222, Sept. 1958.
[8] V. K. Jain, "Statistical error analysis of VLSI architectures: Methodology and a case study," In preparation, 1989.
[9] J. I. Raffel, et. al., "A wafer scale digital integrator," Proc. IEEE Conference on Computer Design, p. 121, 1984.
[10] P. W. Wyatt, and J. I. Raffel, "Restructurable VLSI - A demonstrated wafer scale technology," Proceedings of the International Conference on WSI, January 1989.
[11] R. Frankel, J. J. Hunt, M. V. Alstyne, and G. Young, "SLASH - An RVLSI CAD system," Proceedings of the International Conference on WSI, January 1989.
[12] D. L. Landis, "A self test architecture for reconfigurable WSI," Technical Report No. WSI-T5, University of South Florida, 1989.
[13] P1149 T-Bus Standardization Committee (TBSC) Working Group of the IEEE Computer Society's Test Technology Committee, "Testability Bus Specification, P1149/D7", Draft 7, October 1988.
[14] The Technical Sub-Committee of the Joint Test Action Group (JTAG), "A Standard Boundary Scan Architecture," version 0.4, March, 1987.
[15] A. H. Anderson, "Computer aided design and testing for RVLSI," in Wafer Scale Integration (C. Jesshope editor), pp. 216-222, Taylor Francis Publications, 1987.

Wafer Scale Integration, III
M. Sami and F. Distante (Editors)
Elsevier Science Publishers B.V. (North-Holland)

A Fault Tolerant Wafer Scale Architecture for Vision Processing*

John A. Trotter and Will R. Moore

Dept. Engineering Science, University of Oxford,
Parks Road, Oxford, OX1 3PJ.
England

1. INTRODUCTION

Wafer Scale Integration (WSI) has the potential of providing the end-user with a very large circuit with substantial speed, size and power advantages. However WSI is still predominantly a research activity because of production difficulties which make it hard to obtain an economic yield of good wafers. If these problems can be solved successfully then WSI could yield a very cost effective solution to many large circuit design problems.

One of the major problems with WSI is that of defects which occur during the fabrication process, this problem has to be addressed before cost effective circuits can be made. Because of their increased size, wafer scale circuits are more prone to manufacturing defects and therefore return a very low yield. In addition it is very costly to throw them away so viable wafer scale circuits must include defect tolerance to substantially improve the yield and make them economical.

Vision processing is one application that can benefit from wafer scale integration. Vision processing algorithms require a substantial amount of processing power and map easily into SIMD array machines. The 2-D SIMD array maps well onto a wafer and the architecture described is based on a 2-D SIMD array for vision processing.

2. HISTORY

Several methods of achieving fault tolerance on SIMD 2-D arrays have been proposed and are summarised below.

Restructurable VLSI (RVLSI). A working array is constructed from known good processing elements (PE's). Once the circuit has been fabricated it is tested then configured. A laser is used for this and connects working PE's into the array by joining two layers of metal or cutting a metal line on either level. A 16x16 Fourier transform chip has been configured in this way at MIT Lincoln labs[1].

Spare row/column redundancy. Spare rows and columns of cells have been used successfully in the production of memory chips. If a defect occurs in a memory cell then the row containing the cell is replaced with a spare one. This method is good for memory as the cells

---

* This work was supported by the Science and Engineering Research Council, U.K..

are very small, but when it is applied to a processor array a large area is wasted because of the large number of good processors that remain unused in the excluded row. An extension of this method is to use switching networks which allow individual spares to replace defective cells.

Switching schemes. The switching schemes try and construct a perfect array from the processors and spares available. Switches are used to connect processors so configuring the array and avoiding the faulty cells. A number of schemes have been proposed which use this method [2,3]. First the faulty processors are identified and then the switching network is used to try and connect the working PE's into an array of a desired size. A more sophisticated switching network is more likely to construct an array of a particular size from the given working processors. However delays in the communication are incurred when a complex switch network is used, so to reduce this configuration schemes have been proposed which configure the array with the minimum possible number of switches between two processing elements [4].

The schemes proposed configure a perfect array of processors from the processors and spares available. The switching schemes are a good approach to this configuration problem as they allow reconfiguration if faults occur during operation and do not require extra process steps as in the laser approach. More complex switching networks harvest a higher proportion of the working processors in the array. However there are a number of problems, as well as inter processor communication taking more time the larger area for the interconnect makes it more prone to defects [5]. Also, if the processor size is reduced so that each one is less likely to be effected by a defect, the communication area overhead increases making it still more prone to defects. Even the most sophisticated of these switching networks fail to harvest all the working processors for a given defect distribution, so enough spare processors have to be fabricated to guarantee that an array of a certain size can be configured, consequently a high hardware overhead in the extra cells and switches is required [3]. Configuration of these switches is often non trivial and requires a complex algorithm running on a host processor.

3. THE ARCHITECTURE

This paper describes a fault tolerant strategy for a wafer scale 2-D SIMD array that overcomes many of the problems normally associated with fabrication of wafer scale circuits.

This architecture tackles the problems of the hardware overhead, complexity of configuration algorithm and switching delay. The architecture also degrades gracefully if faults occur during the normal lifetime of the circuit, improving the operational reliability. The fault tolerance scheme can harvest a near perfect number of the working processors with a very low hardware overhead. A novel combination of hardware fault tolerance with algorithmic fault tolerance is used (The ability of certain algorithms to execute even in the presence of noise) to allow low level vision processing algorithms to work.

The architecture is based on a 2-D SIMD array but instead of a perfectly connected array being necessary for vision processing, an imperfectly connected one can be used [6]. This relaxes a number of constraints that are imposed on the system designer.

3.1 The defect tolerance scheme.

The defect tolerance scheme has two parts, a hardware part where the hardware switching network is configured and a mapping part which maps the image data to the processing resources in a special way.

3.2 Hardware defect tolerance

If a faulty processor is detected it is excluded from the array with the switching network shown in Figure 1. Because the defective PE has been bypassed any signals from its South neighbour now go to its North neighbour and any signals from its East neighbour go to its West neighbour. In this way a working processor may still communicate with its four working neighbours, although the array is no longer regular. The switching scheme thus includes every working processor in the array with a very simple switching network that introduces only one switch delay per defective processor. The communication lines are a sensitive part of any 2-D array and the architecture described minimises the possibility of a defect by using the very simple switching and communication network described. In addition the lines and switches would be designed using conservative design rules. In the unlikely event of a communication line being defective the whole of that line can be excluded from the array. This, however, would result in a number of fault free processors not being utilized. This is the essence of the hardware part of the defect tolerance strategy, Figure 2 shows an array where faulty processors have been detected and are excluded from the array.

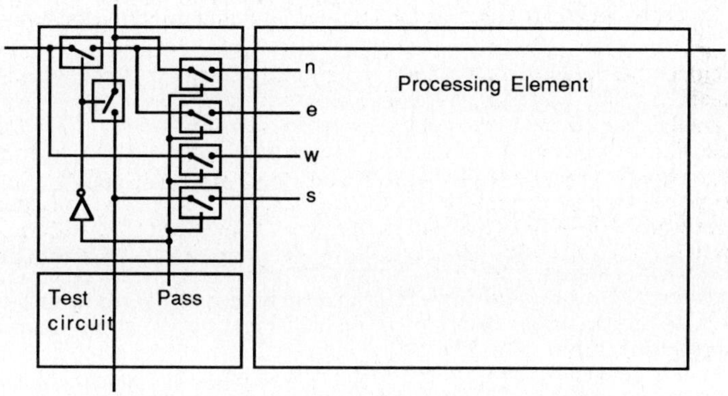

Figure 1. The switching scheme for each processor.

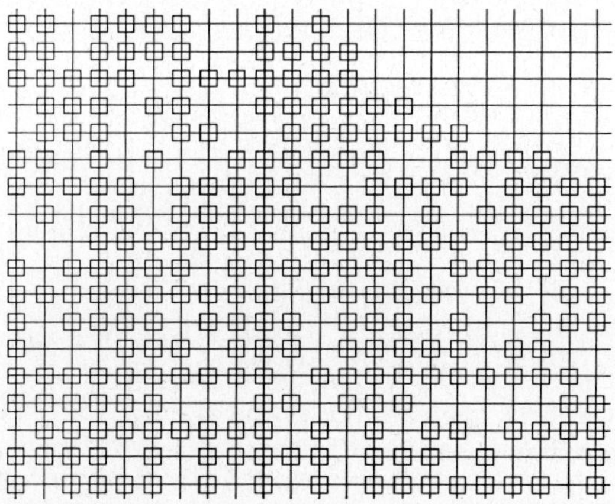

Figure 2. A portion of the array with defective processors not shown.

3.3 Mapping part.

If an image were to be mapped into the array at this stage on a one to one pixel to processor basis then data which would normally have been mapped into the faulty processors would be lost. Clustering of defects can effect a large numbers of processors in a particular area, so instead of discarding this information it is redirected to processors around the defective area. The data can be mapped into the array in this way so that the loss is distributed over the array. This data mapping can be split into two parts so that it has minimum impact on execution time. The first part is a movement of the image area that each processor covers (configuration) and the second an input mapping.

Configuration. The configuration can be thought of as a movement in the conceptual processor positions which alters the area of the image that a processor is responsible for. This configuration is done once at power up. Then the data is projected from the image onto the processors. A number of algorithms can be used for the configuration and we have used one based on minimum energy which minimises the distance between working processors. Figure 3 shows an array configured with this algorithm, the squares representing the conceptual positions of the working processors. We have investigated other algorithms that can be used to configure the array but all are based on the following minimum energy equations:

$$X_{i,j} = [X_{i,j+1} + X_{i,j-1} + X_{i+1,j} + X_{i-1,j}]/4$$
$$Y_{i,j} = [Y_{i,j+1} + Y_{i,j-1} + Y_{i+1,j} + Y_{i-1,j}]/4$$

Where $X_{i,j}$ ($Y_{i,j}$) is the X (Y) position of a working processor and $X_{i-1,j}$ ($Y_{i-1,j}$) is its nearest left hand working neighbour etc..

This is an iterative equation, once it converges the processors will have calculated their conceptual position in the array. Other methods that we have investigated include constraining the movement of processors to discrete and half discrete positions and constraining the movement to one dimension only. The governing equations are the same as before but the processors can only calculate their positions to the nearest processor position or half processor position. Examples of portions of arrays that have been configured with two of these schemes are shown in Figure 4. Configuring the array with these simpler configuration strategies means that the problem space is less evenly covered with processing resources. However the strategies do have advantages for the input and output of data which will be described in the next section.

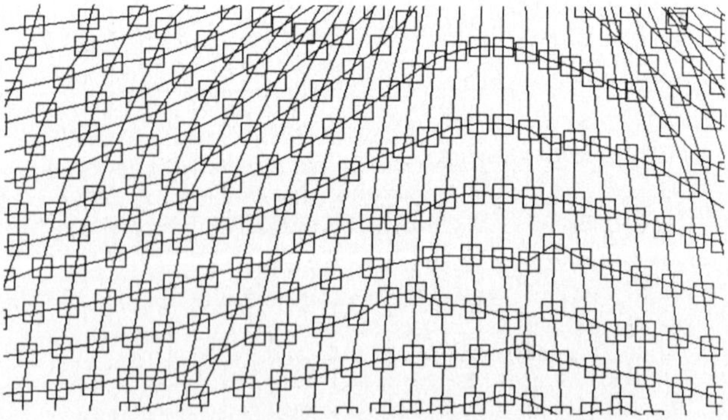

Figure 3. A portion of the array after configuration.

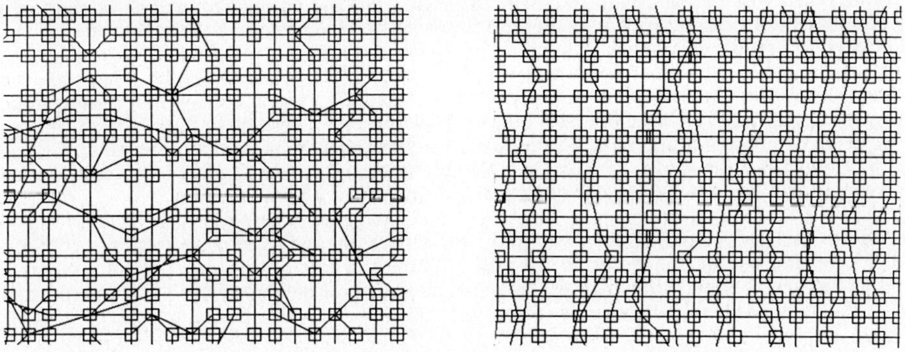

(a) (b)

Figure 4. (a) An array configured with processors allowed to move only discrete positions. (b) An array configured with processors only allowed to move to half discrete positions in one dimension.

Input mapping. Once this configuration is complete the array is ready for use and the data can be input. The input mapping projects the image data onto the array, in a perfect array this would just be a one to one mapping of pixels to processors. A number of input mappings are possible but the scheme we use allows processors to take the data from the image pixel that is closest to the calculated conceptual position of the processor. Data is input to the edge of the array and passed through the array so processors can intercept the data that they need. If the processors have been configured in two dimensions then the data will have to be injected on more than the one row to ensure that all processors are presented with the data that they require. This is because a processor may now be covering an area on a different row of the image than it would have been if the array had not been configured (Figure 5). Once this stage is complete, processors contain data which they can then work on. The particular data that a processor needs can be more easily calculated by the processor if it is restricted in its movement during configuration. If processors have only been configured in one direction then the data need only be made available to one row, reducing the data input problem substantially.

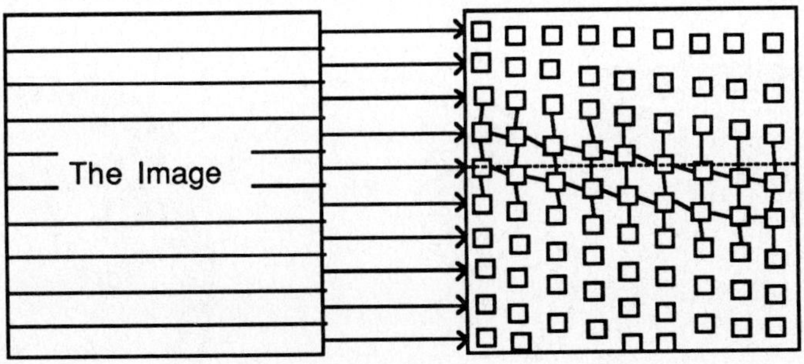

Figure 5. Data input to a configured array

Operation: Now the data is in the array the vision processing algorithms can operate in a normal fashion. Vision processing algorithms, based on nearest neighbour operations can run successfully on the array even though the hardware that they use is not perfectly regular [6]. Figure 6(c) shows the result of a Marr-Hildreth edge detection on a 128 by 128 array, the array has 50% of its processors defective and was configured with the continuous minimum energy algorithm. The edge operator still ran, finding the edge successfully. The algorithm has also been tried on arrays configured with the different strategies, and the results for these can be seen in Figures 6(c) to 6(f). Even though these arrays were covered with a less even distribution of processors the edge detection still executed successfully.

Output: This is done in a similar way to the input mapping, now the processors project their data onto the output image surface. Again a number of algorithms are applicable for this operation, but the one with which we have had most success is that each processor outputs its data to the output pixel position it is closest to.

4. RESULTS

Here we present the results of simulations of the algorithm on a parallel simulation of the architecture. The algorithms were investigated using a combination of defect maps (random and clustered) with a range of configuration algorithms. From the results obtained we can propose a hypothesis which relates the defect density to the effective resolution of an array.

Hypothesis: The resolution of the output image when it has been processed on the array is directly related to the average Inter Processor Distance (IPD) of the processors in the array, that is, the average distance a processor is from its four neighbours.

Practical results have shown the hypothesis to be accurate at predicting the performance of an array with random defects. Figure 7 shows a graph relating the defect density of an array to its effective processing resolution. From this hypothesis it is possible to predict the performance of an array whichever configuration strategy was used. Theoretical results show the one dimensional strategies to be only slightly worse than the original 2-D strategies. With the inclusion of a very simple hardware fault tolerance strategy this can be applied to other array processors which may be able to cope with the loss of a small number of processing elements without much degradation in the final image.

It is possible to consider clustered defects as a combination of different areas of random defects and the same method can be used to successfully predict the processing resolution of such an array.

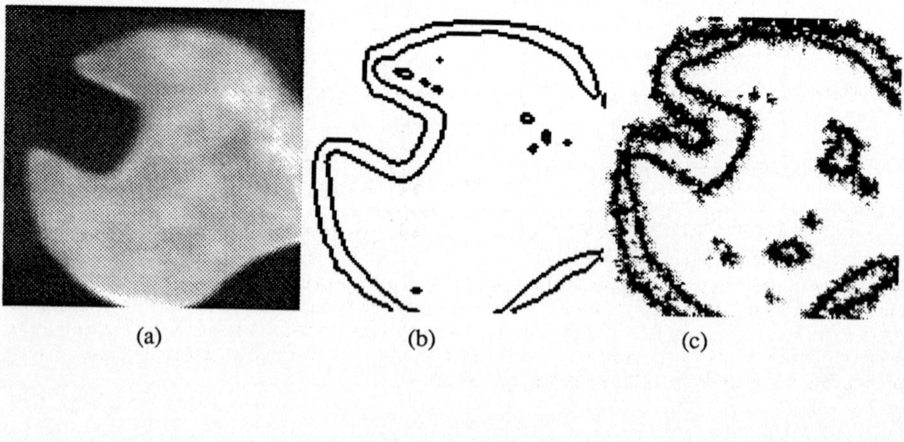

(a) (b) (c)

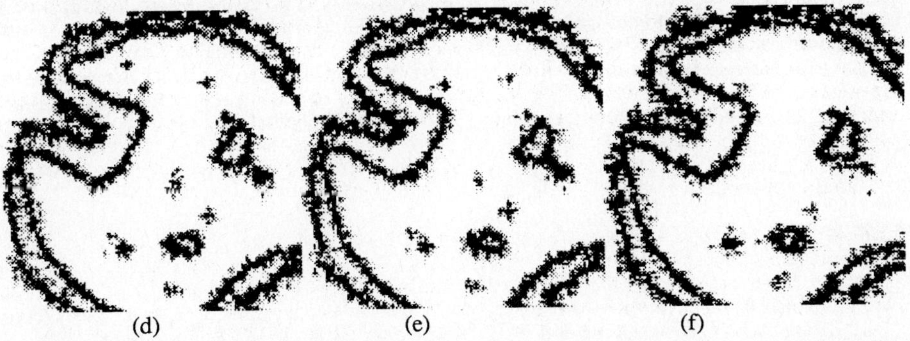

Figure 6. (a) The image. (b) Its edge detection on a perfect machine. And edge detections on an arrays with 50% random defects: Reconfigured with (c) A continuous strategy. (d) A strategy that allows processors to move only half discrete positions. (e) A strategy that allows processors to move only discrete positions. (f) A strategy that allows processors to move half discrete positions in one dimension only.

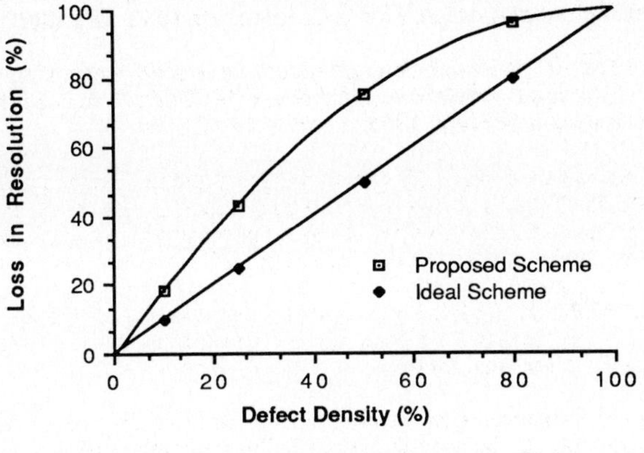

Figure 7. A graph showing the relationship between number of defects and loss in processing resolution for the proposed scheme and a theoretically ideal scheme.

5. SUMMARY

This paper describes an architecture for vision processing that is designed with fabrication and algorithmic issues in mind, these can be combined allowing a complete and cost effective system to be designed. Thus the system uses a fault tolerance scheme that exploits the kind of algorithms the machine runs. The system, unlike previous approaches, does not try to restructure perfect hardware on which any program can run, but sacrifices some of this flexibility by restricting itself to only executing vision processing algorithms. This allows certain compromises in the hardware to be made. The system trades defective processors for processing resolution in the image, distributing the loss over the array. A defect tolerance strategy that can harvest nearly all working processors can now be included, and vision processing algorithms can still be run successfully.

This system approach can lead to a highly cost effective and efficient design as it allows a large area of the silicon to be used for processors and the approach can withstand clusters of faults which would not be tolerated by other approaches. The architecture can be used for vision processing operations and little time overhead is incurred because the switching network is simple. The architecture can degrade gracefully following additional faults, further adding to its cost effectiveness and enhancing it for use in high reliability applications.

REFERNCES

[1] J. Raffell, "The RVLSI Approach to Wafer Scale Integration", in *Wafer Scale Integration,* C.R. Jesshope and W.R. Moore Eds. (Bristol UK: Adam Hilger), 1986.

[2] R. A. Evans J. V. McCanny and K. W. Wood,"Wafer scale integration based on self organization",*Wafer Scale Integration,* C.R. Jesshope and W.R. Moore Eds., Adam Hilger 1986.

[3] M. Sami and R. Stefanelli, "Reconfigurable architectures for VLSI processor arrays", *Proc. IEEE ,Vol. 74, No. 5,* May 1986.

[4] F. Lombardi, D. Sciuto and R. Stefanelli, "An Algorithm for Reconfiguration of Fixed Size Arrays", *IEEE Trans. on CAD of Integrated Circuits and Systems. Vol 7, No. 10,* October 1988.

[5] P. Franzon, "Interconnect strategies for fault tolerant 2-D VLSI arrays", *IEEE ICCD,* October 1986.

[6] J. A. Trotter and W. R. Moore, "An architecture for image processing", in *Parallel Processing for Computer Vision and Display,* P.M. Dew, R.A. Earnshaw and T.R. Heywood, Eds. Addison Wesley, 1989.

Wafer Scale Integration, III
M. Sami and F. Distante (Editors)
Elsevier Science Publishers B.V. (North-Holland)

MODERN ASPECTS OF SILICON TECHNOLOGY FOR MICROELECTRONICS

M. Domenici

DNS Electronic Materials. Viale Gherzi 31, Novara - Italy

1. INTRODUCTION

Today silicon characteristics, as required by microelectronics, are driven by the efforts to increase packing density of integrated circuits. Memory devices have been on the leading edge of scaled integration.

Table 1 shows the development in the last 30 years of the scale of integration in DRAMS together with silicon material parameters, like wafer diameter (last column).

This is only one aspect that modern silicon technology is facing Very Large Scale Integration (VLSI). It implies, beside large wafer diameter, a more uniform dopant and oxygen distribution within the wafer and from wafer to wafer.
The 4th column in the table shows line definition by lithography. The trend to smaller geometries puts stringent requirements on at least three characteristics that silicon wafers must meet with higher emphasis than in the past:
- defect density
- cleanliness
- particulate density

The first two points are addressed nowadays by utilizing the gettering phenomenon, as it will be discussed in 2.

Low particle density has required the construction of quite new production areas, in which ambient particles are conditioned and controlled (clean rooms) by monitoring [1] air, ambient (humidity, pressure, etc.) and particle density to classes like 100, 10 and lower ("class" means number of particles in a cubic feet).
Wafer handling and packaging have been also substantially improved by using robotics and suitable materials (of suitable design) [2] that are in contact with wafers.
In the following sections the impact of VLSI technology on various material properties will be presented.
WSI devices in particular can offer the opportunity to "design" the material properties suitable to a particular device architecture and in turn WSI device architecture can be defined with the latest achievements of the material properties in mind.

On the other hand, WSI devices have the potentiality of anticipating larger capacity memories that will be available, on single chips, only beyond the year 2000 (see fig.1)

1. LARGER DIAMETERS, MORE UNIFORMITY

The two requirements seem to be contradicting each other. This can be overcome if one can handle Czochralski crucible charges of increasing weight or, as a real breakthrough in the crystal growth technology, one makes available an almost "indefinite" crucible charge, by using a continuous feeding CZ equipment: CCZ technology (see fig.2) |3|.

Table II shows a tentative (and conservative) list of the parameters and tolerances achievable by using such a technology (CCZ vs. CZ)
In such a design the possibility of implementing a magnetic field crossing the melt (MCZ) is considered as an optional so that a combination (M-CCZ) is also envisaged, in order to further reduce or suppress thermal convection motions in the melt.

In this case the control of oxygen content will also be more accurate over a wider range. The ultimate goal of this project is to produce large diameter crystals with very uniform bulk properties along crystal axis and diameter as far as oxygen and carrier concentration are concerned. Large diameter wafers seem to be particularly attractive for WSI, since a 200 Mbit memory device on a 6-inch wafer has been announced at the Int. Solid State Circuit Conference in New York, early this year.

2. SUPERCLEAN WAFERS BY GETTERING

Wafer cleanliness, in terms of electrically active impurities concentration, has reached an unprecedented level both at the surface and in the bulk, that can only be detected by very sophisticated techniques (DLTS and NAA for the bulk, SIMS and XPS for the surface). The attention is presently devoted to the problem of maintaining such a cleanliness during device process in the production lines. In fact IC manufacturing processes, especially high temperature cycles, introduce an almost unavoidable amount of contamination. Such a contamination can be disastrous to the device yield of any kind and in particular to WSI device yield:
the faulty elements to be discarded and replaced using switches could overcome any reasonable redundancy concept.

Gettering of impurities away from the active device regions in a wafer has been proved to be very effective in improving device yields.
Existing gettering techniques are of two kinds:

a) <u>intrinsic</u> gettering (IG) is produced by suitable precipitation of interstitial oxygen (contents ranging from 7·10E17 to 1·10E18 at/cm3) approximately 20 microns beneath the front surface with the formation of defect sites for gettering, leaving a defect free zone (Denuded Zone) in the active device area.

Fig.3 shows how the oxygen nucleation and precipitation proceed for three increasing initial oxygen concentration ranges (from 6 to 8.5 10^{17} Oi/cm^3)
The targeting of a particular oxygen concentration range is presently possible thanks to automated pulling equipment: the effect of increased rotation rate on the axial concentration of oxygen in Czochralski crystals is shown in Fig.4. By programming crucible rotation to increase during growth, this effects is used to offset the normal decrease in oxygen concentration along the crystal to produce silicon of closely specified oxygen content.

b) extrinsic gettering (EG) typically is produced on the back surface of the wafers:

b1) by mechanical damage (abrasion, sand blasting, laser-damaging, Ion Implantation etc.)
b2) by deposition of a suitable layer of disordered material (poly-Si [4], α-Si, Si_3N_4, SiO_2 etc)

All the a.m. methods have their advantages and disadvantages:
For istance IG is not a property that is already present in a wafer. It is induced by high temperature treatments that initiate Oi precipitation. It becomes effective midstream the device manufacture.
On the contrary EG is effective from the beginning but it tends to annihilate (damage recovery) during high temperature steps.
So a combination of the two (EG+IG) is sometimes requested. Polysilicon deposition on the back, as compared with mechanical backdamage, is commonly considered as more effective for VLSI applications.

A dramatic view of the trapping capacity of polysilicon is seen in fig.5 where the intentionally introduced Gold is measured by SIMS scanning of the last five microns from the wafer back. An intensity drop of 3 orders of magnitude appears between the poly layer and the bulk.

Again, with increasing packing density in C-MOS devices, one of the problem to overcome, not only from the design point of view, but taking also into account the material, is the latch-up problem. The industry seems to be oriented towards the use of heavily doped substrates, with a higher resistivity-long lifetime epitaxial layer on the top.

CONCLUSION

The "traditional" wafering process (including ingot cutting, wafer lapping and polishing) has been implemented nowadays with a suitable parallel process for the manufacture of crystal-engineered slices (including back side damage and/or polysilicon deposition for external gettering, suitable heat treatments for oxygen precipitation to provide internal gettering and, in some cases, epitaxial deposition of a thin monocrystal layer on the top side). All those characteristics have to be tailored to meet the specification demanded by the device manufacturers. The situation is depicted in fig.6 where a wafer cross-section is shown: the top device architecture is built on a silicon wafer, which is no more a simple substrate but participates actively in the behaviour of the device itself.

REFERENCES

[1] Domenici, M., and Guaglio. A., Monitoraggio Computerizzato dei Parametri Ambientali (Bias, Milano, 1987)

[2] DNS U.S.Patent, 4,718,549

[3] Duncan, C.S., Stickel, W.C., NASA Tech. Briefs (USA) volume 6, n.1, (1981) p. 94.

[4] Borionetti, G., Domenici, M., Ferrero, G.C., Optimum Poly Deposition on Wafer Back for Gettering Purpose, 5th Symp. on Semiconductor Processing ASTM (St.Clara, California, February 3÷5, 1988)

TABLE I: COMPLEXITY OF SEMICONDUCTOR INTEGRATED CIRCUITS

year	class	n'devices bit	litho μm	chip size mm2	wafer size mm
60-68	SSI	2E1-2E7(2-128)	>10	1-15	25
65-75	MSI	2E6-2E12(64-4K)	3-10	10-25	50
72-83	LSI	2E11-2E17(2-128K)	1.5-4	15-50	50-100
80-88	VLSI	2E16-2E22(64K-4M)	0.75-2	25-75	100-125
85-93	ULSI	2E21-2E27(2M-128M)	0.5-1	50-200	125-200
90-99	HSI	2E26-2E32(64M-4000M)	<0.5	100-400	>200

TABLE II

QUALITY PARAMETERS	PRESENT TOLERANCES by CZ		TOLERANCES ACHIEVABLE by CCZ	
1. <u>LONGITUDINAL</u>	TYP	MAX	TYP	MAX
(along crystal axis) variations (%) on 80% of melt pulled				
N-type resistivity	±30	±35	±15	±20
P-type resistivity	±15	±20	±10	±15
High oxygen (8·10E17 At/cc)	±7	±10	±5	±7
Medium oxygen (7.5·10E17 At/cc)	±7	±10	±5	±7
Low oxygen (6.5·10E17 At/cc)	±8	±15	±5	±7
2. <u>RADIAL</u>				
(6 mm from edge to center) Variation (%) On 80% of melt pulled (for 5" crystals)				
N(111) Resistivity	16	25	12	20
N(100) "	8	12	5	10
P(111) "	5	10	4	8
P(100) "	5	10	4	8
Oxygen concentration	5	10	5	7

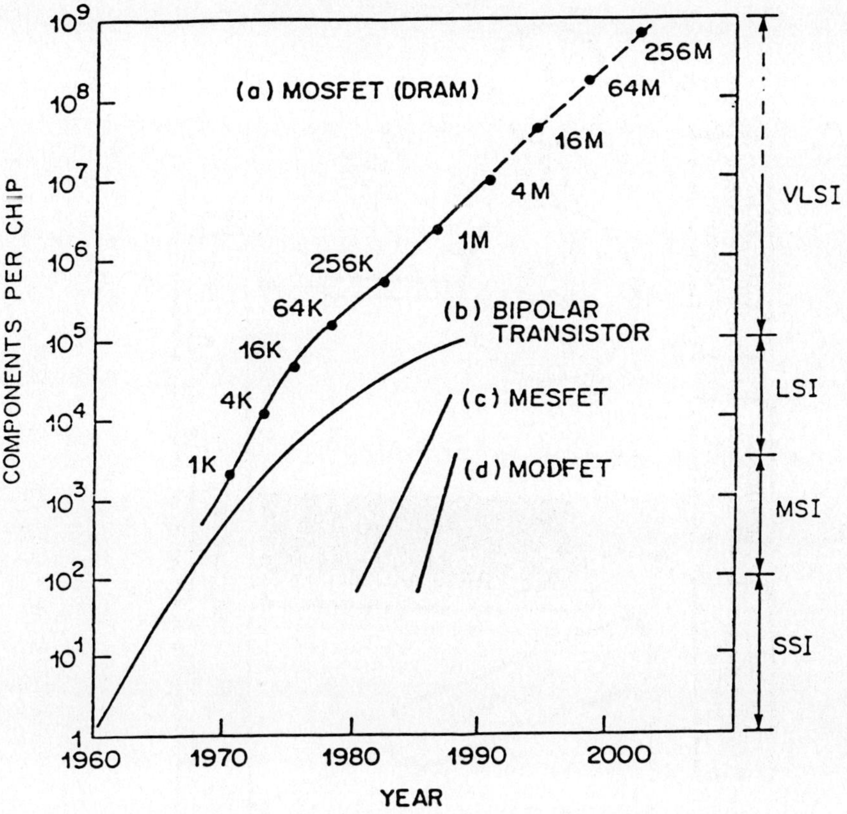

FIGURE 1
TREND IN IC PACKING DENSITY

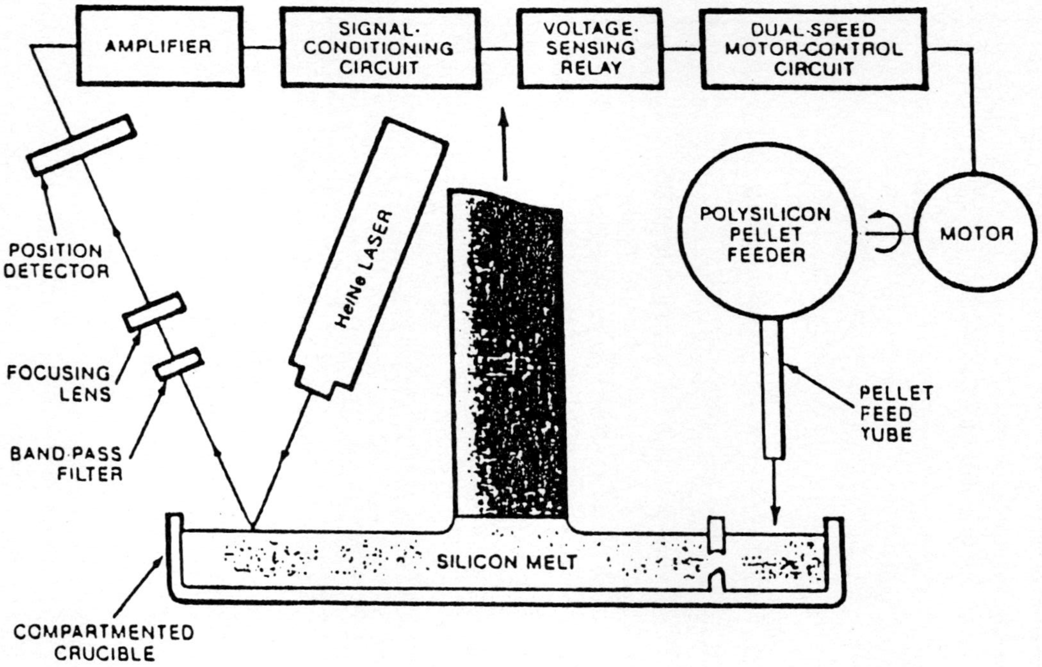

FIGURE 2

SCHEMATICS OF A CCZ PULLING TECHNOLOGY

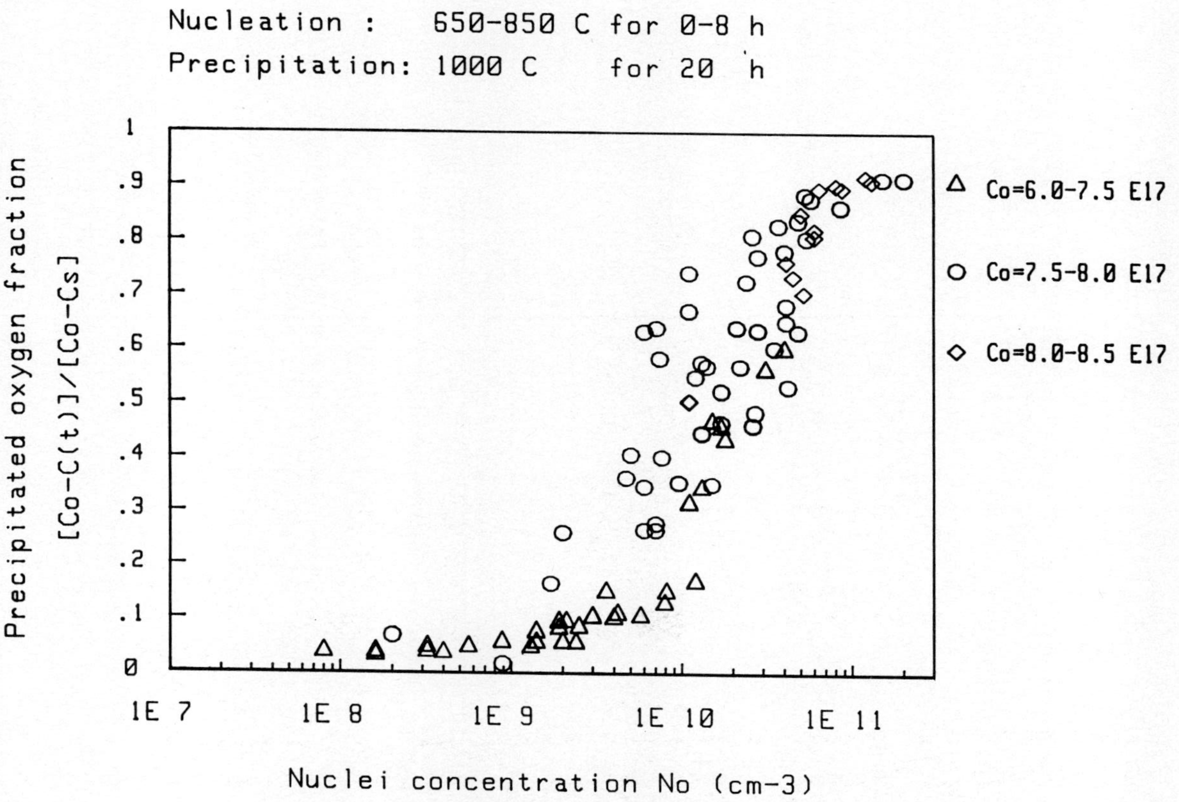

FIGURE 3

OXYGEN PRECIPITATION FOR DIFFERENT INITIAL OXYGEN CONCENTRATION

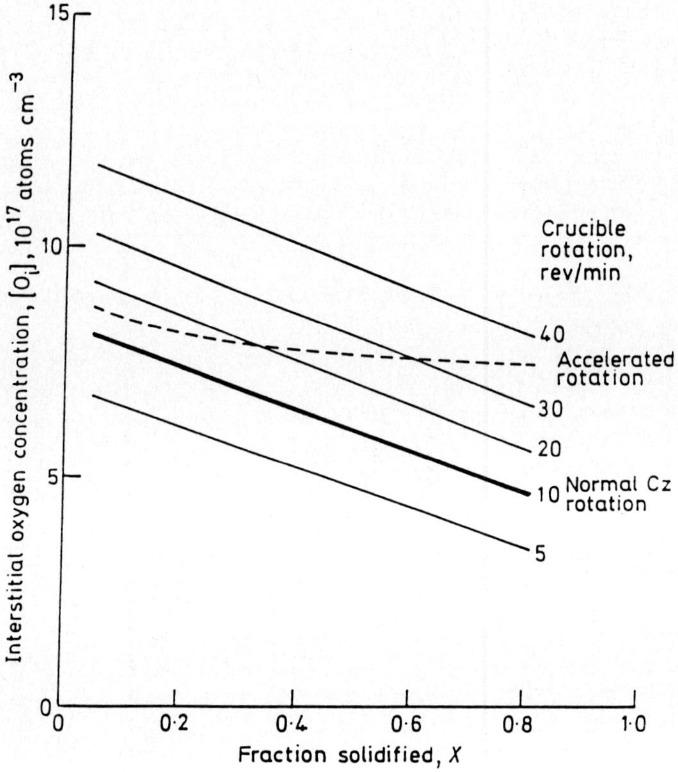

FIGURE 4

EFFECT OF CRUCIBLE ROTATION RATE ON INCORPORATION OF OXYGEN INTO CZOCHRALSKI-GROWN SILICON

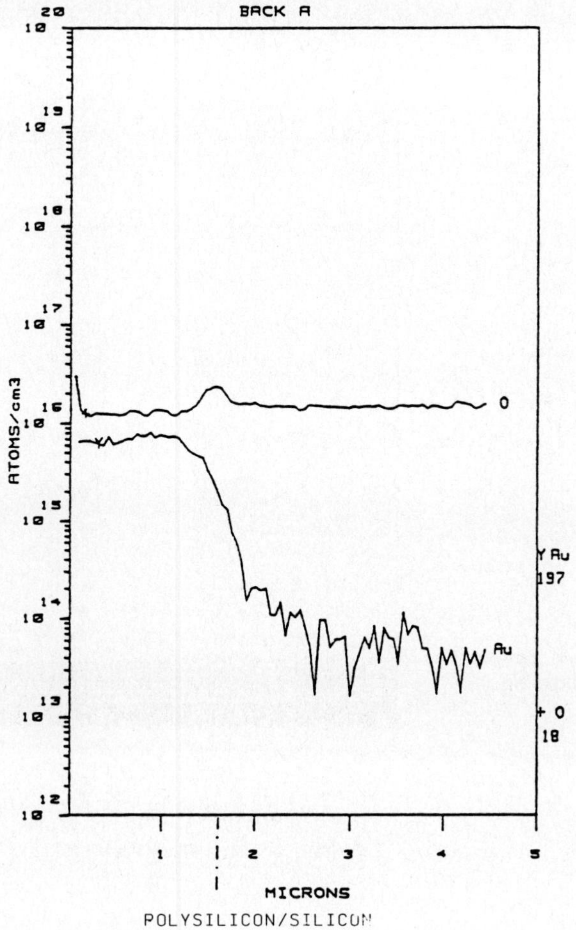

POLYSILICON/SILICON

FIGURE 5

Au DEPTH PROFILE IN POLYSILICON ON THE BACK SAMPLE

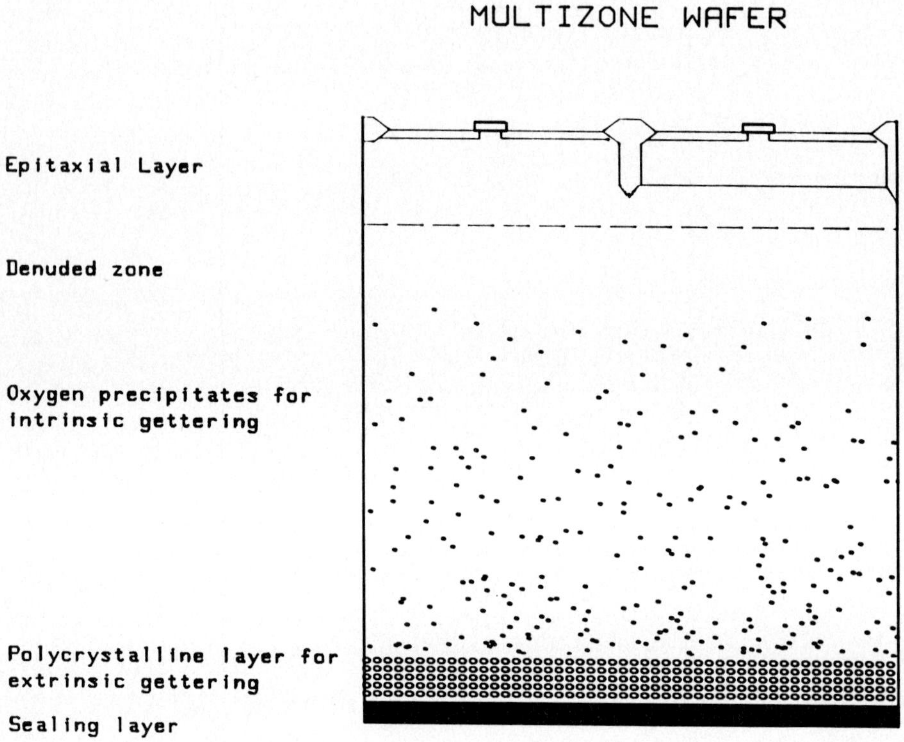

FIGURE 6

CRYSTAL - ENGINEERED WAFER

Wafer Scale Integration, III
M. Sami and F. Distante (Editors)
Elsevier Science Publishers B.V. (North-Holland)

DESIGN OF MOS POWER SWITCHES FOR SHORT-CIRCUIT PROTECTION IN LARGE-AREA INTEGRATED CIRCUITS

Y. SAVARIA*, N.C. RUMIN+, B. MANDAVA+ and S. MADHYASTHA+

* Department of Electrical Engineering
Ecole Polytechnique de Montréal
Montreal, Canada

\+ Department of Electrical Engineering
McGill University
Montreal, Canada

Several aspects relevant to the design of circuit breakers for protection against V_{DD} to V_{SS} shorts in large area devices are presented. Solutions based on parasitic bipolar transistors and MOS transistors only are analyzed and compared. Serious objections against the use of parasitic bipolar transistors are raised. A set of electrical design rules is proposed for circuit breaker solutions applicable to bulk CMOS processes. A design method for large "waffle-iron" MOS transistor with Manhattan geometry is also presented. An experimental validation of the design method is presented, and the behavior of the breakers in the presence of defects is assessed. This work supports evidence that MOS circuit breakers are feasible.

1. INTRODUCTION

Even though large-area integrated circuits (IC) are designed with redundant modules for defect tolerance [1,2,3,4], their yield can still be unacceptably low if they are not protected against defect-induced shorts between the supply and ground. The reason is that the power distribution network is shared by a number of modules and, therefore, a single short between the supply (V_{DD}) and ground can disable all those modules. Disconnecting by laser cutting was the usual solution [1] until some of us proposed an automatic circuit breaker as a feasible alternative [5]. Figure 1 shows the initial proposal, which assumed a bulk CMOS p-well process. The circuit breaker is composed of a low-impedance n-channel MOS transistor and a control circuit. In normal operation, the size of the power transistor is sufficient to guarantee that its voltage drop is low, thus providing the module with an adequate virtual ground. In the presence of excessive supply current, the drop in the transistor increases and eventually sets the latch in the control circuit. Setting this latch turns off the power transistor by forcing its gate low. The mechanism which is automatic at power-up can later be used to disconnect good but unused modules in order to limit power dissipation. Since the mechanism is automatic, even a module with a solid short-circuit does not prevent testing.

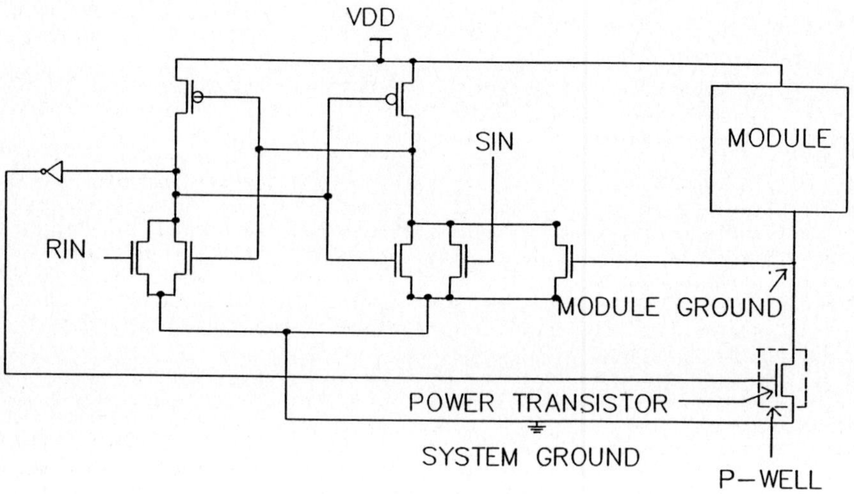

Figure 1 - Schematic of the circuit breaker.

More recently, Carley and Maly [6] proposed the use of parasitic bipolar transistors, which occur in every bulk CMOS process, as an alternative to our solution. Very recently, the need and the feasibility of using circuit breakers was demonstrated in a 200 Mbit wafer-scale memory [2]. One of the purposes of this paper is to compare Carley and Maly's approach with ours. It will be seen from this comparison that it is unlikely that the designers in [2] followed Carley and Maly's approach. We also present some important rules which govern the design of such circuit breakers. Experimental results for circuit breakers of various sizes are given. Finally, the effects of defects in the breaker are assessed.

2. MOS VERSUS PARASITIC BIPOLAR BREAKERS

Although it is generally true that bipolar transistors exhibit a lower dynamic resistance than MOSFETs of similar feature size, parasitic bipolar transistors may offer no real advantage in the design of circuit breakers. Without loss of generality, let us assume a p-well process to simplify the discussion. Carley and Maly [6] offer two solutions: a vertical n-p-n for breaking V_{DD} and a lateral n-p-n for breaking V_{SS}.

The lateral n-p-n solution is to be compared with a nMOS power transistor solution shown in Fig. 1. The lateral n-p-n V_{SS} breaker is fatally flawed by the existence of a vertical n-p-n which, according to Carley and Maly's results, injects from 2 to 3 times as much current into the substrate as it does into the modules. That would multiply the power dissipation by a factor of 3 to 4, which is significant since a wafer can easily dissipate more than 100 W. In addition, this current would have to be collected very efficiently to avoid latchup. If the collection guard ring is close to the transistor for maximum efficiency, one would most certainly produce a hot spot problem when large currents are handled by the switch.

Let us now consider the vertical n-p-n V_{DD} circuit breaker. Carley and Maly mention that the power distribution network would have to be separated from the substrate bias distribution network. Otherwise, interrupting V_{DD} is pointless since a short remains through all substrate connections, which would most probably induce latchup and destroy the circuit by excessive dissipation. However, how could one guarantee that the V_{DD} to V_{SS} short does not occur in the substrate bias distribution network? Since there is no practical solution to this problem, the substrate bias network, which occupies a large area, remains unprotected against a short to V_{SS}. Thus the second solution is only a partial solution, and it introduces a significant area overhead not accounted for in [6].

Thus, if one is limited to parasitic bipolar transistors, they cannot be considered as a practical solution for implementing circuit breakers, despite the potential area savings associated with the better transconductance. Note also that the transconductance of lateral transistors appears to be 3 times smaller than that of vertical transistors [6].

3. DESIGN AND TESTING OF BREAKERS

The following design considerations apply to any circuit breaker in a bulk CMOS process, even where the breaking mechanism is permanent in the form of laser-cutting, fuses or anti-fuses. Unless otherwise indicated, a p-well process is assumed.

a) Only the well bias potential can be safely interrupted.

Figure 2 shows the consequence of not following this rule. Cutting the power to the defective module is ineffective since a short through the substrate remains. Separating the power and substrate potentials is obviously an incomplete solution. In CMOS, this rule means that a p-well process uses a nMOS switch, and a n-well process uses a pMOS switch. A double-well process will normally have a low conductivity substrate which defines the well in an unambiguous manner. Naturally, the area overhead associated with n-well processes will tend to be 2.5 times larger.

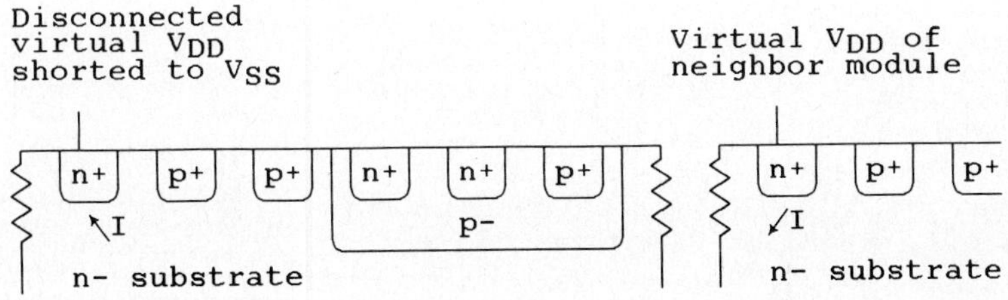

Figure 2 - Structure which demonstrates the possible existence of shorts through the substrate.

b) Connecting transistor drains outside a module is not permitted

Figure 3 illustrates the type of connections which have to be avoided, namely tying together diffusions in different modules. The buffers driving the bus contain n-channel transistors. Each is in a different module and, therefore, has a different virtual ground. If one of the breakers is turned off because of a supply-to-ground short, its virtual ground will float up to V_{DD} due to the short. Hence the bus will be clamped at one diode drop below this potential, due to the voltage drop across the drain- to-well junction of the n transistor.

The above situation can not occur if the connections between the modules are via transistor gates, as illustrated in Figure 4. Therefore, when a bus is required, the solution to the problem in Figure 3 is to locate the bus and the associated drivers in a separate module, as shown in Figure 5.

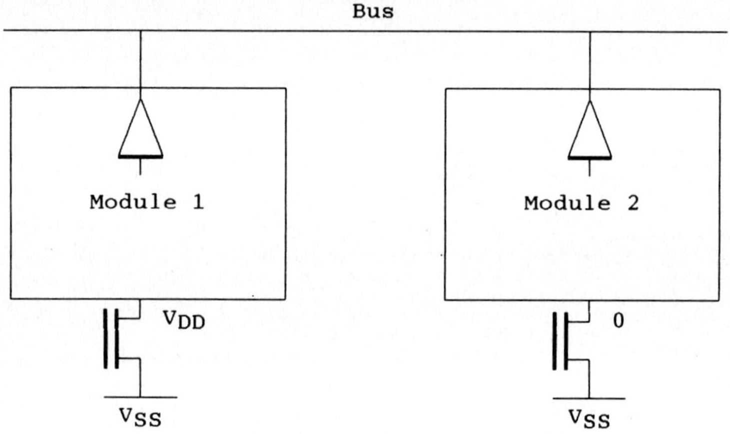

Figure 3 - Diffusion-interconnected modules which result in bus clamping.

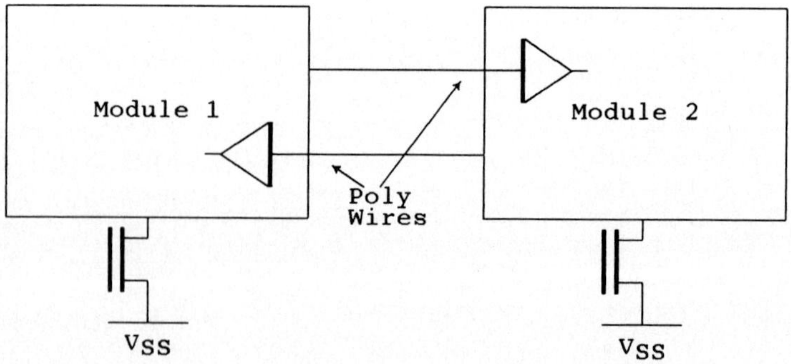

Figure 4 - Poly-interconnected modules.

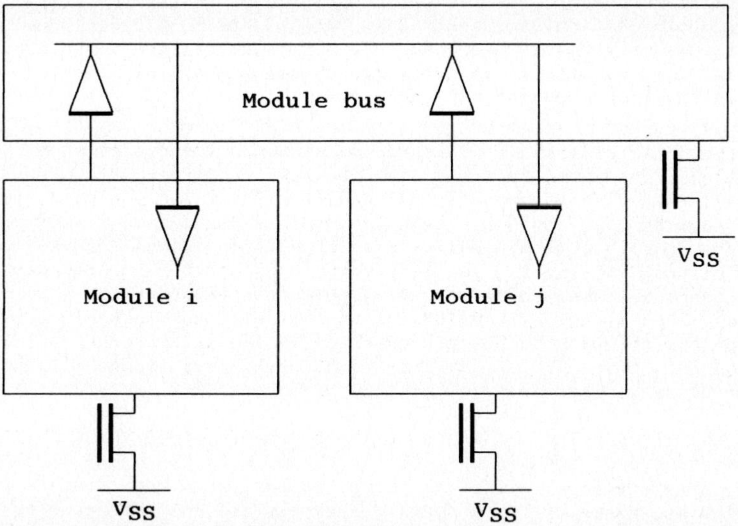

Figure 5 - A bus implemented without connecting diffusions in separate modules.

c) The maximum peak (dynamic) voltage drop across the breaker must be limited to 700 mV.

The presence of dynamic storage nodes, as depicted in Figure 6, is a factor limiting the peak breaker drop. Whenever there is a large dynamic node where the capacitance is to the common V_{DD} substrate, any switching noise temporarily raising the virtual ground potential above 700 mV can

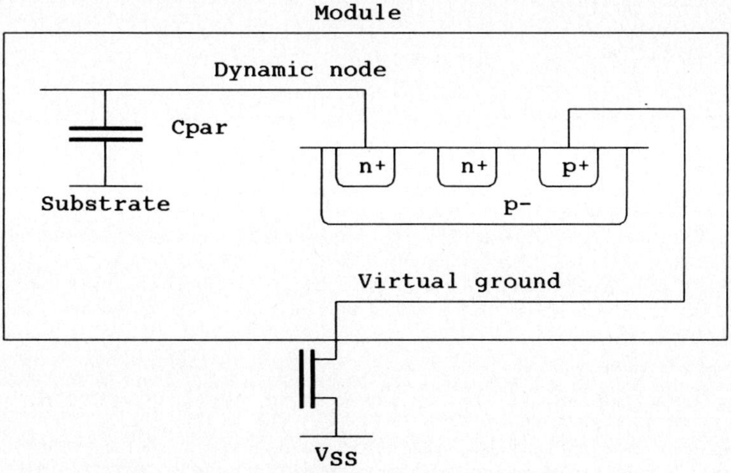

Figure 6 - A module with large dynamic nodes.

trigger latchup if the parasitic capacitance to the substrate is large enough. Note that this circuit does not sustain latchup due to the capacitive nature of the load, however, once triggered who can predict what may happen. Laying out all large dynamic nodes on top of a well reduces the likelihood of this phenomenon.

Even if 700 mV of virtual ground noise does not induce latchup, it can inject current into p-wells. Any pair of n-diffusion is also a latent lateral n-p-n. Thus, if one of these diffusion is connected to a dynamic node storing V_{DD}, while its neighbor is a dynamic node storing 0, the corresponding lateral bipolar transistor will provide a leakage path. A dynamic node storing V_{DD} can leak down to almost 0 V in some situations. The presence of dynamic nodes thus imposes the earliest limit on the peak virtual ground noise. Considering that significant leakage could result well before the base to emitter junction of the lateral n-p-n is fully turned on, the use of dynamic nodes would thus impose a peak virtual ground noise smaller than 700 mV (500 mV for example).

Interestingly, if the modules contain no dynamic nodes and if they are connected through transistor gates as shown in figure 4 and 5, they could safely be operated with virtual ground potentials that differ by more than 700 mV. Peaks of 1.5 V for a 5 V supply are not unthinkable.

Evidently, there is a direct trade-off between the size of the breaker transistor and its maximum drop under normal conditions. Designing for a drop of 300 mV guarantees a good noise margin without incurring an unacceptable area overhead. This assumes $V_{DD} = 5$ V. At lower supply voltage, 300 mV may be too large. Larger voltage drops may of course be tolerable if their duration is short in comparison with the clock period.

d) The maximum peak current must be computed.

Clearly the design of the breaker can not be optimized without a knowledge of the maximum peak current associated with the module. Predicting this current is a separate problem which is receiving a growing amount of attention, but which did not constitute part of the present work [7].

e) After testing a module, turning off power via the breaker is a latchup hazard

Returning to Figure 6, one remarks that, if a large dynamic node is set to 0 during testing, and the module containing the node is later turned off via the breaker, latchup can be triggered when the virtual ground potential rises. The phenomenon is inherently transient and is only harmful if latchup spreads beyond a module's boundary. Assuming that the parasitic capacitance is not extremely large and that each module is surrounded by guard rings to prevent spreading, that mechanism should not be a problem.

If it does happen to be a problem, a robust solution exists. This consists of ensuring that, after testing, the system is turned off via the main supply for a time sufficient to allow all the internal nodes to discharge. The system is repowered with all breakers in the off state, and only the modules which will be part of the final configuration are turned on. This sequence guarantees that no dynamic node can charge to 0 in a module that will not be used. In summary, testing can be followed by a cold restart rendering the described latchup induction mechanism impossible.

4. DESIGN OF LARGE MOS TRANSISTORS

The equivalent width of the required MOS transistor can be several millimeters. Using an efficient layout method such as the "waffle-iron" [8] reduces the overhead associated with the breaker. Figure 7 shows a typical "Manhattan-style" waffle-iron transistor. The power transistor is built by replicating a cell composed of a contact surrounded by four gate segments.

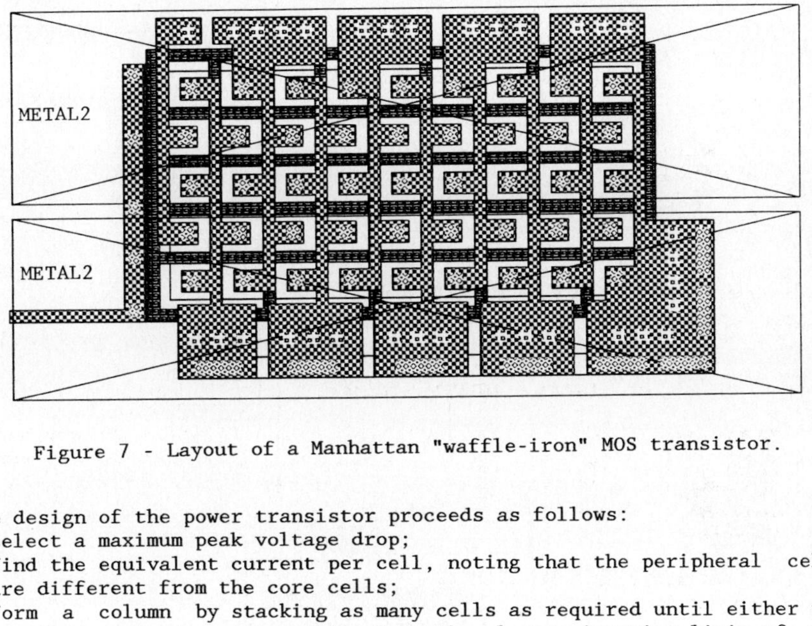

Figure 7 - Layout of a Manhattan "waffle-iron" MOS transistor.

The design of the power transistor proceeds as follows:
- select a maximum peak voltage drop;
- find the equivalent current per cell, noting that the peripheral cells are different from the core cells;
- form a column by stacking as many cells as required until either the required peak current is achieved or the electromigration limit of the metal wire is attained;
- for large currents, several columns will have to be grouped in an array;
- if the array becomes too large, stack several arrays as required (electromigration becomes a problem at this level too).

5. TEST CHIPS AND EXPERIMENTAL RESULTS

Two test chips were designed, fabricated and tested. The first one which was designed according to Figure 1 contains a 100 mA @ 300 mV (nominal) device. The second test chip was designed according to Figure 8 where the same sensor controls two power transistors of different ratings (10 mA and 50 mA @ 300 mV). This second design would be of interest when some portions of a circuit are less sensitive to ground noise. A good example is the peripheral drivers as compared to internal logic with dynamic nodes. Figure 9 is the layout of this second test chip fabricated in a 3 μm double metal p-well CMOS process (actually double-well with n-substrate).

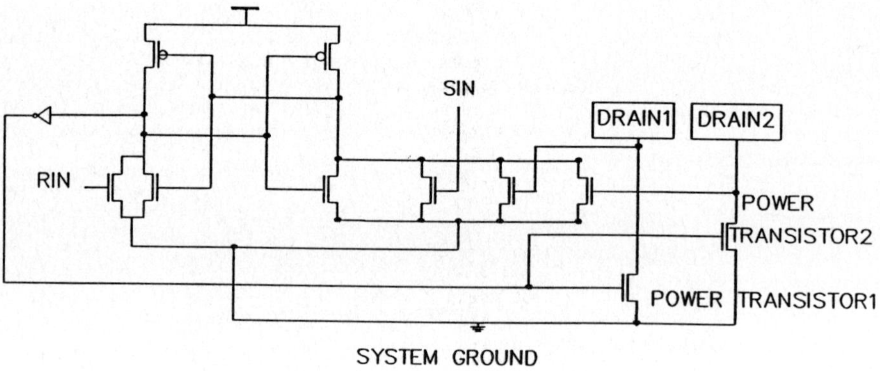

Figure 8 - Schematic of a modified circuit breaker.

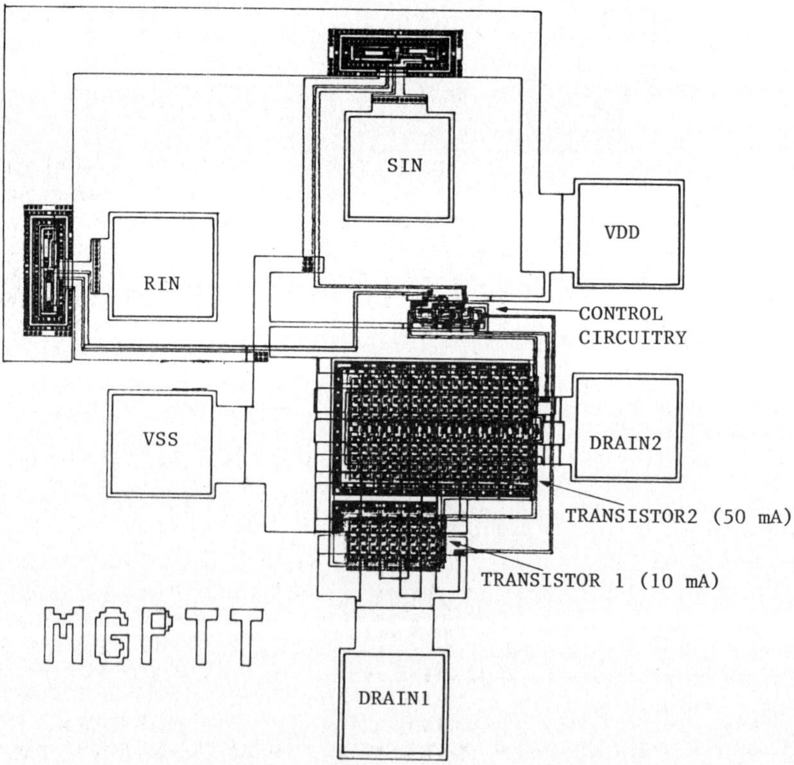

Figure 9 - Layout of the test circuit.

Up to 10 samples of each device were measured to confirm the design method. Table 1 shows a comparison of measurements with predicted results. It includes data from [6] on a parasitic bipolar breaker. As can be seen, there was a good correlation between the measured and the expected currents when parasitics are included. Note the significance of parasitics for the 100 mA device which degraded its expected current to 72 mA.

Table 1
Sizes and measured currents of various structures

	transistors			
	10 mA [1]	50 mA [1]	100 mA [1]	10 mA vertical n-p-n [6]
dimensions (μm)	151*92	270*187	389*246	
transistor area (μm^2)	13900	50500	95700	7300
control area (μm^2)	8000	8000	8000	
total area (μm^2)	21900	58500	103700	18000
relative area [2]	0.224	0.6	1.06	0.183
expected currents [3] (mA @ 300 mV)	10.5	39.5	72	
measured currents (mA @ 300 mV)	13	50	82	

1 nominal current ratings (neglecting parasitics)
2 compared to a standard library I/O pad 234 μm * 420 μm (I/O-limited design)
3 includes all parasitic resistances (diffusion, metal, contacts, etc.)

The control circuit was designed to switch when the voltage drop on the power transistor is around 2 V or roughly 7 times the nominal rating. Testing showed that a minimum size device in the control circuit is sufficient to turn off the largest power transistor in less than 1.5 μs. This switching time is rather long but perfectly adequate considering the normal rise time of a power supply and the amount of heat that would be dissipated in less than 2 μs, even in the presence of a dead short.

An examination of Table 1 is quite revealing. First of all, the 10 mA n-transistor was 1.9 times larger than the 10 mA vertical n-p-n device proposed in [6]. Considering that the MOS transistor was actually a 13 mA @ 300 mV device, whereas the bipolar transistor was a 10 mA @ 250 mV (dynamic drop) device, the comparison is a 1.75 factor in favor of the bipolar device. This is to be compared with the factor of 7 predicted in [6]. It is also interesting that, for small transistors, the size of the control circuit becomes significant. Due to the smaller size of our control circuit the advantage of the bipolar solution reduces to a mere 13%.

A meaningful figure of merit for a circuit breaker is to compare its size to a normal I/O pad. Interestingly, the 82 mA device has an area roughly equal to one of our library I/O pads. If the area devoted to circuit breakers is significant in a large area circuit, the relative area would clearly be much smaller than what is currently devoted to pads at the VLSI level of complexity, where circuits with more than 100 pins are common.

6. DEFECT SENSITIVITY

The size of the power switch raises a question with respect to its defect sensitivity. Indeed, if 1% of the area of a large chip devoted to circuit breakers is worthwhile from an area overhead standpoint, it may not be if the complete breaker is intolerant to defects ("kill" area). The yield of the breakers could become a first-order factor in reducing the yield of the complete device.

One important feature of our breakers is that they are fail safe to most defects. Any "stuck-at" defect is indeed tolerated since the breaker is either permanently on or off. When permanently off, a module is lost and when permanently on the protection is lost. Loosing the protection is of consequence only if there is also a short in the corresponding module, and partitioning can be used to ensure that this event has a low probability.

The power transistor is inherently tolerant to a break in the poly or in the metal as well as to missing contacts. Shorts in the polysilicon gate is of no consequence since it is an equipotential. Shorts in the metal could cause the breaker to be permanently on as already discussed. A drain-to-well excessive leakage current would have a similar effect. Thus, none of the above mentioned defect types is capable of failing a system using circuit breakers.

There is one failure mechanism, however, which could defeat the system. A metal-to-poly short between the drain and gate would put a breaker in a medium impedance mode where the gate voltage is above the threshold of the n-device, but is not high enough to provide a good virtual ground at the drain. The waffle-iron layout, which superimposes the drain metal wiring and the gate enhances the probability of this defect mode. Trying to turn off the breaker would only work if the module is designed to draw essentially no current when inactive, since the breaker would be self-polarizing. This solution is feasible but would impose additional precautions in the design of a module to avoid unpredictable and potentially malicious behavior. A more robust solution would consist of fragmenting the power transistor into a multiplicity of independent transistors, each of which is individually controlled by independent drivers (see Fig. 10). The same drain to gate short would then simply amount to an increased dynamic resistance in a breaker that cannot be turned off.

7. CONCLUSION

Several aspects relevant to the design of circuit breakers for protection against V_{DD} to V_{SS} shorts in large area devices have been presented. A solution based on MOS transistors was analyzed and compared to a second

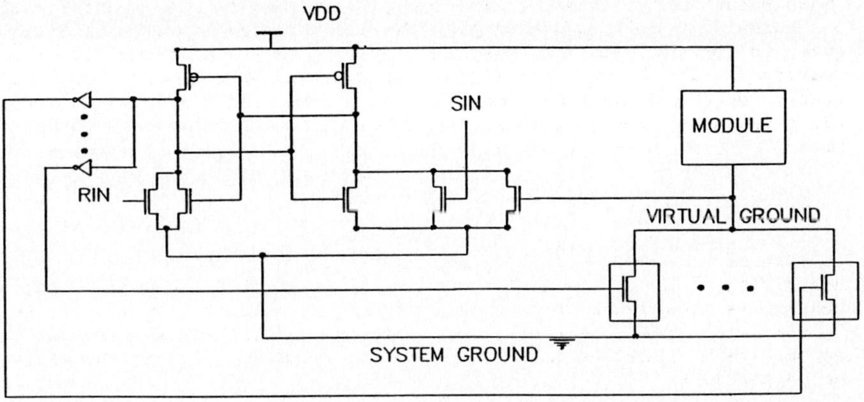

Figure 10 - Circuit breaker with partitioned power transistor.

solution where parasitic bipolar structures are used to implement low impedance transistors. Serious objections against the latter were raised. A set of electrical design and test rules was proposed for circuit breaker solutions applicable to bulk CMOS processes. The design method for a large waffle-iron MOS transistor with Manhattan geometry was also presented. An experimental validation based on two test chips that were fabricated in a 3 μm CMOS process was also presented, and the behavior of the breakers in the presence of defects was assessed. This work supports evidence that MOS circuit breakers are feasible and that they are an effective means of implementing large area devices tolerant to shorts between V_{DD} and V_{ss}.

ACKNOWLEDGEMENTS

This work was partly supported by individual operating grants as well as and by a Strategic Grant from the Natural Sciences and Engineering Research Council of Canada, by an "Actions structurantes" grant from the Ministry of Higher Education and Science of Quebec, and by the Canadian Microelectronics Corporation.

REFERENCES

[1] R.O. Carlson, and C.A. Neugebaur, "Future Trends in Wafer Scale Integration", Proceedings of the IEEE, vol.74, no. 12, dec. 1986, pp. 1741-1752.

[2] N. MacDonald, G. Neish, A. Sinclair, F. Baba, T. Tatematsu, K. Hirawa, K. Miyasaka, "200 Mb Wafer Memory", 1989 IEEE International Solid State Circuit Conference, p. 240-241.

[3] R. Nasreddine, E.F. Kouka, Y. Wang, D. Marron, and J. Trilhe, "A Reconfigurable SRAM 4.5 Mbits WSI Memory", International Workshop on Defect and Fault Tolerance in VLSI Systems, Amherst, October 1988, pp. 6.2-1, 6.2-14.

[4] J.-L. Patry, E.-F. Kouka, and G. Saucier, "Defect Tolerance in Wafer Scale Array for Image Processing", International Workshop on Defect and Fault Tolerance in VLSI Systems, Amherst, October 1988, pp. 8.2-1, 8.2-13.

[5] V.K. Agarwal, E. Cerny, N.C. Rumin, Y. Savaria, and P. Schvan, "Techniques for Implementing Large Area Devices", ICCD PortChester N.Y., October 1986, pp. 220-224.
[6] R. Carley and W. Maly, "A Circuit Breaker for Redundant IC Systems", IEEE 1988 Custom Integrated Circuit Conference, Rochester N.Y., May 1988, pp. 27.6.1 - 27.6.6.
[7] F. Najin, R. Burch, P. Young, and I. Majj, "A Current Estimator for CMOS Circuits", IEEE Conference on Computer-Aided Design, Digest of Technical Papers, pp. 204-207, Nov. 1988.
[8] L. Glasser, and D. Dobberphul, The Design and Analysis of VLSI Circuits, Addison-Wesley 1985.

Wafer Scale Integration, III
M. Sami and F. Distante (Editors)
Elsevier Science Publishers B.V. (North-Holland)

Organic Polymer Optical Interconnections for Photonics in WSI and WSHP

H. T. Lin, S. Dabral, and J. F. McDonald

Center for Integrated Electronics
Rensselaer Polytechnic Institute
Troy, NY, 12181, USA

The limitations of metal interconnections in WSI are examined with respect to their bandwidth and coupling. Comparison is made with the alternative of employing optical interconnections. Some of the technical issues which lie ahead to make photonic WSI possible are explored. Certain organic materials are identified as promising candidates for implementing optical interconnections. These are found to be interesting because of their low deposition temperatures, and because they lend themselves to fabrication by techniques that are already finding acceptance for metal interconnections.

1. Introduction

1.1 Limitations of Metal Interconnections

Metal interconnections between die sites in WSI and WSHP play the same role as metal traces on a printed circuit board, or the wiring on a hybrid substrate. Their behavior in high speed applications must be quite different from that of the metal lines used inside of a chip. Metal lines inside a chip have traditionally been operated as diffusion lines or distributed resistor capacitor RC charging lines. For short lines this is acceptable. For long lines the delay experienced by these RC lines during charging and discharging can become excessive. This is because both the resistance of a line and its capacitance grows with the line length. Therefore, the delay for an RC line grows as the square of the length of the line. Also, since the delay for an RC line largely consists of the rise or fall time of its exponential charging behavior, the slope of this charging curve can become very low leading to instabilities of an analog type in high gain circuits.

Hence, long interconnections must be fabricated in such a way that they behave as LC or weakly lossy LRC lines [1]. Signal propagation on such lines is at the speed of an electromagnetic wave in the dielectric medium surrounding the metal lines. For lines the size of a wafer, the difference between RC and LC type delays can be several orders of magnitude! Techniques exist for fabricating metal lines which exhibit LC transmission line behavior for wires for the length typically found in wafer scale. However, there is another important limitation on metal lines, namely the skin effect. At sufficiently high frequencies the current in the wires is confined by inductive effects to travel along the first few microns of the surface of the metal. As a result the lines become very lossy as frequencies increase, and the sharp rise times of the ideal transmission line become limited by the loss at these higher frequencies. Rise times on the order of 200 ps might be observed for such signals. These might seem adequate by today's standards, but they are actually already limiting systems employing high speed silicon bipolar and GaAs MESFET circuits. It will not be long before this limit is felt in submicron CMOS design.

For faster rise times one could cool the substrate substantially. Aluminum at liquid nitrogen temperature is roughly 10 times as conductive as it is at the normal temperatures for electronic equipment. This decreases the skin depth still further, but the net sheet resistance through the more conductive skin would actually decrease by the square root of this factor, or by about 3.33. Alternatively, one could pursue the new high T_c superconductor materials in an attempt to drive down the R component of the RC wire until it behaved as an LC line. The skin effect of ideal superconductors is absent, and the bandwidth of superconductivity should theoretically extend into the IR frequencies. However, the currently deposited thin film high T_c superconductors behave as chains of weak link Josephson junctions due to undesirable grain structure, and their bandwidth appears not to be any better than Cu cooled to the same temperature. Also, most T_c's for these materials are only slightly above liquid nitrogen temperature, and it is necessary to operate well below T_c to achieve good bandwidth even in the ideal superconductor (possessing ideal grain structure). Barring some further breakthrough it would appear that cooling the substrate is awkward to implement, and does not bring large practical improvements in performance. However, this is a rapidly evolving field and should be monitored for such breakthroughs.

Coupling, particularly the rise time sensitive component of coupling, constitutes another limitation on metal interconnections. This problem becomes aggravated when the separation between lines is reduced due to packing requirements. Homogeneous dielectrics can be utilized to help suppress rise time sensitive coupling. However, this effect applies only to coupling between one dimensional transmission lines. By this we mean that the two lines encounter no variation in their environment along their length. Any disturbance to this one dimensional character can begin to bring back the rise time sensitive component of the coupling. For example, if the two coupled lines cross over an array of orthogonally oriented lines, or if the ground plane is perforated with holes to permit outgassing during fabrication then the effectiveness of the use of a homogeneous dielectric is reduced in suppressing rise time sensitive coupling. The presence of such changes in the electromagnetic environment also causes variations in characteristic impedance which act to reduce rise time.

If the operation of WSI at reduced temperatures is undesirable, or if in fact very fast rise times are desired, then the only alternative would be the use of optical interconnections. In such a case we would have photonic WSI.

1.2 Optical Interconnections

The primary requirements for optical interconnections are low attenuation, and the availability for means to modulate and demodulate the light. Traditionally the materials studied for this purpose have been glasses, Gallium Arsenide (GaAs), and the crystalline ceramics such as Lithium Niobate. Even a newcomer to the field of photonics can see that these materials are incompatible with each other to varying degrees, and furthermore they are incompatible with silicon, which is the basis of today's VLSI.

The origins of the current selection of photonic materials has been dictated by the desire to generate light, which requires GaAs technology, or at least compound semiconductors whose merit is the indirect band gap. Glasses have achieved extremely low attenuation in long haul fiber optics, and it is natural to see them considered for micron sized optical waveguides. Finally, Lithium Niobate is unique because it alone among the inorganic crystalline ceramics has the highest linear electro-optic effect coefficient (-32 pm/V). Hence, with this material it is possible to change the effective dielectric constant by applying a field across it in the right orientation. This can be profitably employed to slow down or speed up the light wave and hence alter its phase. This permits modulation of the light beam by interference techniques. Devices to do this even on Lithium Niobate are often several millimeters in length since even with this material the strength of the linear electro-optic effect is not as strong as one might like. In spite of this, wafer sized optical crossbar switches have been routinely fabricated on wafer substrates of Lithium Niobate [2]. In effect, this is the earliest manifestation of photonic WSI.

However, these materials all vaporize at high temperature, and almost any method of depositing them on a common substrate would lead to incompatibilities due to damage of the substrate or previously deposited materials. Unfortunately, for the moment these remain interesting but separated technologies with little chance of integration.

There has been some limited amount of work depositing GaAs on silicon [3]. This might eventually lead to the integration of GaAs light sources on silicon substrates, but even this seems a distant possibility, as there appears to be a reliability problem with respect to these devices. As a result, one might conclude that the possibility for photonic WSI would appear dim. However, recent breakthroughs in optically active polymers have created some new approaches which look viable. Since these do not produce the light, but can only provide modulation, it will be necessary to solve the light generation problem with an off-wafer source which will act like a power supply for photons. This will be in the form of a nonmonolithic solid state laser source and fiber optic coupling to the wafer surface. This would also remove the heat dissipation associated with light generation to a point removed from the wafer. We assume monolithic light amplification, which requires GaAs technology, will also be absent, and therefore all optical paths must be kept relatively fan-out free, and as short as possible to minimize attenuation.

2. Waveguiding Polymer Materials

The most important goal is the fabrication of the photonic materials on the wafer in the most innocuous manner possible, including low deposition temperatures, and low radiation damage from etching and metallizations. Confinement of the light (usually operating at wavelengths in the near IR) to the guide can be by two methods. The first is by the index variation method, and the second is by cladding with metal. In the index variation method the index of refraction inside the guide must be higher than that outside the guide by a margin sufficient that all light launched into the guide will reflect perfectly from the guide walls. In the metal clad enclosure for comparison (see Figure 1) the electrically conducting material at optical frequencies behaves like a lossy dielectric rather than a metal in the conventional sense. At wavelengths larger than one micron the reflection even at normal incidence of IR from smooth, clean Cu or Au surfaces is extremely high (see Figure 2) so this approach is viable for the short lengths of guides on a wafer surface. However, passivation of these surfaces is mandatory to prevent corrosion.

PMMA, an electron resist material, is unable to crystallize, and forms an extremely clear amorphous polymer which has been successfully used to make short fiber optic cables. Its attenuation can be made as low as 0.03 dB/cm by simply purifying the material [4]. There is no molecular mechanism for absorption, only scattering sites in the material which can deflect light out of the guide. The amorphous nature of the material greatly reduces crystallite boundaries which can scatter light. However, if ebeam resist chemistry is needed elsewhere in the process there will be a conflict.

Some colorless polyimides are also promising materials for transmission of light [5]. However, there have been complaints about certain of these polyimides taking up water (which can absorb in the IR), and the tendency of some to melt at low temperatures. They can also be more crystalline than PMMA.

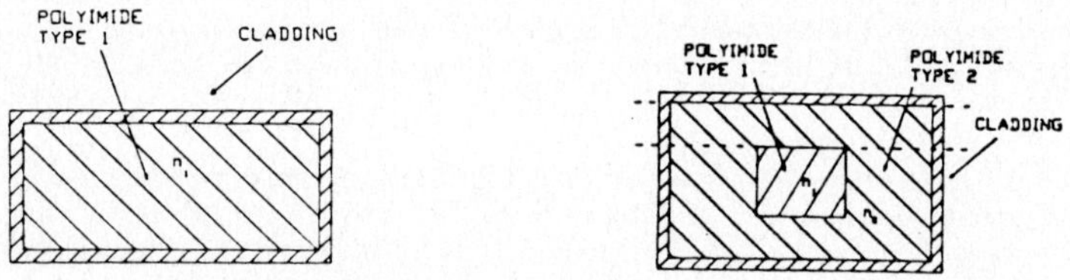

Fig. 1: Possible Optical Waveguide Structures for Confinement of Light.

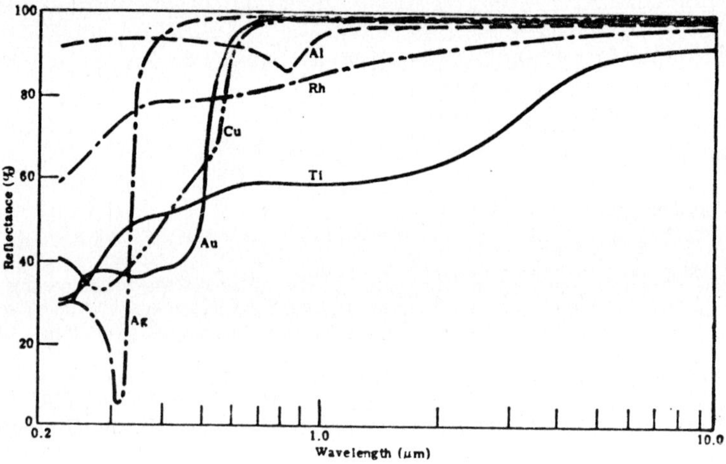

Fig. 2: Reflectivity of Clean, Smooth Metal Surfaces at Normal Incidence.

Another very promising group of polymers is the class of parylenes. These are vapor depositable with the wafer held at room temperature. From the outset workers have though these might be suitable for clear optical waveguides [6]. One of these parylenes (the E version) also cannot crystallize and appears to be extremely satisfactory for this application. The material has been used as an encapsulant in the pellet fusion program where light scattering would be undesirable.

3. Electro-optically Active Polymers

3.1 "Guest-Host" and "Backbone-Pendant" Polymers

There now exist polymers which exhibit the same kind of linear electro-optic effect as Lithium Niobate. That is, by applying a low voltage across the material its effective dielectric constant at optical frequencies can be affected. These are the so called NonLinear Optical (NLO) polymer systems.

The key is to find a linearly electro-optically active material which can be embedded in a clear transparent waveguiding polymer. Promising candidates for these materials have been found. Their great advantage is that they can be applied to the surface of a wafer at low processing temperatures, making them at once low in stress, and compatible with other electronic materials. These come in two varieties, the doped "guest-host" solution, and the "backbone-pendant" polymer [7]. Currently these polymer matrices are spin-cast onto a wafer in a solvent like polyimides, or resists, with some advantageous planarizing effects. After driving off the solvent and curing the polymer the film must be oriented or poled.

In either case the embedded electro-optically active material is a molecule one end of which acts like a donor, and the other an acceptor. Electrons displaced from the donor towards the acceptor end form a dipole. To obtain a net macroscopic dipole moment the molecules need to be ordered. One way to accomplish this is by applying a poling field while the embedding polymer is heated above its so-called glass transition temperature, T_g, (see Figure 3). This is the temperature at which small molecular motions just begin to occur within the polymer such as rotation or change of orientation of a part of the polymer chain. For the polymers of interest here the glass transition temperatures are all low ($T_g \leq 200^o$ C). The applied poling field is held fixed while the temperature of the embedding polymer matrix is cooled below T_g.

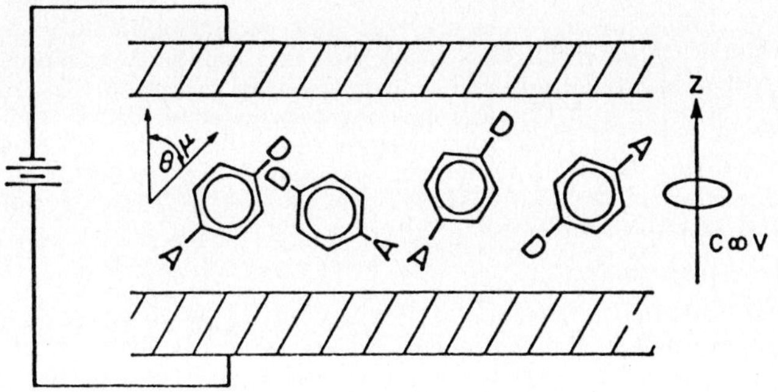

Fig. 3: Poling of the initially disordered electro-optic material.

Poling of the as-deposited polymers must contend with the random polymer chains that arise when curing is from the spin cast state. Poling then must contend with the stress associated with imposing order on the disordered chains. As a result the frozen in order has a tendency to decay with time. For the doped guest host polymer matrix the decay can be all the way to zero. For the backbone-pendant approach the decay usually approaches some asymptotic value over time. In an attempt to impose order on the polymers as they are forming some have attempted to apply the poling field as the polymer cures. This has led to some improvements. However, the mobility of the molecules in the spin cast state is not high.

It is our contention that only certain techniques for vapor deposition of these polymers in the presence of a substrate bias field to orient the molecules will have a chance to substantially reduce this problem. These techniques are the Ionized Cluster Beam codeposition of organic polymeric materials [8] with the accelerating ion bias field serving to perform orientation. It is important in this work to avoid introduction of scattering sites in the deposited material. So it is likely that the embedding material tend towards the amorphous state, while the electroactive material becomes oriented in a manner similar to that found in liquid crystal polymers.

3.1.1 Doped Guest-Host Systems

Workers at Bell Laboratories [9] have experimented with doped guest-host solution systems. For example, the dye Red Disperse-I has been dissolved in the host PMMA which as we have mentioned is a clear polymer suitable for waveguide material.

3.1.2 Backbone-Pendant Systems

Workers at American Hoechst Celanese [10] have experimented with backbone-pendant molecules possessing a PMMA-like backbone, and having electro-optic pendant molecules attached directly onto the backbone polymer (see Figure 4). In some cases this has led to the alternation of more than one type of electro-optically active compound along the backbone polymer as examined by workers at Kodak [11]. This permits study of interaction between two or more types of pendants. One reason for studying this effect is that a kind of superhyperpolarizabilty effect has been observed in alternating Langmuir-Blodgett multilayers of different electro-optically active compounds [12]. The effect of the combination is in this case greater than the sum of the separate contributions toward electro-activity.

Finally, a recent study by workers at IBM has led to the discovery that a polymer employing a backbone based on polyethylene and a pendant based on p-NA called PPNA (see Figure 5) has shown low attenuation, and an electro-optic activity near the level of Lithium Niobate right after poling, with a relaxation after poling to an asymptotic strength approximately only 35 % less than its maximum [13]. This is perhaps the most encouraging result to date, because it appears that this material is similar enough to Styrene that it may vaporize by breaking bonds primarily along the backbone in such a way that repolymerization might occur on deposition upon a cold substrate retaining the pendant intact.

One must approach much of the published literature with caution in interpreting the results. Most papers report second harmonic generation (SHG) efficiencies, these are related to but not identical to the currently more useful linear electro-optic effect, which is the one we wish to employ. The SHG conversion process, which is an optical to optical frequency interaction, is relatively easy to measure. Consequently one sees in the literature many papers on the SHG effect of materials. On the other hand, the linear electro-optic effect which is a mixing of RF and optical frequencies in the

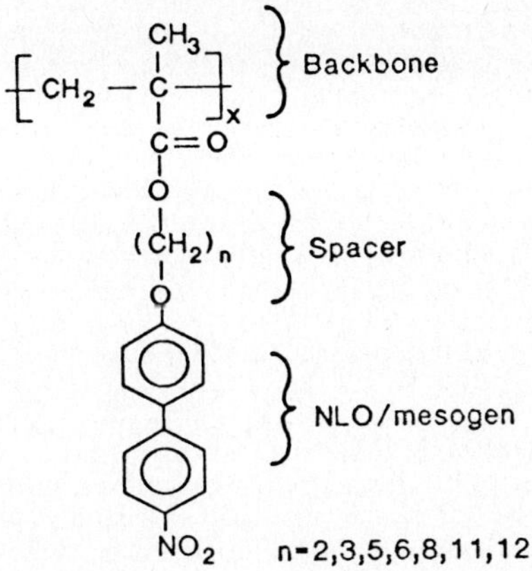

Fig. 4: Backbone-Pendant polymer of the type examined by Hoechst-Celanese. This one has the NLO active pendant of 4,4'oxynitrobiphenyl attached through a variable spacer Alkyl chain to a methacrylate backbone.

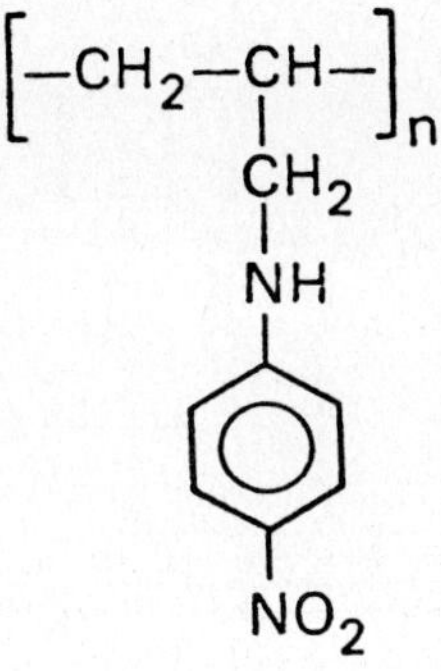

Fig. 5: Backbone-Pendant PPNA developed at IBM Almaden.

material nonlinearity, requires building a device structure to measure the effect directly. Consequently it is often the case that the reader is left only with SHG information, which is indicative, but not directly usable for our purposes.

4. Organic Electro-optic Devices

There are a number of interesting candidates for devices for use in photonic WSI [14]. To be useful operating with the current generation of microelectronics they must be efficient in converting electrical signals into modulated light beams. Furthermore, they must be fabricatable with, or on top of, silicon devices. One device which is suited for this purpose is the Mach-Zehnder (MZ) interferometer.

4.1 The Mach-Zehnder Interferometer as a Modulator

The Mach-Zehnder interferometer consists of two identical parallel paths transmitting coherent light. The light for these two paths can be derived by equally splitting light from a single guide by a proper design taking into account the modal structure of the light in each section. Both of these paths have the same electro-optically active waveguiding material, and both legs of the device have electrodes placed across the walls of the paths in such a way that a voltage across these plates can change the velocity of light through the guide (see Figure 6). One plate is then grounded, and the other is driven by an electrical modulating voltage. If the voltage is large enough, the light in the driven section will be shifted in phase by exactly π. The light from the grounded leg, and the driven leg can then be recombined in such a manner as to promote their destructive interference, thereby extinguishing the light as it proceeds toward the output waveguide of the device. In reality the light does not disappear but scatters out of the waveguide and is prevented from propagating forward along the guide, so a means to absorb the scattered light must be provided.

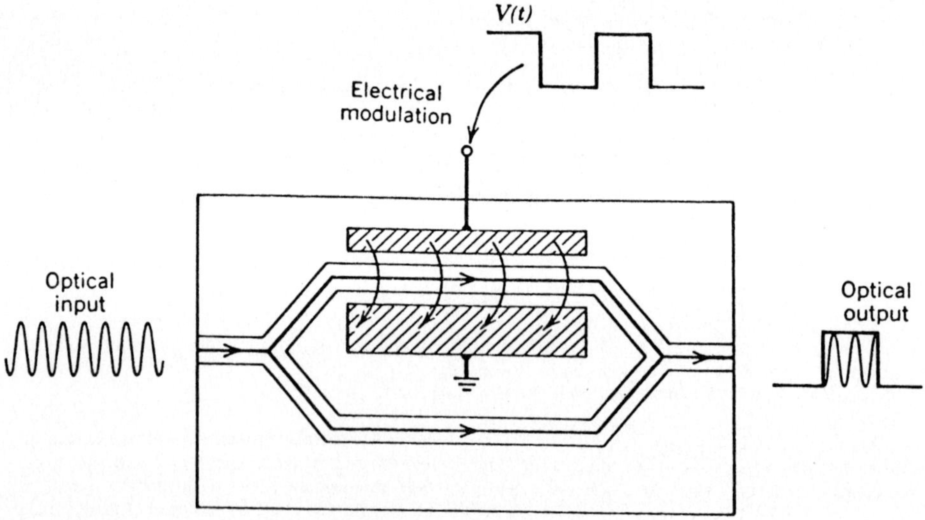

Fig. 6: Mach-Zehnder Interferometer for light modulation.

Now, electrically the MZ interferometer looks like a simple RC charging network to its driver, which can be implemented in a silicon die (i.e. GaAs is not essential). One immediate advantage of the polymer based electro-optic materials is that their typical dielectric constant at RF frequencies (which might be different from that at optical frequencies) is low (in the range 3-4). This is in comparison with Lithium Niobate which has a high dielectric constant at RF (about 64). Hence, for identical linear electro-optic coefficients the polymer will be easier to drive at high frequencies than the Lithium Niobate by a factor of 16 because the devices will have equal length and perhaps even equal width (note, however, the difference in modal structure in the two cases).

At very high frequencies it is desirable to treat the electrodes on the legs of the MZ device as transmission lines, and terminate them with the characteristic impedance of that line. This permits the RF to enter at one end of the electrode and travel toward the terminator without reflections. Provided the velocity at optical frequencies in the electro-optic material is the same as that at RF frequencies the light and modulating RF travel down the leg in synchronism and an appropriately modulated light signal emerges at the end of the leg. The result is called a "Traveling Wave" MZ interferometer [15].

4.2 Light Detection

As is conventional in WSI we can fabricate devices in that substrate, and in particular we can fabricate silicon based IR photodetector devices. These can be in the form of high speed photoresistors or phototransistors which must be extremely small to minimize the parasitics they entail. Although the speed limitations of such silicon based devices are obvious, it is clear that a photonic interconnection technology based on an all silicon and polymer approach, and involving no mandatory monolithic integration of GaAs devices is possible. Furthermore, at the WSI level it is possible to have a completely planarized self-aligned fabrication process for the optical devices and interconnections.

5. Fabrication Considerations

Most of the organic polymers considered here can be patterned by conventional techniques. For example, most polymers consisting only of C, O, H, N are readily etched by reactive ion etchers, tools which have found successful application in metal interconnection structures. The patterning of electrodes is similar to metal interconnections in WSI.

6. Micro-optical Fixtures

In addition to the means for modulating and detecting light which is generated off of the wafer a means must be found for fabricating many of the other fixtures needed in optical systems on a microscopic scale. These include gratings, dome lenses and mirrors. In this section we will limit our attention to mirrors.

Index variation confined guides have difficulty changing direction. It might take many millimeters to execute a right angle turn without having the light radiate out of the guide [16]. Although light waveguides could be fabricated in the same plane at right angles to each other (Figure 7) thereby reducing the need for vias, any vertical turns requiring such large radii of curvature would lead to impractical thicknesses for the material on the wafer. In such cases vertical mirrors could be required using metal surfaces.

Consequently, it is necessary to etch patterns with laterally varied dwell times. In the case that there are many varied surface features like this structure which must be fabricated on a given layer the focused ion beam (FIB) is a useful tool for ion milling the polymers [17]. A SEM photograph for an ion beam milled mirror surface in polyimide is shown in Figure 8. Note the extraordinary smoothness of the milled polymer surface. This is an essential prerequisite for a smooth metal surface to be deposited onto the polymer surface.

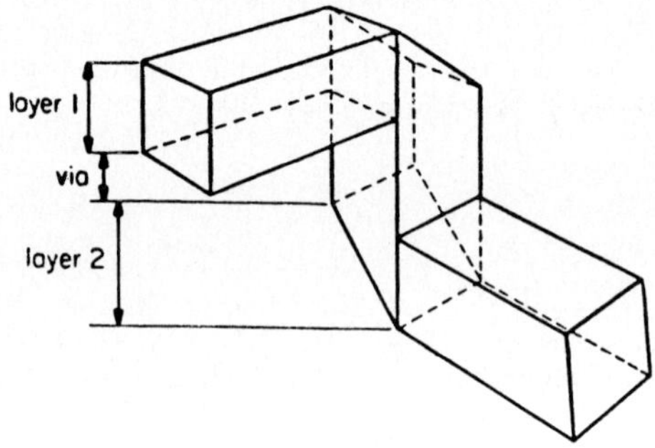

Fig. 7: Vertical optical mirrors used in a "periscope" optical via.

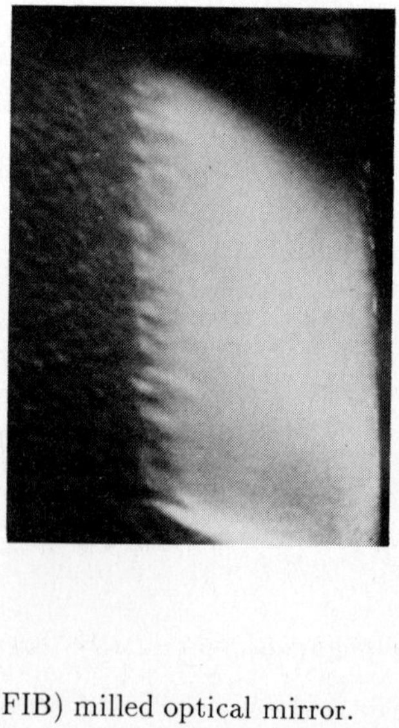

Fig. 8: Focused Ion Beam (FIB) milled optical mirror.

4. Conclusion

It has been suggested that a photonic WSI technology based exclusively on silicon substrates and linear electro-optically active polymers may be possible. Key technologies to pursue are development of vapor depositable versions of these polymers, development of polymers with giant hyperpolerizabilities, and ICB methods for deposition of organics.

Acknowledgement

The assistance provided by Eastman Kodak, IBM, Raytheon and DEC over the years is acknowledged.

References

[1] Jaesok, Kim et al, Transient and crosstalk Analysis of slightly lossy interconnect lines for WSI and WSHP - weak coupling case, IEEE Trans. on Circuits and Systems, Vol. CAS- 35(#11), Nov 1988, pp 1369.

[2] Alferness, R.C., "Waveguide Electrooptic Switch Arrays," IEEE J. on Spec. Topics in Commun., Vol. JSAC-6(#7), 1117-1130, Aug. 1988.

[3] Tsaur, B.Y, et al, "Low Dislocation Epilayers Grown on Ge-coated Silicon Substrates by Means of Lateral Epitaxial Overgrowth," Appl. Phys. Let., Vol. 41, 347-349, 1982.

[4] Williams, D.J., Ed., "Nonlinear Optical Properties of Organic and Polymeric Materials," American Chemical Symposium Ser. #233, 1983.

[5] Selvaraj, R., et al, Integrated Optical Waveguides in Polyimide for Wafer Scale Integration," IEEE J. Lightwave Tech., Vol. LT-6(#6), 1034-1044, June, 1988.

[6] Majid, N. et al, "The Parylene- Al Multilayer Interconnection System," IEEE JEM vol-18(#2), March, 301-311, 1989.

[7] Bowden, M.J., and Turner, S.R., Eds., "Electronic and Photonic Applications of Polymers," Advances in Chemistry Series # 218, 1988.

[8] Usui, H., et al, "Anthracene and Polyethylene Thin Film Deposition by Ionized Cluster Beam," J. Vac. Sci. Technol., A., Vol. 4(#1), 52-60, Jan./Feb., 1986.

[9] Singer, K.D., et al, "Orientationally Ordered Electro-optic Materials," in Nonlinear Optical and Electroactive Polymers, by Prasad, P.N., and Ulrich, D.R., Plenum Press, 189-204, 1987.

[10] DeMartino, R., et al, "Nonlinear Optical Polymers for Electrooptical Devices," in Nonlinear Optical Properties of Polymers, Heeger, A.J., Orenstein, J., and Ulrich, D. R., MRS Symposium Proceedings, Vol. 109, 65-76, 1988.

[11] Meredith, G.R., et al, Macromolecules, Vol. 15, 1385, 1982.

[12] Neal, D.B., et al, Electron. Lett., Vol. 22, 460, 1986.

[13] Eich, M. et al, "Corona Poling and Real Time SHG Study of a Novel Covalently Functionalized Amorphous NLO-Polymer," IBM Technical Report # 6663 (64238) 1/24/89, IBM, Almaden Research Center, 38 p., 1989.

[14] Lee, D.L., Electromagnetic Principles of Integrated Optics," John Wiley & Sons Publishers, 1986.

[15] Nishihara, H., et al, "Optical Integrated Circuits," McGraw Hill Publishing, 1985.

[16] Marcatili, E.A.J., "Bends in Optical Dielectric Guides," BSTJ, Vol. 48, 2103-2132, 1969.

[17] McDonald, J.F., "Applications of Focused Ion Beam for Testing, Repair, and Fabrication of Metal and Optical Interconnections in Wafer Scale Integration," Proc. 1st IEEE Int. Conf. on WSI, ICWSI89, San Francisco, 161-172, Jan., 1989 (IEEE Cat. # 89CH2680-7).

Wafer Scale Integration, III
M. Sami and F. Distante (Editors)
Elsevier Science Publishers B.V. (North-Holland)

The WASP Project : Bridging the Gap Between Hypothetical Wafer-Scale Architectures and Engineering Reality

I.P.Jalowiecki, S.J.Hedge, K.D.Warren[1] and R.M.Lea

Aspex Microsystems Ltd.
Department of Electrical Engineering and Electronics,
Brunel University,
Uxbridge, Middlesex, UK.

INTRODUCTION

The initial hypothesis upon which an innovative wafer-scale architecture is founded must have sufficient scope and vision to excite the imagination of researchers and funding bodies alike. However, the establishment of architectural, physical and electrical models, no matter how detailed, can never expose the full scale of the engineering problems which must be overcome.

Eventually, the only way to progress from the hypothesis is to take the strategies derived from these initial models and apply them to a practical series of experiments. Designed to answer detailed feasibility questions, these experiments provide the crucial engineering perspective upon which to base the evolution of the initial hypothesis.

At Brunel University, research has been underway for several years to assess the architectural, electrical and physical benefits and constraints of the WASP wafer-scale Associative String Processor (ASP). This is intended to implement a massively parallel processor entirely within the medium of WSI.

Progress in such a major endeavour must be made with caution, through a series of carefully planned test chips and demonstrators. This paper outlines the WASP research project, detailing the progress of the programme. It includes a summary of the outcome of a series of test chips, and describes the encouraging results emerging from testing of the latest wafer-scale demonstrator, WASP 1, and the design of WASP2, the next major milestone in the project.

It is a feature of the WASP programme that fundamental engineering constraints have been actively addressed throughout. The architecture has been developed to derive the maximum benefit from the massive speed and bandwidth opportunities of WSI. Strategies for a hierarchical approach to defect-tolerance, signal and power distribution, as well as selective power isolation have been investigated. This paper explores the lessons learnt from attempting to implement these architectural and implementation strategies within practical engineering constraints.

Studies, based on the results of the current series of experiments, suggest that WASP devices will achieve significant decreases in system size, weight, production cost and power dissipation over an equivalent VLSI system, combined with improvements in performance and reliability. Furthermore, the results to date are helping to shape the future direction of the WASP experimental research programme - represented by the WASP3 and WASP4 functional demonstrators.

---

[1] **Now with Anamartic Ltd, Cambridge UK.**

WSI ASSOCIATIVE STRING PROCESSOR (WASP)

For a detailed description of the ASP architecture, inappropriate here, the reader is referred to [1], [2], [3] and [4].

The implementation of a complete WSI ASP system, known as WASP (WSI Associative String Processor), is the ultimate goal of the present research programme. Through the introduction of defect-tolerant design techniques, ASP structures comprising many thousands of Associative Processing Elements (APEs) can be integrated on a single undiced silicon wafer.

The ASP system architecture is mapped into a WASP wafer as a two dimensional array of modules, where each row comprises a complete ASP substring. System composition is by stepping the identical sized modules onto the wafer, maintaining electrical connectivity through exactly butting module interfaces across the reticle 'stitching' boundary. Only three module types are employed to make an entire WSI system as shown in Figure 1.

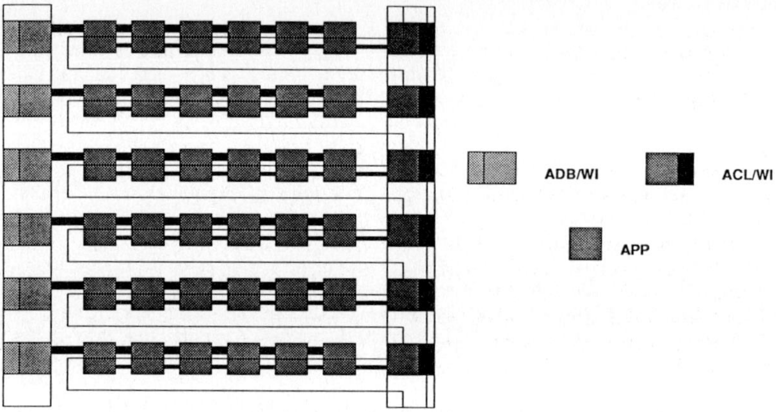

Figure 1 : WASP architecture

A definition of the WASP component modules is given below.

(1) APP : Associative Parallel Processor

The processing kernel of WASP, comprising a number of APEs plus relevant local control logic and defect-tolerance, which is cascaded to form an ASP substring.

(2) ADB/WI : ASP Data Buffer/Wafer Interface

Comprising secondary storage for data, exchanging with the local APE primary data registers via a high speed, high-bandwidth bus. Furthermore, stored data may be exchanged with the host system (via an external Data Interface) or with other ADB modules.

(3) ACL/WI : ASP Control Logic/Wafer Interface

Supporting a defect-tolerant ASP control interface, plus support for inter-substring communications, substring test and reconfiguration.

This generic WASP structure is designed to facilitate hierarchical testing and defect/fault-tolerance and clock, signal and power distribution, as described in [1].

WASP DEVELOPMENT PROGRAMME

The initial involvement of Brunel University in WSI research dates back to 1978, when work began on a parallel programme to the ongoing investigation of VLSI ASP devices.

Currently, the WASP team at Brunel University is attempting to demonstrate the feasibility of a range of WASP variants, the size of which will depend on the available packaging technology. These might comprise a ULSI ASP (i.e. a single monolithic ULSI device, with the structure described above, having an APE harvest of 2048) or a full WSI ASP (comprising four ULSI devices, with a projected APE harvest in excess of 8192).

WASP1 and WASP2 are the WSI technology demonstrators of the UK Alvey programme. Two further, functional, demonstrators are funded by a 3 -year US ONR contract. This contract also includes research into the application of the WASP architecture to iconic to symbolic image processing tasks, based on a wide range of image processing benchmarks [5].

The following sections outline the experimental devices which form the basis of this highly practical research project.

WSI test chips

The first experimental iterations of the WASP architecture follow substantial preparatory experimentation, which has included test chips intended to establish the fundamental constraints. The following list outlines the nature of these test chips.

(1) WSI optimised low-power CMOS Content Addressable Memory (CAM) designs for APE local storage. Comprising four CAM designs, namely a standard VLSI pseudo-static, a VLSI static, a WSI static and a WSI self-isolating CAM, this test chip established the design fundamentals of designing WSI CAMs. CAM (or associative memory) comprises the key element of the WASP processing strategy, support address independent data storage and parallel data manipulation.

(2) Defect-tolerant reconfiguration networks for APE strings.
Comprising four variants of intra-module defect-tolerant bypassing logic, this device was designed to validate the library cells and provide an experimental verification of the optimum bypassing strategy for reconfiguring the APE string in the event of failure of one or more APEs.

(3) Defect-tolerant PLA design and power isolation test.
A key feature of the WASP methodology is reliance on standard CMOS technologies. This experimental device investigated the 'soft' electronic repair of faulty PLAs and provided a practical test of power isolation strategies.

(4) Signal distribution tests.
A critical factor in the design of successful cost-effective WSI is the minimisation of overheads (i.e. bussing, reconfiguration logic) while maximising the usable processing logic. This experiment attempted to establish the quality of bus interconnect, upon which was ultimately based the decisions as to whether to adopt area-consuming, defect-tolerant bus design methodologies.

(5) Reticle 'stitching' experimental wafer.
In keeping with the motivation of applying standard CMOS fabrication techniques, a major milestone of the project was the demonstration of reliable inter-module connection and communication. This is implemented by selective reticle shuttering and exposure at the stepper phase, providing connections between die-size modules.

ULSI and WSI technology demonstrators

Following evaluation of the WSI test chips, the experimental programme progressed to a planned series of full scale technology and functional demonstrators.

A major constraint on these demonstrators is the requirement, in order to minimise masking costs, to use only a single 13 x 13 mm reticle per demonstrator. This, together with a requirement for four module types per demonstrator, leads to an arbitrary, and certainly non-optimal choice of module size of 6.5 x 6.5mm. The WASP 1 quadrature reticle, based on these constraints, is illustrated in Figure 2 below.

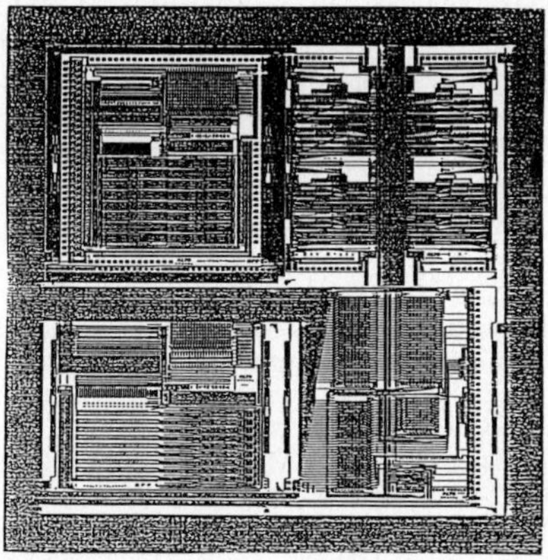

Figure 2: WASP1 reticle

WASP1

WASP1 is a technology demonstrator, providing a suitable experimental vehicle for the investigation of some key aspects of WASP implementation. The objectives of this design are to demonstrate and assess the following areas.

(1) ASP defect-tolerance, namely
  (a) intra-APP (fine-grain defect-tolerance within APP modules)
  (b) inter-APP (coarse-grain defect-tolerance bypassing failed modules).

(2) The power distribution strategy, benefitting from suitable low-power CMOS design techniques, hierarchical power distribution network, and selective dynamic (i.e. electrically controllable and reconfigurable) isolation of modules from the power distribution network.

(3) The clock distribution strategy, involving realistic and achievable clock zoning and synchronisation design techniques.

WASP1 organisation

The following section outlines the functions of major modules integrated onto WASP1.

(1) APP : Associative Parallel Processor

Demonstrating the key elements of a WSI APP, this module integrates 36 APEs.

Intra-module fault-tolerance is provided by the IOI (Input-Output Interface, which supports reconfiguration and test scan-paths, as well as the logic to bypass all inter-processor communications links and disconnect an unrecoverable module from the APP Global Bus (AGB). The AGB is a 132-bit highway for the Data, Activity and Control busses, which are common to all APP modules.

The APP module must clearly contain a complete 'framework' for the support of the APEs, namely, an ASP global bus, power routing and isolation, control sequencer and data routing logic. Once these structures were in place in the module layout, the remaining area was filled with as many APEs as possible, namely 36. This number is a consequence of the constrained 6.5 x 6.5 mm APP module size, and is wholly unimportant in the context of the WASP 1 experiment. In a practical WASP however, the fixed size APE support framework would be amortised over many more APEs, implemented in a much larger APP module.

Table 1 provides APP module statistics. It is noteworthy that the intra-module and inter-module defect-tolerance area overheads incurred are very low.

module area	42.25 square mm
#.transistors	44K
power (worst case)	100mW-220mW
clock rate	12MHz
relative areas	
AGB global bus	16%
power defect-tolerance and routing	10%
control	29%
APE array	
APE active logic	25%
intra-APE defect-tolerance	3%
inter-APP defect-tolerance (IOI)	1%

Table 1: APP module statistics

(2) CC/WI1 : Communications and Control/Wafer Interface

This is a module designed solely to provide a control and testing interface to each WASP1 substring. The module was designed in semi-custom technology, and will not be reused in future WASP generations. Its design was therefore a necessary overhead in order to achieve the main experimental objectives; namely the test and measurement of the APP modules.

(3) TRUNK : Trunk (Defect-tolerant bus trunking experiment)

An experimental device to assess the effectiveness of defect-tolerant bus networks for WSI devices.

The WASP1 module layout designs are based on 2-micron design rules. The composition of WASP 1, fabricated by Plessey (Roborough) is illustrated in Figure 3.

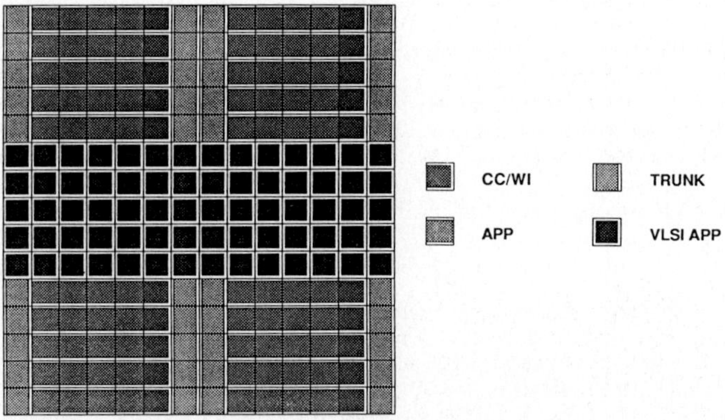

Figure 3: WASP1 wafer composition

For the purposes of experiment, the partitioning, structure and overall composition of these modules is suggestive but not representative of a definitive WASP. The composition of the wafer comprises a number of individual ASP substrings, each configured with four APP modules and a CC/WI substring control and interface module as shown in Figure 3.

In addition, provision is made for a number of independent 64-APE VLSI ASP test chips, ultimately intended for comparative VLSI/ULSI/WSI and HDI (High Density Interconnect e.g. ceramic hybrid and Silicon-on-Silicon assembly) cost-performance analysis.

Results

WASP1 was fabricated in 3Q88. Several key experiments were performed on the wafer. These comprised

(1) Selective power isolation
Employing the test and reconfiguration scan paths, individual modules were successfully and repeatedly attached to and isolated from the power network. The current drain into active processors proved erratic and difficult to quantify. Estimated current values were exceeded by several times, although higher than expected leakage currents were suspected.

(2) Selective module isolation and bypass (inter-module defect-tolerance)
Employing the test and reconfiguration scan paths, individual modules were successfully and repeatedly attached to and isolated from the global busses, and bypassed by the communications highways.

(3) Signal distribution
Validation of global clock distribution strategies was verified by successful scanning of module test and reconfiguration signatures across neighbour site boundaries. In addition syncronous clocking bypassing multiple modules was demonstrated.

(4) Control distribution
Successful demonstration of broadcast instructions across four APP modules, with correct instruction capture and decoding at individual modules, at up to 10MHz, (the maximum frequency supported by the test equipment employed). Validation of correct behaviour of the individual modules was by examination of the output from the

module scan paths, which were successfully clocked back through the substring control. Furthermore, these tests consolidated the earlier test results on bus quality.

Additional testing was carried out on the 64-APE VLSI ASP test chips integrated on the same wafer. These chips were tested by excitation of all scan paths, plus demonstration of basic instruction capture and decoding and functional testing. All of the above tests were carried out using manual wafer probing techniques. The results of testing are illustrated in Figure 4, which shows an example wafer map.

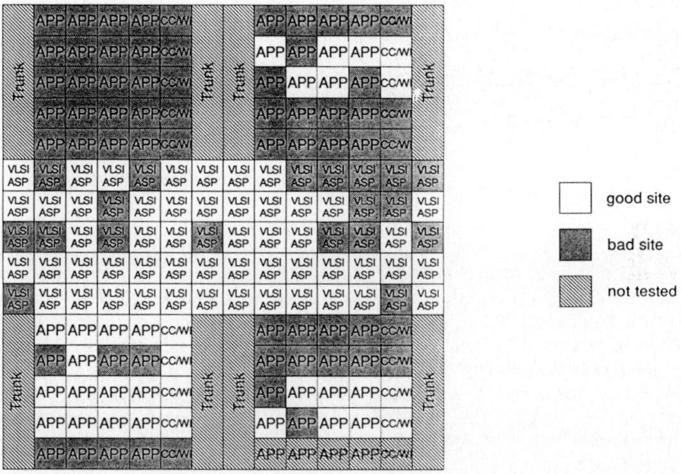

Figure 4: WASP1 wafer map example

Die sites which successfully passed all of the above tests are clearly marked. Although indications of 40-50% harvest may be inferred from these limited results, the level of testing is still relatively coarse in terms of validating full internal functionality. Additional data is still being accumulated, including work to validate the intra-module fault-tolerance, by exercising the processing logic internal scan-paths and inter-APE communications network in bypassed and unbypassed modes.

Furthermore, caution must be exercised in interpreting these results, since WASP1 designs do not adopt any defect-tolerance in the CC/WI module, which was included merely to provide a test interface to the APP modules. Consequently a failure in this module will reflect as a complete failure of the substring, several of which may be observed in Figure 4.

The ultimate availability of these test results is expected to provide data to validate and enhance the yield models being developed as part of the WASP research programme.

WASP2

WASP2 is a more elaborate technology demonstrator, which includes many of the structures which are to be expected on a WASP functional demonstrator. It comprises a dynamically reconfigurable parallel processing structure of communicating ASP sub-strings, each supported by an ASP Data Buffer (ADB/WI), and controlled by ASP Control Logic modules (ACL/WI), which also support inter-substring communication and reconfiguration.

The experimental objectives of WASP2 are outlined below.

(1) Demonstrate inter-branch defect-tolerance.
(2) Demonstrate inter-branch communications.
(3) Demonstrate realistic array power distribution (comprising dynamically controlled power isolation combined with separate power nets per substring).
(4) Demonstrate on-chip array data exchange, illustrating the capacity of WSI to support high speed/bandwidth processor-memory interfaces.
(5) Identify the constraints of a realistic defect-tolerant wafer interface.

WASP2 organisation

This section outlines the function of the current WASP2 modules.

(1) APP : Associative Parallel Processor

This design is identical to that employed in the WASP1 demonstrator.

(2) ADB/WI : ASP Data Buffer/Wafer Interface

This module is a buffer memory for I/O operations which may be concurrent with normal ASP substring activity. It comprises a double-buffered RAM module, defect-tolerance and reconfiguration logic plus a Wafer Interface. Defect-tolerance in this module is restricted to selective power isolation and whole module isolation and bypass (i.e. inter-module defect-tolerance). However, it is anticipated that provision of fault-tolerant RAMs will be a logical extension of this methodology. In the case of WASP2, the severe area constraints artificially imposed by the quadrature reticle approach, precluded inclusion of this option, even though it would only have required a modest area increase.

(3) CC/WI2 : ASP Control Logic/Wafer Interface

A wafer interface module, providing defect-tolerant array control interface plus local substring testing logic. The control interface of the WASP2 device is distributed across multiple CC/WI2 modules, which are fully interconnected via a 116-bit Control Global Bus (CGB). There is a minimum requirement for only four working modules in order to fully configure the control interface of the WASP2 device. Partially working modules may also be effectively used.

In addition, the module supports the inter-substring communications and defect-tolerance networks, which allow selective bypassing of whole failed substrings.

WASP 2 is currently being fabricated in relaxed 2-micron design rules at Plessey Roborough. Two distinct variants are being fabricated from the basic three module types, and will be available 3Q88. These WASP2 variants are

(1) WASP2A
The structure of the target WASP2A composition is shown in Figure 5, and comprises four independent WASP devices, each integrating 864 APEs.

The ULSI devices are accompanied by the 64-APE VLSI ASP chips, which have been re-introduced as the fourth module type to provide further data for the monolithic vs. hybrid WSI cost-effectiveness comparison. The design of a hybrid WASP employing these chips has been completed and will be manufactured by ICL as part of the Alvey collaboration.

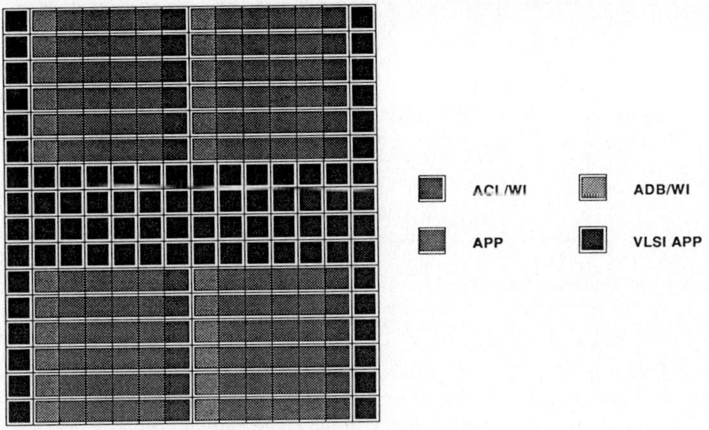

Figure 5: WASP2A wafer composition

(2) WASP2B
The WASP2B composition comprises an array of 180 APP devices, integrating some 6480 APEs on a whole 6 inch wafer (Figure 6). This device is being fabricated to analyse the properties of whole wafer signal and power distribution.

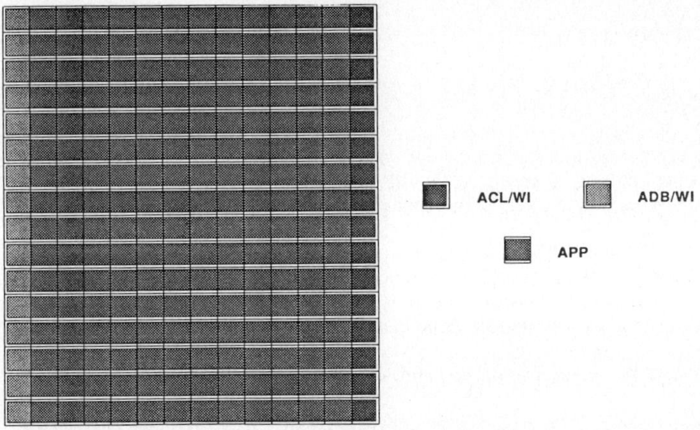

Figure 6: WASP2B wafer composition

	WASP2A	WASP2B
#.APEs	864	6480
device area	3.9cmx3.9cm	9.1cmx9.8cm
#.transistors	1.26M	8.43M
power	5.2-8.0W	29.6-51.2W
maximum internal bond wires	420	1050
#.external control signals	132	132
external clock	6MHz	6MHz
internal clock	12MHz	12MHz

Table 2: WASP2 statistics

Considering the characteristics of the projected WASP2 devices (Table 2), the dangers of misinterpreting results are clear. These figures are derived from wafer-scale compositions using experimental modules. At first glance, neither of these compositions appear to achieve the APE density promised by WSI technology. However, this is attributable to the constraints of small module size, imposed by employing a single (quadrature) reticle to fabricate all of the modules.

Detailed modelling work is underway to assess the effectiveness of this and future architectures, once unrealistic masking constraints are removed. In addition, the work to date has concentrated on relatively old 2-micron technology. Future work will begin to extend this architecture and design methodology to finer feature size processes.

CONCLUSIONS

The WASP (WSI Associative String Processor) device has been proposed as a highly-versatile, cost-effective, parallel processor. WASP hardware and software modules have demonstrated the potential to achieve step-functions in cost-performance over a wide range of computer vision tasks [6].

Outstanding experimental results to date have confirmed the feasibility of the WASP architecture, and as a consequence, the programme will continue along its planned course. The WASP1 experiment has demonstrated

(1) Inter-module defect-tolerance
(2) Power distribution and module power isolation
(3) Inter-module (substring) control, clock and signal distribution at up to 10MHz, the limit of the test equipment used to date.

Work is continuing to validate the intra-module defect-tolerance.

Furthermore, the WASP2 development has expanded in scope, and devices are currently being fabricated to validate ULSI device characteristics, as well as address the significant issues of whole wafer (6 inch) monolithic WSI. The future WASP 3 and 4 devices will provide a functional demonstration of the WASP architecture.

ACKNOWLEDGEMENTS

The authors gratefully acknowledge enthusiastic contributions from past and present members of the WASP team (namely M.B.E. Abdelrazik, D. R. Bean, H.S. Bolouri, S.R. Jones, D.K. Koshal, R.D. McKirdy and N. G. Sheridan), from A. Krikelis and the financial support of the UK Alvey initiative and US ONR. Thanks also to Mars Electronics for donating the 'Gazelle' test equipment used in this work.

REFERENCES

[1] R.M. Lea, The ASP: a cost-effective micro-computer, IEEE Micro, October 1988, pp 10 - 29.

[2] R.M. Lea, A WSI image processing module, Wafer-Scale Integration (eds. G. Saucier and J. Trilhe), Elsevier Science Publishers B.V. (North-Holland) 1986, pp. 43 - 58.

[3] R.M. Lea, WASP: A WSI Associative String Processor for Structured Data Processing, Wafer-Scale Integration (eds. C.Jesshope and W.Moore), Adam Hilger, 1985, pp. 140 - 148.

[4] R.M. Lea, WASP: a reconfigurable WSI Associative String Processor for highly parallel information processing, Wafer-Scale Integration (ed. R.M.Lea), Elsevier Science Publishers B.V. (North-Holland) 1988, pp. 68 - 81.

[5] A.Krikelis and R.M.Lea, Performance of the ASP on the DARPA architecture benchmarks, Proc. of Frontiers of Massively Parallel Computation, George Mason University, Fairfax, Virginia, October, 1988.

[6] R. D. McKirdy and R. M. Lea, A comparative appraisal of VLSI, ULSI and WSI Associative String Processors, Proc. Electronic Design Automation conf. (Wembley, England), 1987, pp. 291-301.

Wafer Scale Integration, III
M. Sami and F. Distante (Editors)
Elsevier Science Publishers B.V. (North-Holland)

AN ARCHITECTURE FOR AN INTELLIGENT MULTI-MODE WSI MEMORY

K.E. Grosspietsch

Gesellschaft für Mathematik u. Datenverarbeitung
Postfach 1240
D-5205 St. Augustin 1, W. Germany

1. Introduction

Wafer scale integration implies the potential to efficiently implement also new architectures which formerly could not be realized because of technological restrictions like pin limitation, too small bit capacities etc. In this context, especially approaches appear interesting which are to eliminate the 'bottleneck' between processor and data: In conventional von Neumann machines, every time when data are to be manipulated, they have first to be fetched from memory and transferred to the processor. The result of the processing operation has to be stored back into memory. So, for many of these operations, most of the execution time is caused by data transfers.

This principle in the beginning of computing technology of course was dictated by the very high price of the hardware for the processing logic. Today, at the current state of hardware integration technology, this situation has changed. So, today an alternative to the processor/memory bottleneck could be to integrate more logic directly into the memory structure, i.e. to make the memory more intelligent.

Such 'intelligent memory' architecture can especially provide support for non-numerical data processing fields like database management, logic programming, pattern recognition, image processing, CAD graphics etc., and find here a growing field of applications.

In this context, as one interesting step towards smarter

memories, also systems for associative (meant here always as a synonym for content-addressable) storage and processing of data can win importance again: Content-addressable memories (CAMs) have been in the discussion of system developers already for a long time. The benefit of such structures for speeding up search operations is unqestionable, but for a long time the state of technology allowed only very small CAM capacities. It is now the first time that the state of art in hardware integration , i.e. the current VLSI technology and the advent of wafer scale integration (WSI), promises to enable the implementation of such memories with reasonable capacities and cost/bit ratio. Moreover, the extension of the principle of content-addressing from storage and retrieval of data also to processing of data has been proven /5/ to imply many interesting features for various fields of parallel processing, e.g. again image processing, CAD support etc.

2. The proposed approach

Several interesting approaches on CAMs and coontent-addressable processor systems have been reported in the last years /5, 7, 9/. The approach discussed here is mainly based on the ideas of /5/; it tries to extend this solution with regard to the following objectives:

- extend the flexibility of the logic elements
- combine processor cell arrays with ordinary CAM and RAM parts
- modify the resulting architecture for testabilility and for fault tolerance features; the latter objective thereby implies not only structural redundancy (by spare components) but also functional redundancy in the sense that the function of a more complex component can be stepwise degraded to less comfortable functionality.

2.1. Basic principles and requirements

This goal is to be achieved by inclusion of the new features into an existing RAM structure. RAM, CAM and a content-addressable processor/register array (CAPRA) together form a kind of storage hierarchy where the main part can consist of

an ordinary RAM (see Fig. 1). The additional 'smarter' components are included as additional memory segments within one uniform physical memory space; thereby their individual capacities can arbitrarily be tailored to the specific needs of the application.
Compatibility between these steps of a hierarchy is achieved in the sense that 'smarter' parts also provide the full functionality of the simpler ones. So, the CAM parts are requested also to be operable as RAMs; the CAPRA is implemented by some functional extension of the CAM architecture.
With respect to more flexible logic features, the following architectural properties were felt promising to support more flexible use of content-addressable memory/ processor systems:

- (a) Extension of the usual very simple hit mechanisms of present CAM implementations (equivalence between data, potentially modified by some kind of masking) to more flexible solutions which, apart from the mentioned principle allows more sophisticated hit evaluation (e.g. by thresholds for the number of eqivalent bits, similarity metrics between search pattern and data in storage etc.).
- (b) With relatively low hardware effort the usual comparison logic of CAM cell can be extended to provide an entire set of one bit Boolean operations.
- (c) Extension of arithmetics elements from sequential one bit adder elements per word cell to at least 4 bit adder elements; together with data paths to such ALUs in the two nearest neighbour word cells, parallel exchange and processing of data can be provided: This could, e.g., especially support the processing of pixels with 16 grey levels in image processing.
- (d) test of the mentioned architecture is to be supported by implementing the possibility for multi-access write operations and parallel evaluation of test outcomes in the word cells by means of the extended logic mentioned above.

In contrary to other architectural approaches, the described architectural features just fit into the usual programming paradigm of the von Neumann machine, the control-flow. Therefore, integration of the discussed structure into a conventional system can be performed relatively easily; thereby, our architecture is planned to work as a co-processor of a conventional processor; the instructions of the co-processor can be modularly added to that ones of the processor.

2.2. The resulting architecture

Our approach specified to fulfil the mentioned requirements has the following characteristics:

- the CAM has also an additional RAM access mode so that it can, e.g., be loaded or read like ordinary RAM cells.
- in the CAPRA (see Fig. 2), as the base cells we again have RAM bit cells. Moreover to every bit cell a simple logic block is associated which enables Boolean one bit operations between two one-bit operands (thereby including, as one of these operations, the compare operation needed to fulfil the CAM functionality); this enables bit-parallel and word-parallel execution of a Boolean operation on all words of the CAPRA.
- to every word cell in the CAPRA, a simple 4-bit adder/shifter unit is assigned; so, an arithmetic operation can be performed on all the, e.g. 32-bit, words of the CAPRA segment in parallel in about 8 cycles.
- an additional flag bit is associated to each bit cell of the CAPRA; this enables us to flexibly define arbitrary 'activity patterns' for the cells of the array; so it is possible to process data not only on all processor elements, but instead on a previously defined arbitrary sub-pattern of processing elements.
- For the RAM access, a simple extension of the memory decoder /7/ is provided by means of an additional mask register; by setting bits of this mask register to 1, an arbitrary part of the address bits can be declared to be 'don't care' bits; thus, it is possible to implement concurrent access to word cells which in their addresses have common address bit sub-patterns (see Fig. 3).

The operation of the CAPRA structure is as follows (see Fig. 4):
Data can be written into a word cell - controlled by the corresponding word line emanating from the memory decoder - via the MDR and read/write lines. For every bit of a word cell, its contents stored in the storage flipflop SF can be combined with the contents of the read/write line by a Boolean operation in the functional block BOOL; the result of this operation is latched in the intermediate flipflop IF. From there it can be propagated further (control line TRANSFER in Fig. 4) either to the adjacent ALU (line GLOBAL/LOCAL=1) or to be memorized in the storage flipflop SF (GLOBAL/LOCAL=0). These transfers are performed either unconditionally (control line UNCOND=1) or dependent on the status of the cell which is memorized in the activity flag AF (control line COND=1, if AF stores a '1'). As a third sink for the bit transfer from the intermediate flipflop IF, setting of the activity flag is possible (control line SET FLAG). This is performed either unconditionally (control line UNCOND = 1) or dependent on the present status of AF (control line $\overline{\text{COND}}$ = 1); in the latter case, setting of AF is possible only if AF has not yet been set, i.e. if AF is storing a '0' (control line $\overline{\text{COND}}$ then equals 1).
The intermediate flipflop can not only receive a data bit from the functional block BOOL, but, alternatively, also from the adjacent ALU (control line REC=1).

3. State of the implementation

With regard to classical CAM structures, some promising approaches have been evaluated and simulated and might be ripe for implementation in the near future /6/; moreover we included studies to derive architectural support for testability and fault tolerance in CAMs /1,2,3/.
With regard to the building block of a CAPRA, a first approach has been under investigation. The resulting basic cell in CMOS implementation turned out to comprise about 60

transistor functions. The layout of one CAPRA bit cell takes an area of about 120 × 100 μ; this cell is still in the state of optimization. For the ALU slices, we consider a simple and concise structure as proposed in /4,8/. This ALU structure can perform 16 Boolean and 16 arithmetic operations; correspondingly, it needs 4 control line for operation selection and 1 control line for selection of Boolean/arithmetic mode (see Fig. 5).

Moreover, we are evaluating the structure of an interface of our proposed system to existing computing systems; our system is specified to work as a co-processor to a conventional wide-spread processor family (Motorola 68000 family). As a first step to this adaption, a coprocessor interface has been realized by one VENUS standard cell chip.

4. Conclusion

The described architecture is an ideal candidate for utilizing WSI technology: Internal coupling lines within word lines and between adjacent word cells in a multi-chip realisation probably would cause considerable pin limitation problems; it is just by avoiding these restrictions in WSI that for the regarded structure reasonable capacities can be achieved.

By developing different, but compatible building blocks for the RAM, the CAM, and the CAPRA, with regard to the final memory hierarchy we are not confined to fixed fractions of these components within the entire memory capacity. In contrast, it is possible to vary these fractions on the wafer, dependent on the individual planned application.

5. References

/1/ Grosspietsch, K.E., Huber,H., Müller,A., The Concept of a Fault-Tolerant and Easily-Testable Associative Memory, in: Proc. Int. Conf. Fault-Tol. Computing FTCS-16 (1986), pp. 34-39

/2/ Grosspietsch, K.E., Huber, H., Müller, A., The VLSI

Implementation of a Fault-Tolerant and Easily-Testable Associative Memory, in: Proebster, W. (ed.), Proc. COMPEURO '87 (1987), pp. 46-50

/3/ Grosspietsch, K.E., Concepts for Fault Tolerance and Testability in WSI Associative Memories, in: Lea, M. (ed.), Proc. IFIP Workshop Wafer-Scale Integration (North Holland 1988) pp. 135-144

/4/ Kaiser, J., Fehlerdiagnose in einem hochzuverlässigen Einprozessorsystem, (GMD, St. Augustin (W. Germany), 1980)

/5/ Lea, M., ASP: A Cost-Effective Parallel Microcomputer, in: IEEE Micro, October 1988, pp. 10-29

/6/ Qiu Yulin, Design of a Testable Content Addressable Memory, in: Preprints Proc. IFIP Working Conference 'Design Methodologies for VLSI and Computer Architecture'(1988)

/7/ Tavangarian, D., Ortsadressierbarer Assoziativspeicher, in: Elektronische Rechenanlagen (1985), pp. 264-278

/8/ Texas Instruments, The TTL Data Book

/9/ Waldschmidt, K., Associative Processors and Memories: Overview and Current Status, in: Proebster,W. (ed.), Proc. COMPEURO '87 (1987) pp. 19-26

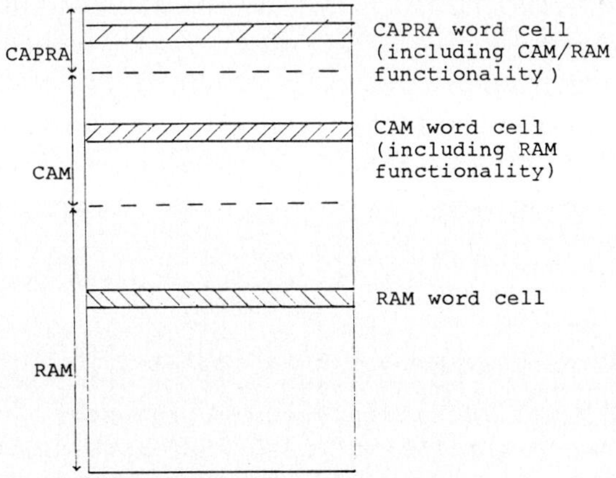

Fig. 1: Schematic structure of the memory hierarchy

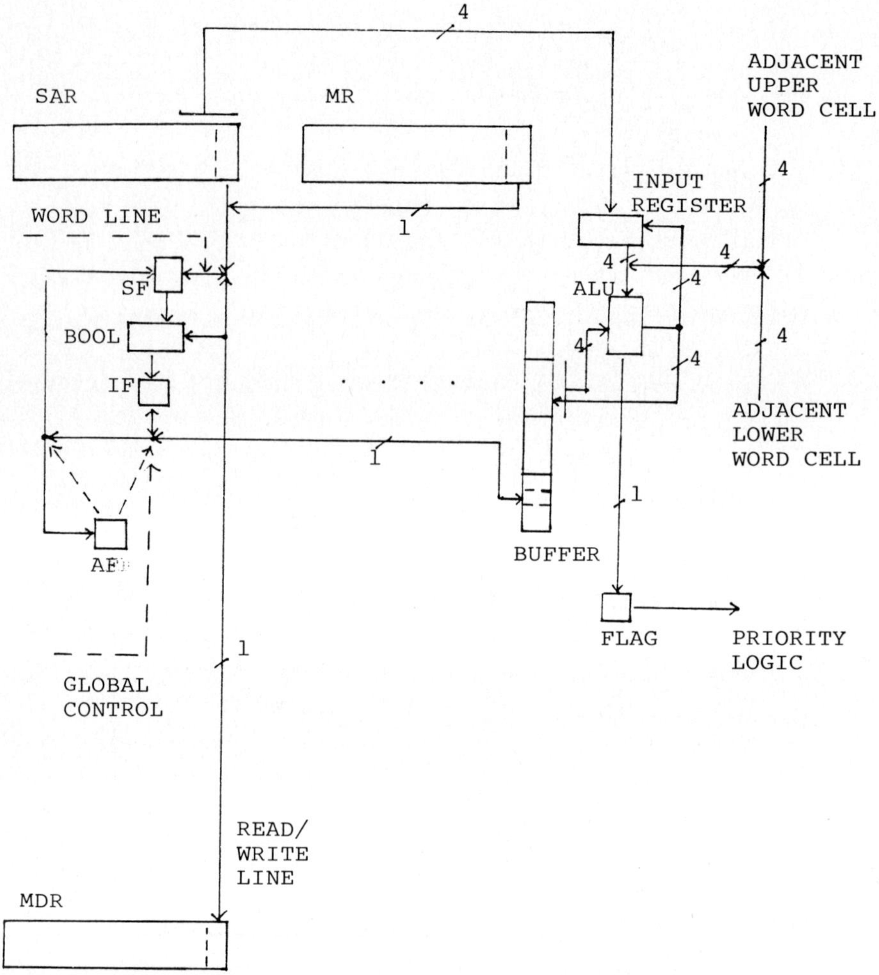

Fig. 2: Scheme of a one bit storage cell and the 4 bit ALU stage within a CAPRA word cell

Abbreviations:

AF	activity flag
BOOL	logic block for Boolean operations
BUFF	ALU input buffer
IF	intermediate flipflop
MDR	memory data register
MR	mask register
SAR	search argument register
SF	storage flipflop
⟶	data lines
- - →	control lines

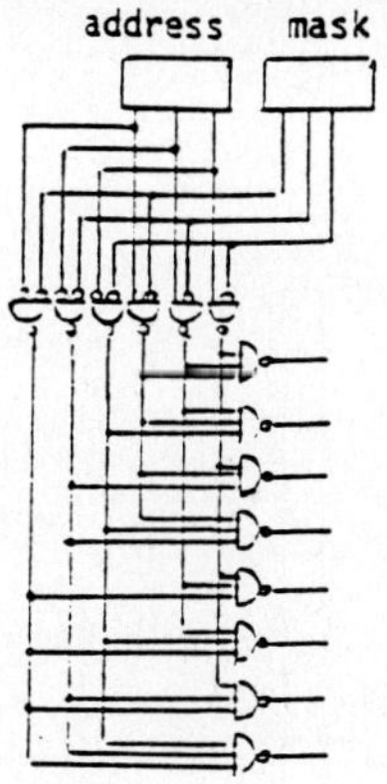

Fig. 3 Extended (masked) memory decoder /7/

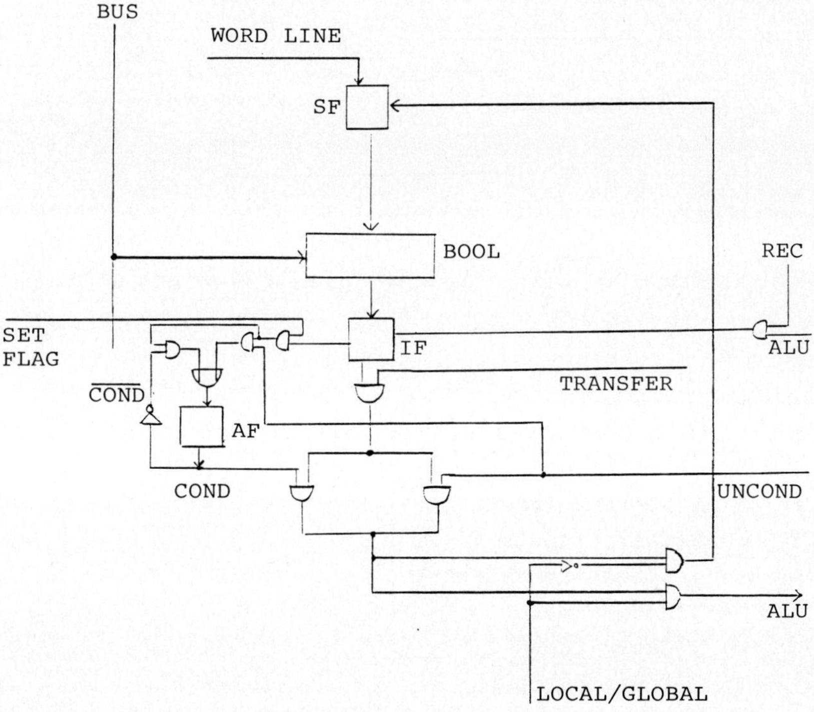

Fig. 4: Logic level structure of a bit cell

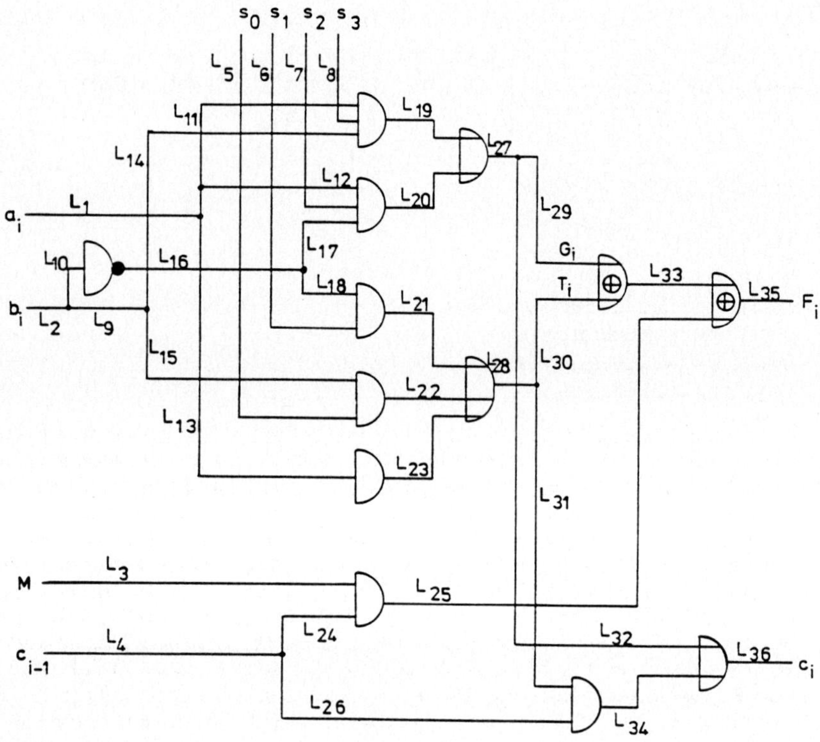

$$F_i = [(a_i\, b_i\, s_3 + a_i\, \overline{b}_i\, s_2) \oplus (\overline{b}_i\, s_1 + b_i\, s_o + a_i)] \oplus Mc_{i-1}$$

$$c_i = G_i + T_i\, c_{i-1}$$

$$G_i = a_i b_i s_3 + a_i \overline{b}_i s_2$$

$$T_i = \overline{b}_i s_1 + b_i s_o + a_i$$

a_i, b_i : Operand inputs
c_{i-1} : Carry inputs
s_o,..s_3 : Operation selection
M : Mode input
c_i : Carry input
F_i : Function output

Fig. 5 Structure of an ALU slice /4,8/

Wafer Scale Integration, III
M. Sami and F. Distante (Editors)
Elsevier Science Publishers B.V. (North-Holland)

ON REPAIRABILITY/UNREPAIRABILITY DETECTION AND YIELD ENHANCEMENT OF WSI REDUNDANT MEMORIES*

Yi-Nan Shen and Fabrizio Lombardi

TEXAS A&M UNIVERSITY
Department of Computer Science
College Station Tx 77843-3112

*This research supported in part by grants from AT&T and NATO.

ABSTRACT.

In this paper, new techniques for repairability/unrepairability detection of WSI memory chips with redundancy are presented. Initially, an heuristic, yet efficient approach is proposed. This first approach is based on finding a very good approximation to the minimum covering set. An algorithm, which executes in quadratic time with respect to the largest dimension of the memory, is presented. This algorithm is executed off-line i.e. when the memory has been fully diagnosed. New conditions for detection are presented and fully analyzed. These are based on a more accurate estimation of the regions of repairability and unrepairability. The second approach to repairability/unrepairability detection is based on a more complex covering relationship, namely the generalized leading element. The main benefit of these algorithms is their practicality with respect to fast execution time and the improved ability of diagnosing a WSI redundant memory prior to the generation of the repair-solution. This characteristic proves that repairability/unrepairability detection is a cost effective solution for increasing the throughput of memory repair. A new repair algorithm which utilizes a ternary tree approach, is also presented.

1. INTRODUCTION.

Today's technologies, such as VLSI and WSI, have had a great impact on the manufacturing of large density memories. Capacity of a single memory chip has increased by many orders of magnitude [1,2]. One of the major problems encountered in the manufacturing of WSI memories is the inability of producing absolutely fault free devices. Redundancy in the form of additional (spare) rows and columns of cells, has been extensively used [3,4,5] to enhance the yield. Yield is defined as the percentage of devices which can be shipped to custumers over the whole number of manufactured devices [12]. Spares are used to replace either the rows and/or the columns where faulty cells lie [11]. This process is referred as *repair*. Repair can reclaim functional all those devices in which the spare rows and columns can be used to replace (or to cover [6]) all faulty cells (usually a degradation in access time is found to accomodate extra decoding due to the redundancy). It has been proved that repair by row/column deletion is an *NP*-complete problem [7] as equivalent to the logic minimization problem.

Repair algorithms have been proposed in [6,7,8,9,10,14]. These algorithms have different characteristics. Some algorithms have a fast execution time at the expense of the the ability of always generating a repair-solution, whenever one exists (suboptimality) [6]. Other algorithms strive for optimality, but at an exponential complexity (perfect repair) [7,9,10]. The nature of repair with respect to the testing phase has also been analyzed [8]. In [6], a novel approach for repairability/unrepairability detection has been proposed. The objectives of this paper are to extend the approach for repairability/unrepairability detection and diagnosis found in [6], such that a more efficient heuristic (polynomial time execution) technique can be formulated. This also results in a new and more efficient data structure for repairing memories.

2. PRELIMINARIES AND REVIEW.

Prior to present the review of the approach of [6], definitions and notation which are used in the analysis, are presented.

Definition 1. A faulty cell which does not share the row or column with any other faulty cells, is referred as a *single faulty cell.* A faulty cell is said to be $R(C)$-single faulty cell if the faulty cell is single only along the row (column).

Definition 2. A row (column) is referred as a *covered* row (column) if it is selected for deletion in the repair process. This row (column) is repaired by employing a spare row (column).

Definition 3. A faulty cell is referred as a covered cell if it is on a covered row or column.

Definition 4. A spare row (column) is referred as a *covering* row (column) if it used to repair a row (column) where faulty cells lie. A covering row (column) has the same number of the row (column) which is being covered with.

Definition 5. A row (or a column) in which at least a faulty cell lies, is referred as a faulty line.

The following Lemma of [9] is included for completeness. It characterizes the *line covering principle.*

Lemma 1 [9]. A faulty line (column or row) with k faulty cells can only be covered by a spare column (or row), or by k spare rows (or columns) without redundantly using spare columns (or rows).

The memory consists of M rows and N columns. Redundancy consists of R_A rows (out of the original M rows) and C_A columns (out of the original N columns). Columns (rows) are numbered from 0 to $N-1$ ($M-1$). These are referred as the physical indices and are denoted as PC_i, $0 \le i \le N$-1, and PR_j, $0 \le j \le M$-1. The number of single faulty cells is denoted by S_F. It will be assumed throughout this paper that without loss of correctness $R_A = C_A \neq 0$. The number of faulty cells is denoted by N_F. The results of the memory test are stored in a diagnostic matrix $A = \{a_{ij}\}$; a_{ij} is 1 (0) if the memory cell with physical coordinates (i,j) is faulty (fault free).

Definition 6. The leading element (LE) of A (say a_{kh}) is the first nonzero entry in A such that the kth row is the first row (i.e. the lowest row index) with a nonzero entry and (k,h) entry is the first nonzero entry of row k.

The *repair-solution* is denoted by two sets: the set R whose elements denote the numbers of the rows to be deleted and the set C whose elements denote the numbers of the columns to be deleted. A fault counter is associated with each row and column of the memory. The content of the fault counter of row i (column j) is denoted to as $C(R_i)$ ($C(C_j)$) for $i = 1,M$ ($j = 1,N$). The total fault count is

$$N_F = \sum_{i=0}^{M-1} C(R_i) = \sum_{j=0}^{N-1} C(C_j) \qquad (01)$$

The following characteristics are assumed in this paper:

I. In the case of a large WSI memory chip, redundancy is applied to the whole unpartitioned array.

II. Only faulty cells affected by hard defects are considered [12].

III. Faulty cells appear in the die according to different distributions. This depends on the technology used[12].

Repair of redundant memories by row/column deletion can be analyzed according to the following criteria.

I. The repair-time should not be excessive. A fully exhaustive search for an optimal repair-solution becomes realistically viable only if its execution is not computationally intensive.

II. Early abort conditions should be used to promptly identify unrepairable devices, thus reducing the time spent on finding the repair-solution.

III. The repair technique should be compatible with a hardware implementation in a defect analysis system.

These criteria have been accomplished in [6] by using a new approach to memory repair. This approach consists of establishing repairability/unrepairability of a device prior to finding its repair-solution. *Repairability* (unrepairability) is defined to as the probability of *correctly diagnosing* a memory as repairable (unrepairable) *prior to* the execution of the repair algorithm and finding the repair-solution. A description of the approach of [6] will follow as applicable to the description of the proposed approach. The first stage of the repair algorithm of [6] is similar to Repair-Most, because it establishes repairability detection using fault counters. Repairability detection consists of calculating the total fault count [10]. The *maximum number of faulty cells* that permit repair, is given by

$$F_t = (NR_A + MC_A) - R_A C_A \tag{02}$$

F_t can be compared with the total fault count. If this limit is not exceeded, repair can take place. Detection can be extended to repairability and to repair. (02) represents an obvious condition, that can be intuitively obtained by using the approach of [4] also. It should not be mislead with repair, because the content of a counter does not account for the locations of faulty cells. An alternative solution for establishing repairability is one of the algorithms of [8,9,10]. These methods are based on the exhaustive search out of the binary tree made of all possible repair solutions, hence yielding an exponential complexity [7,8]. A significant computational overhead is therefore, encountered.

It is evident [6] that unrepairability of large memory devices should be established before the repair-solution is found, because the number of records required for repair grows with the number of faulty cells. However, the process of detection should not preclude at a later stage an efficient repair process of those devices which can be repaired using the provided redundancy. Uniqueness conditions in the allocation of spares for memory repair have been presented in [6].

It has been proved [6] that occurrence of faulty cells can determine a repair process that in most cases can be uniquely described by the above criteria [6]. This characterization is very important, because repair is independent of the locations of all faulty cells. This can be used to define the regions for repairability, unrepairability and uncertainty of a redundant memory. These regions are found using an approximation to the repair-solution in polynomial time complexity [6]. This is required for not incurring into a large and rather wasteful execution-time if the memory is unrepairable. This approximation is given by the *repair-set*. The repair-set is denoted by S and can be constructed as follows. Initially, the leading element of A is selected. All entries along either the row and/or column of the leading element are replaced with zero entries. This process is repeated until the diagnostic matrix is a zero matrix.

Selection of a leading element is based in general on spare availability [6]. In this paper, however, $R_A=C_A$ by assumption, hence the criteria for spare allocation of [6] are not directly applicable. For a leading element a_{ij}, i is added to the repair-row set S_R, while j is added to the the repair-column set S_C, where $|S| = \max\ \{|S_R|,\ |S_C|\}$. In [6], criteria for repairability/unreparaibility detection have been established as function of the cardinality of the repair-set and the number of spare rows and columns. These are as follows.

1. $|S| \leq \max\ \{R_A, C_A\}$; the memory is R_A/C_A repairable (repairability region).

2. $|S| > R_A + C_A$; the memory is R_A/C_A unrepairable (unrepairability region).

3. $\max\{R_A, C_A\} < |S| \leq R_A + C_A$; the memory cannot be diagnosed (uncertainty region).

The uncertainty region is analyzed in [6] using the condition of *complement*, while at the same time generating the repair-solution. For a leading element a_{ij}, a column complement is defined as a nonzero element, say a_{ik}, such that the kth column $C_k \notin S_C$. Similarly, a nonzero element a_{tj} is a row complement of a_{ij}, if the tth row $\notin S_R$. The set of rows and columns required to cover all faulty cells (i.e. the repair-solution) can be found by using the repair-set found previously and complementing all of its elements. Rows and/or columns in S which cover faulty cells that can be all covered by other elements of S also, can be deleted from the repair-set. The repair-solution is therefore given by a refinement of S. The unrepairability region at completion of the complementing process is defined by the condition $|S| + d > R_A + C_A$, where d is the number of complements. In [6], other conditions of repairability/unrepairability detection, as applicable to other repair schemes such as totalizers [10], are also proposed.

The main advantage of the approach of [6] is the low order of complexity of the detection and repair processes: the repair-set can found in $O(MN^{0.5})$ (without using early abort criteria as the worst case), while the algorithm for finding the repair-solution (including the complement generation) is quadratic with respect to the provided redundancy.

3. LOGICAL TRANSPOSITION.

The main disadvantage of [6] is that diagnosis is not always possible. This is caused by the very simplistic criterion (i.e. the leading elements), which is used in the generation of the repair-set. This heuristic condition however, has the advantage that a reduced execution complexity is possible. This characteristic is reflected in the presence of the uncertainty region and in the strict, but limiting relationship between the elements of the repair-set and those of the repair-solution.

The problem of detection can be formulated as follows: does it exists a mapping from the repair-set to the repair-solution such that a redundant memory can always be diagnosed using a suitable repair-set ? The nature of this mapping function should be obviously one-to-one onto, because a repairable (unrepairable) memory must be always mapped in the repairability (unrepairability) region. Only in these circumstances the uncertainty region would be null. It has been proven that in general this problem is NP-complete for functions with finite set size [13]. This is equivalent to state that the detection and diagnosis problem can be solved correctly with 100% confidence (no uncertainty region) only when the fully exhaustive search for finding the repair-solution has been completed. However, some more sophisticated, yet heuristic criteria can be added to the generation of the repair-set by leading elements to improve the repairability/unrepairability detection of redundant memories without incurring in a fully exhaustive search. In this paper, the technique of logical transposition is proposed. The diagnostic matrix A can be thought of as a table with a faulty pattern which is fully defined by the physical column and row indices. This pattern can be modified if a faulty line (with a given physical index) can be logically moved (or transposed) to a new position.

Let a logical column (row) index be represented as LC_j (LR_i). The process of transposing a column with physical index PC_j to a column with logical index LC_k is denoted as $PC_j \rightarrow LC_k$ if $j < k$, or $LC_k \leftarrow PC_j$ if $k < j$ (similar notation applies for a row transposition). Note that this notation also implies that the logical indices of all columns and rows which lie between the physical line and the new logical line, must be modified accordingly. Consider as example, the fault pattern shown in Figure (1), $R_A = C_A = 1$. The algorithm of [6] finds two LEs: (0,0) and (1,1). The transposition process consists of $PC_0 \rightarrow LC_9$. The new diagnostic matrix has only one LE, namely the cell with physical indices (0,1). A reduction in the cardinality of the repair-set is therefore, accomplished. A new diagnostic matrix results from this transposi-

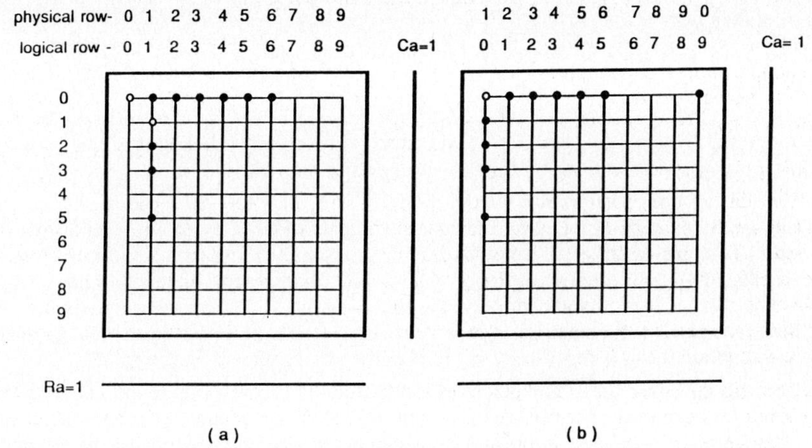

Fig. 1 (a) before transposing s = 2

(b) after transposing s = 1

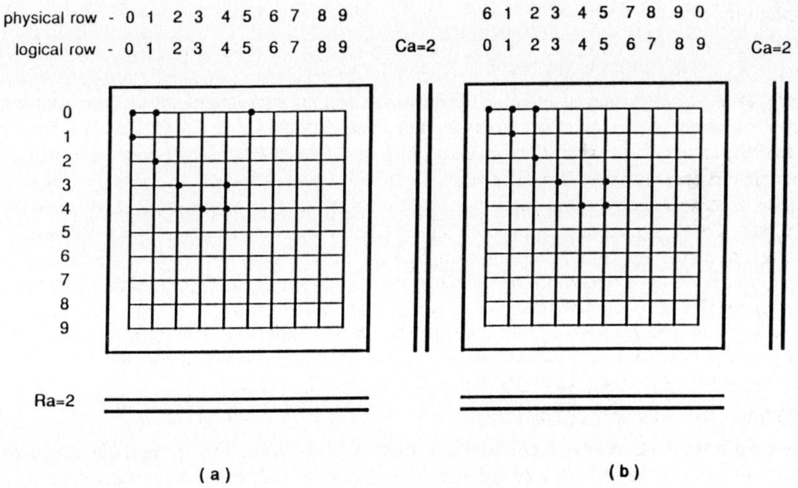

Fig.2 (a) before transposing s = 4

(b) after transposing s = 5

tion. This new diagnostic matrix A_{new} has the same basic properties as A, but the fault pattern has been rearranged.

Definition 7. A faulty line is said to be complete if it has cells which are either all R or all C-single faulty cells.

Theorem 1. Let a_{ij} be an LE in A and a C-single faulty cell. Let a_{ik} be faulty, a_{mk} be a R-single faulty cell, where $k > j$ and $m > i$, and either all cells corresponding to a_{ih} be fault free ($i < h < k$), or a_{mk} be an LE in A. If $PC_j \to LC_{k+1}$, then $|S_{new}|=|S|-1$.

Proof: The LE a_{ij} can cover prior to the transposition of C_j only the faulty cells along R_i. This follows from the definition of C-single faulty cell. a_{mk} is covered by C_k in S (following the definition of R-single faulty cell). Note that this is an LE in A using the algorithm for repair-set generation of [6]. The transposition of $PC_j \to LC_{k+1}$ still results in covering these faulty cells on R_i and a_{mk}. It results also in only one LE (a_{ij} is not a LE any more by definition of A_{new}). The execution of the repair-set algorithm on A_{new} results in a new repair-set S_{new} which has an element less than the repair-set S of A. The Theorem is therefored proved. □.

Note that under the same initial scenario of Theorem (1) the transposition $LC_{j-1} \leftarrow PC_k$ would result in the an equivalent situation. Figure (1) shows an example of column transposition. Theorem (1) can be used to prove the following covering characteristic of the algorithm for repair-set generation by leading element selection of [6].

Theorem 2. The repair-set S generated by the covering algorithm of [6] is not unique, i.e it is dependent on the selection of the starting LE.

Proof: Omitted due to lack of space. □

The following Theorem characterizes the inverse transposition and its effect on the cardinality of the repair-set.

Theorem 3. Let a_{ij} be an LE in A. Let a_{ik} be a C-single faulty cell and a_{mj} be an LE and a R-single faulty cell. If $LC_{j-1} \leftarrow PC_k$, then $|S_{new}|=|S|+1$.

Proof: Omitted, it is similar to the proof of Theorem (1). □

Note that the properties of column transposition follow equivalent rules as for row transposition in Theorems (1) and (3). Consider the fault pattern shown in Figure (2) [6]. The generation of the repair-set using the algorithm of [6] results in $|S|=4$; the transposition $LC_0 \leftarrow PC_6$ generates a repair-set such that $|S_{new}|=5$.

Corollary 1. The transposition of all C-single faulty cells into the last columns of a new diagnostic matrix A_{new} results in a repair-set S_{new} such that $|S_{new}| \leq |S|$, where S is the repair-set of A and S has been generated using the algorithm of [6].

Proof: Omitted due to lack of space. □

Corollary 2. The transposition of all C-single faulty cells into the first columns of a new diagnostic matrix A_{new} results in a repair-set S_{new} such that $|S_{new}| \geq |S|$, where S is the repair-set of A and S has been generated using the algorithm of [6].

Proof: Omitted due to lack of space. □

The above results can be used in defining new algorithms for generating new diagnostic matrices such that $|S_{new}|$ can be either decreased, or increased. A decrease (increase) in $|S_{new}|$ results in an improvement of the condition of repairability (unrepairability), as the uncertainty region is correspondly reduced. Algorithm (1) finds the minimum value of the repair-set to detect the repairability of a redundant WSI memory.

Algorithm 1: Reduction of the cardinality of the repair-set.

for i:=0 to N-1 do $LC(i)=i$

```
for i:=0 to N-1 do
        begin if C(C_i)=1 then
                begin tmp:= LC(i);
                  for k:= i to N-2 do
                          LC(k):=LC(k+1);
                LC(N-1):= tmp;
        end
for j:=0 to M-1 do LR(j)=j
for j:=0 to M-1 do
        begin if C(R_j)=1 then
                begin tmp:= LR(j)
                  for k:= j to M-2 do
                    LR(k):=LR(k+1);
                LR(M-1):= tmp;
        end
```

Get Repair Set; Use the algorithm of [6] for repair-set generation.

Algorithm (2) find the maximum value of the repair-set to detect the unrepairability of a redundant memory.

Algorithm 2: Increase of the cardinality of the repair-set.

```
for i:=0 to M-1 do LC(i)=i
for i=M-1 step -1 to 0 do
        begin if C(C_i)=1 then
              begin tmp:= LC(i)
                    for k:= i step -1 to 1 do
                      LC(k):=LC(k-1);
                LC(0):= tmp;
        end
for j:=0 to N-1 do LR(j)=j
for j=N-1 step -1 to 0 do
        begin if C(R_j)=1 then
              begin tmp:= LR(i)
                  for k:= j step -1 to 1 do
                      LR(k):=LR(k-1);
                LR(0):= tmp;
```

Get Repair Set; Use the algorithm of [6] for repair-set generation.

The above algorithms have an execution complexity of $O\{D^2\}$, where $D=\max\{N,M\}$, because they try to transpose each faulty line of the memory (rows and columns) once at a time. This operation also involves the shifting of the indices of all lines which lie between the lines involved in the transposing, hence, the quadratic order with respect to the greatest dimension of the memory. Note that the generation of the repair-set using the algorithm of [6] is linear with respect to the number of cells in the array (the worst case of no successfull early abort).

4. REPAIR USING TERNARY TREES.

One of the most desiderable features of a repair algorithm is its capability to find the optimal repair-solution (or perfect repair) [10], i.e. the solution which utilizes the smallest number of redundant rows and columns (whenever one exists) [7]. This problem is an *NP*-complete problem; the fault line covering principle as well as a branch and bound approach have been proposed for optimizing this process [7,9]. These approaches generate repair records using a binary tree as a basic data structure. The nature of the binary tree is caused by the choices which are demanded for the non redundant utilization of spare rows and columns: the covering of a faulty row (column) requires the utilization of either a redundant row (column) or a number of redundant columns (rows) equal to the number of faulty cells on the faulty row (column) (as given in Lemma (1)). This introduces an exponential dependency for the number of records to be generated in the repair process on the number of faulty cells [7,9].

A different criterion can be used to accomplish perfect repair. This new criterion does not rely on the overall number of faulty cells in the memory, but on the cardinality of the repair-set as generated previously for repairability/unrepairability detection. Let a_{ij} be an *LE*, where $C(R_i)= n$ and $C(C_j)= m$. It is obvious that repair is meaningfull (while still preserving the optimality of [9]) only if the following three options are executed in the generation of the records.

Rule 1: Use a spare row and a spare column.

Rule 2: Use m spare rows.

Rule 3: Use n spare columns.

Therefore, at each iteration of the repair algorithm the above three rules must be applied, as shown in Figure (3). This corresponds to the requirement of a ternary tree data structure for the generation of the repair records. The repair algorithm follows the basic structure of [9]: initially, Algorithms (1) and (2) are executed for repairability/unrepairability detection. Provided that the WSI memory under test is repairable, the repair-set with minimum cardinality found using Algorithm (1) is stored. The process for finding the repair-solution can be started. Forced Repair is executed. Forced Repair covers only those lines which can be covered using only one spare line (Criterion 1). Then, the remaining faulty cells are covered using the above three rules. However, single faulty cells and complete lines are not covered until all other faulty cells are repaired (Criterion (2) and (3) respectively). Note that for Criterion (2) the constraint $R_A'+C_A' \geq 1$ must be satisfied, where R_A' and C_A' are the number of unutilized redundant rows and columns. New early abort conditions in excess to those found in [9,10], can be also used. These conditions follows from Criterion (3), i.e. the late covering of the complete rows. Let k complete rows exist (including single faulty cells). Let the number of faulty cells on these lines be m_1, $m_2,...,m_k$ respectively. Assume without loss of generality that $m_1 \leq m_2 \leq ... \leq m_k$. If either of the following two conditions is satisfied, the memory is repairable, else it is unrepairable.

1. $R_A' \geq k$

or

2. $R_A'=k-r,\ C_A' \geq \sum_{i=1}^{r} m_i \ (1 \leq r \leq k).$

where R_A' and C_A' are the number of unused spare rows and columns. The complete columns can be used in a similar way.

As an example, consider the fault pattern of a WSI memory shown in Figure (4) (this is the same pattern as presented in [7]). The exhaustive repair process for the algorithm of [7] is shown in Figure (5); Figure (6) shows the exhaustive repair process using the proposed

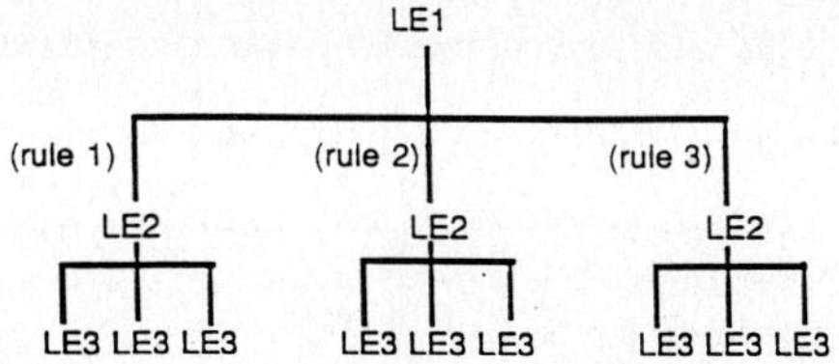

Fig.3 search using ternary tree

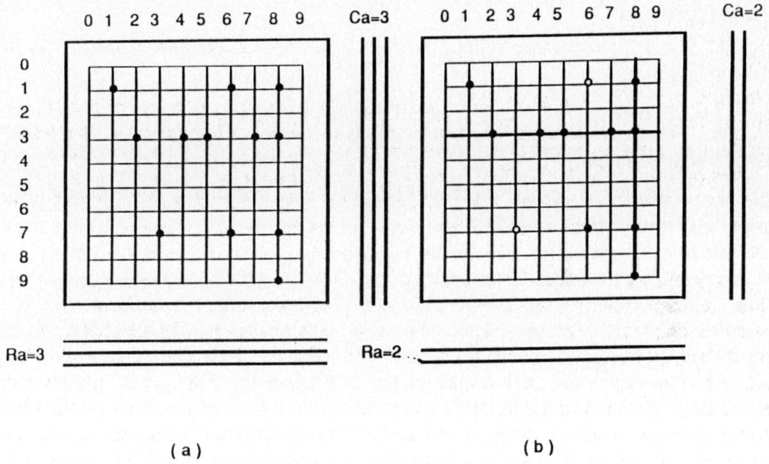

Fig.4 an example of repairing a RRAM

approach. Shaded entries are those records which can be ignored using the criteria of the branch and bound method of [9]. It is evident that the proposed approach requires a significant less number of records than the approach of [7]. The complexity of the proposed approach to the repair of redundant memories is given by the following theorem.

Theorem 4. The maximum number of records required in the repair process is $3^{|S_{min}|}$, where $|S_{min}|$ is the cardinality of the repair set using Algorithm (1).

Proof: Omitted, it follows from previous discussion. □

The above theorem characterizes a repair process that does not strictly depend on the number of faulty cells; in the presence of heavy clustering (as for WSI [9]) the proposed approach can show significant improvement over existing approaches, because in these circumstances the number of *LE* is very small (the number of faulty cells can be however, very large). The following theorems compare the proposed approach with the Fault Driven approach [10] and FLCA [9] (and indirectly the algorithm of [7]). The proofs are omitted due to lack of space.

Theorem 5. The maximum number of records required by the proposed approach is less than the number of records required by the Fault Driven Approach.

Theorem 6. The maximum number of records required by the proposed approach is less than the number of records required by FLCA.

Note that it has been proved in [9] that FLCA requires no more records than the branch and bound approach of [7].

Definition 8. After logical transposition, the set of *LE*s with maximum cardinality is referred to as the maximum *LE* set.

Definition 9. The sum of the number of spare rows and columns is referred to as the number of spare lines. The repair-solution which uses the least number of spare lines, is referred to as the optimal repair-solution.

Theorem 7. The cardinality of the maximum *LE* set is s_{max} if and only if the number of spare lines in for an optimal repair-solution is s_{max}.

Proof: The necessity condition will be initially proved, i.e. if there is an optimal repair-solution, the number of required spare lines is s_{max}. The *LE* set can be constructed as follows. Along every deleted row, it is always possible to find a faulty cell as a candidate for *LE*, such that this cell is not located on any deleted column. This is always possible, because the deleted row is not redundant [9] (which in turn would contradict the definition of an optimal repair-solution). Assume there exist two candidates for *LE* in two rows R_i and R_j which share the same column C_k. These two candidates are not both *LE*s by definition. It is possible to find another faulty cell on either R_i or R_j as a candidate, such that this cell is not located on any deleted columns as well as C_k. Therefore, two *LE*s are constructed. This cell must exists, else it is possible to add to the repair-solution a C_k in place of R_i and R_j. This results in a decrement by one in the number of required spare lines. This is obviously a contradiction. Similiarly, it is possible to construct the *LE* for every deleted column. The sufficiency condition can be proved as follows. Let the maximum *LE* set with a cardinality s_{max} exist. Assume that it is possible to find an optimal repair-solution which utilizes s_m spare lines, where $s_m > s_{max}$. Using the method described previously, it is possible to construct a *LE* set with cardinality $s_m > s_{max}$. This is a contradiction. Obviously, the least number of spare lines for any *LE* set cannot be less than the cardinality of the optimal *LE* set. This proves the theorem. □

This theorem is very important because it relates the problem of detection of repairability/unrepairability to the general modelling approach of [15]. The cardinality condition of the repair-set of Theorem (7) is equivalent to the characterization of the repair process in [14] as determining the critical set in the repair-solution.

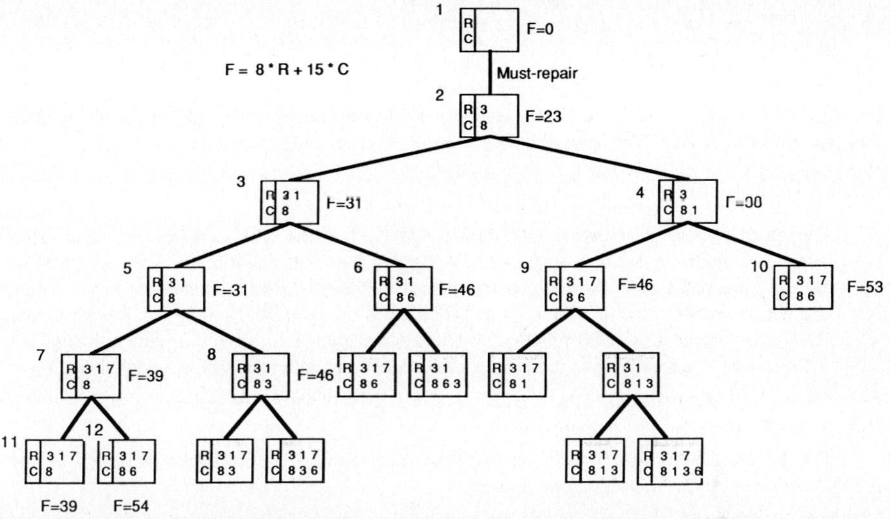

Fig.5 Binary tree search ; if using the branch-bound algorithm the shadowed records can be avoided

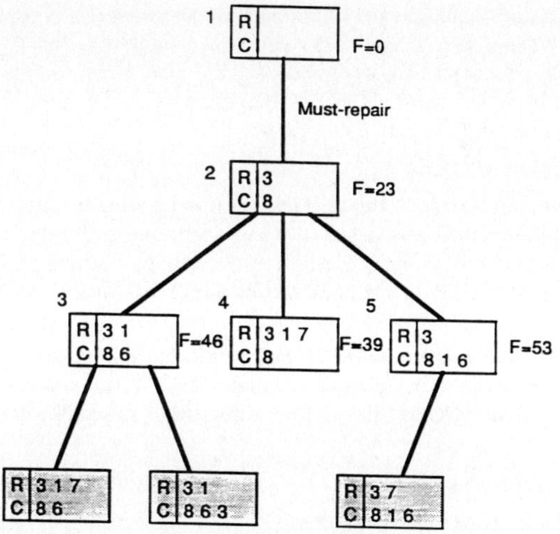

Fig.6 ternary tree search; if using the branch-bound algorithm, the shadowed records can be avoided

5. THE GENERALIZED LEADING ELEMENT.

The process of transposition of rows and columns presented in the previous section does not always allow to detect the repairability/unrepairability of a memory with redundancy. Consider for example the faulty pattern [6] shown in Figure (7), $R_A = C_A = 4$. In this case, $|S|=|S_{new}|=8$ always. A new, but more complex approach can be employed to further resolve this type of ambiguities. This approach is based on the following definition.

Definition 10. Let the ith row and jth column of A have at least one faulty cell each. Then, a_{ij} is referred to as the generalized leading element (*GLE*).

The basic difference between a *LE* and a *GLE* is that the *GLE* a_{ij} does not necessarily correspond to a faulty cell in memory. Consider again the faulty pattern shown in Figure (7). a_{83} is a *GLE*, but not a *LE*. Let the repair-set made of only *GLE*s be denoted as S_{GLE}. Then, it is possible to restate the condition of repairability and unrepairability of [6] by interchanging $|S|$ with $|S_{GLE}|$ without loss of correctness. However, the algorithm of [6] for generation of the repair-set must be drastically changed to accomodate the generation of *GLE*s, rather than *LE*s.

Definition 11. A k-crosspoint k_{ij} is a *GLE* of A if there at least a C-single faulty cell in row i and a R-single faulty cell in column j of A.

For the fault pattern of Figure (7), k_{83} is a k-crosspoint. The following Theorem proves a very important covering property of a *GLE*.

Theorem 8. If a_{ij} is a C-single faulty cell and is chosen as a *GLE*, then the number of faulty cells covered by a_{ij} is less or equal to the number of faulty cells covered by any other *GLE* in row i.

Proof: Omitted due to lack of space. □

The k-crosspoints can be used in connection with the unique covering property of Theorem (8) to further reduce the cardinality of the repair-set as well as providing an early abort condition. To further highlight the importance of Theorem (8), consider the fault pattern shown in Figure (7). The first C-single faulty cell is a_{87}, while the first R-single faulty cell is a_{43}. Hence, a_{83} is a k-crosspoint. a_{94},$a_{10,5}$ and $a_{11,6}$ are also k-crosspoints. This implies that $|S_{GLE}|=4$ which is correct for repairability of the memory. The algorithm for finding the minimum repair-set using *GLE*s is given as follows.

Algorithm 3: Generation of GLEs.

Step 1: Find a row in which there exists at least one C-single faulty cell. If unsuccessfull, go to Step 2. If successfull, assume this is the ith row. Then, find a column in which there exists at least one R-single faulty cell. If unsuccessfull, go to step 2. If successfull, assume this is the jth column. Choose a_{ij} as a *GLE* delete all 1 entries along the ith row and jth column and go to Step 3.

Step 2: Find a row in which the number of faulty cells is maximum. Assume this is the f row. Find a column in which the number of faulty cells is maximum. Assume this is g column. Choose a_{fg} as *GLE*. Delete all 1 entries along the f row and gth column in the diagnostic matrix.

Step 3: Add the newly found *GLE* to S_{GLE}. If A is a non zero matrix go to Step 1. Else, S_{GLE} has been found.

Algorithm (3) is based on a very simple principle: find the single faulty cells, else find the k-intersection of that row and column with the largest number of faulty cells. This is obviously a heuristic criterion which closely resembles the fault counting algorithm [10]. The implications of Algorithm (3) is that the number of *GLE*s is much less than the number of *LE*s (by almost a factor of 2), thus enhancing the detection of repairability.

The complexity of Algorithm (3) can be calculated as follows. Step (1) requires $O(NM)$ operations to find a C- or R- single faulty cell. Step (2) requires $O(\log \max\{M,N\})$ steps to

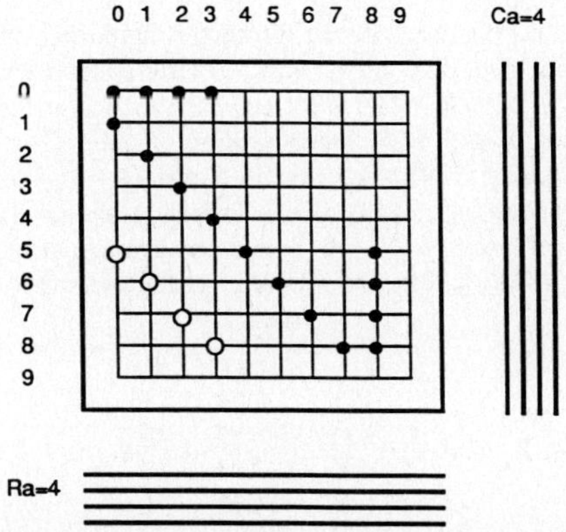

Fig.7 a repairable RRAM

find the maximum line. The loop in Algorithm (3) must be executed (R_A+C_A) times using early-abort conditions. Hence, Algorithm (3) is $O((NM+\log \max\{M,N\})(R_A+C_A))=O(NM\max\{R_A, C_A\})$.

6. SIMULATION RESULTS.

The proposed approaches of Sections (3) and (5) have been simulated. The simulated devices consists of multiple batches of 1M bit square memories (unpartitioned) with $R_A=C_A$. Each batch has been simulated 1000 times with F faulty cells in each device for $1 \leq F \leq 3R_A$ (this is in accordance with expected values for current WSI technology [16]). Faulty cells have been injected both randomly and in a cluster fashion. Tables (1) and (2) show the statistical characteristics of the simulated WSI devices: in Table (1), N_C (N_R) denotes the number of columns (rows) with at least one faulty cell. Table (2) shows the average number of LEs found by the algorithm of Section (3) (LE_{av}) and the average number of GLEs found by the algorithm of Section (5) (GLE_{av}) versus the number of faulty cells in the memory.

Table 1.

N_C	N_R	F
4.73	4.69	5
8.85	8.87	10
12.06	11.99	15
14.85	14.79	20
17.31	17.19	25
19.44	19.15	30

Table 2.

GLE_{av}	LE_{av}	F
2.90	4.43	5
4.68	7.71	10
6.09	9.56	15
7.31	11.25	20
8.61	12.52	25
9.27	13.71	30

Table (3) shows the percentiles of devices found repairable, unrepairable and diagnosed uncertain using for detection logical transposition, the GLE approach of Section (5) and the LE technique of [6], $R_A=C_A=6$. The last row of Table (3) shows the actual percentiles of devices which can be either repaired or discarded using an exhaustive repair algorithm to find the repair-solution [9].

Table 3.

Approach	Rep.	Unrep.	Uncert.
Logical Trans.	31.922	33.357	34.721
GLE	40.000	32.778	27.222
[6]	24.333	32.778	42.889
[9]	60.222	39.778	0.0

Table (4) shows the average number of records in the binary tree search using the exhaustive approach of [6] (R_b) and the ternary tree approach of Section (4) (R_t).

Table 4.

R_b	R_t	$R_A=C_A$	F
4.8	2.8	6	5
36	8	6	10
355	38	6	15
430	84	6	20
464	92	6	25
566	108	6	30
137	29	5	15
134	39	5	20
153	40	5	25
194	48	5	30
42	15	4	15
42	16	4	20
49	17	4	25
66	21	4	30

The following conclusions may be drawn from the above tables.

1. The proposed method achieves a significant improvement in repairability/unrepairability detection at modest increase in execution complexity. This is very important in an on-line repair system, as it can abort the testing of a WSI device once it is found that it is unrepairable.
2. The ternary tree data structure utilizes less records than the commonly used binary structure [7,9]. This improvement is considerable when the number of faulty cells is larger than R_A+C_A, i.e. the memory can not be guaranteed a priori to be repairable [6]. This feature can be readily accomodated in a defect analysis system using fault counters [6] for the generation of the records using ternary trees.
3. The simulation results in Table (4) do not account for the reduction in records due to early abort conditions such as those provided by the branch and bound method of [7]. Note that this can be readily incorporated in the proposed repair algorithm as pruning of the tree is possible at no loss of significant information.
4. The average number of *LE*s found using the proposed algorithm seems to suggest that it exists an asymptotical value as F increases. The existence of this value is crucial as it would imply that WSI memories with a large number of faulty cells would not require on average a large repair time. This implication is very important for the economical feasability of repair for yield enhancement of WSI devices.
5. The throughput for repairing memories has been defined in [8]. The results of Table (3) prove that the throughput can be significantly increased if repairability/unrepairability detection is used prior to the repair algorithm. This improvement is significant because unrepairability detection is rather effective and a large majority of unrepairable devices can be discarded very early in the detection process. The improvement due to logical transposition is also effective in reducing the uncertainty region. This implies that only the status of a small number of devices must be established by the repair algorithm. Only the devices which are diagnosed uncertain, are required to be analyzed by a repair algorithm. The memories which have been diagnosed repairable, can be repaired using the *GLE* or *LE* set in lace of the repair-solution. Note that in this case the solution is not necessarily optimal, but it still meet the requirements of the provided redundancy.
6. The approach of Section (5) results in a considerable improvement of repairability detecion. This is achieved by reducing the uncertainty region at a larger order of complexity than previous detection techniques.

7. REFERENCES.

[1] Benevit, C.A. et al, "A 256k Dynamic Random Access Memory," *IEEE Journal of Solid-State Circuits,* Vol. SC17, No. 5, pp. 857-862, 1982.

[2] Cenker, R.P. et al, "A Fault Tolerant 64k Dynamic Random Access Memory," *IEEE Trans. on Electron Devices,* Vol. ED26, No. 6, pp. 853-860, 1979.

[3] Okajima, Y. et al, "64kb ECL RAM with Redundancy," *Proc. IEEE ISSCC,* pp. 48-49, 1985.

[4] Evans, R.C., "Testing Repairable RAMs and Mostly Good Memories," *Proc. IEEE ITC,* pp. 49-55, 1981.

[5] Smith, R.T. et al, "Laser Programmable Redundancy and Yield Improvement in a 64k DRAM," *IEEE Journal of Solid-State Circuits,* Vol. SC16, No. 5, pp. 506-514, 1981.

[6] Wey, C-L and F. Lombardi, "On the Repair of Redundant RAMs," *IEEE Trans. on CAD,* Vol. CAD6, No. 2, pp. 222-231, 1987.

[7] Kuo, S-Y and W.K. Fuchs, "Efficient Spare Allocation in Reconfigurable Arrays," *Proc. ACM/IEEE DAC,* pp. 385-390, 1986.

[8] Dahbura, A.T. and R.W. Haddad, "Increasing the Throughput of the Testing and Repair of RAMs with Redundancy," *AT&T Bell Labs,* Internal Report, Murray Hill, 1986; also in Proc. *IEEE ICCAD,* Santa Clara, 1987.

[9] Huang, K.W. and F. Lombardi, "Approaches for the Repair of VLSI/WSI RRAMs by Row/Column Deletion," *Proc. Fault Tolerant Computing Symposium,* pp. 342-347, 1988.

[10] Day, J.R., "A Fault-Driven Comprehensive Redundancy Algorithm for Repair of Dynamic RAMs," *IEEE Design and Test of Computers,* Vol. 2, No. 3, pp. 33-44, 1985.

[11] Fitzgerald, B.F. and E.P. Thoma, "Circuit Implementation of Fusible Redundant Addresses of RAMs for Productivity Enhancement," *IBM J. Res. Develop.,* Vol. 24, pp. 291-298, 1980.

[12] Stapper, C.H., "On a Composite Model to the IC Yield Problem," *IEEE J. of Solid State Circuits,* Vol. SC10, pp. 537-539, 1975.

[13] Gary, M.R. and D.S. Johnson, "Computers and Intractability," Freeman, San Francisco, 1979.

[14] Hasan, N. and C.L. Liu, "Minimum Fault Coverage in Reconfigurable Arrays," *Proc. FTCS18,* pp. 348-353, 1988.

[15] Hasan, N., Cong, J. and C.L. Liu, "A General Model for Fault Covering Problems in Reconfigurable Arrays," *Proc. IEEE Workshop on Defect and Fault Tolerance in VLSI Systems,* 1988.

[16] Carlson, R.O. and C.A. Neugenbauer, "Future Trends in WSI," *Proc. IEEE,* Vol. 74, No. 12, pp. 1741-1752, 1986.

Wafer Scale Integration, III
M. Sami and F. Distante (Editors)
Elsevier Science Publishers B.V. (North-Holland)

WESTINGHOUSE WSI DEVELOPMENT PROGRAM

Joe E. Brewer and Dr. Michael E. Michael

Westinghouse Electric Corporation
P.O. Box 746 (MS 5210)
Baltimore, MD 21203
(301) 765-1247

ABSTRACT Westinghouse has initiated a program to devise and to mature wafer scale integrated (WSI) circuit technology for the purpose of building complete signal and data processors. The program treats the development of a variety of types of WSI silicon based circuits as well as provides novel packaging which allows exploitation of the density and reliability potential of the technology.

1 INTRODUCTION

The Westinghouse WSI effort which began in 1985 is evolutionary in nature, and seeks to build on existing silicon integrated circuit technology and packaging techniques. It is expected that as the program proceeds the proven technology base will be expanded, and new and unique capabilities will be merged into the established baseline.

A fabrication approach for silicon wafer scale integrated circuits is being combined with advanced three dimensional packaging to create a producible means for building processor subsystems using WSI components. Because of component count reductions and the elimination of large numbers of interconnections, resultant WSI processors are expected to enjoy more than an order of magnitude improvement in serial failure rate over state of the art VLSI technology. The savings inherent in the chosen WSI fabrication approach should cut processor costs by at least a factor of two. Significant gains are also expected in physical parameters and performance capability.

2 WAFER SCALE INTEGRATION CONCEPTS

A driving motivation for integrated circuit technology development is the achievement of ever increasing levels of circuit integration. Increased density per monolithic device is recognized as providing both reduced cost and increased reliability. It also can be a key for the achievement of improved performance.

Three ways of improving density are commonly recognized: use smaller feature sizes, employ improved layout schemes (i.e., circuit cleverness), and allow the use of larger circuit substrates. In simplest terms, wafer scale integration is the culmination of the drive to achieve higher density by using larger substrates.

A suggested [1] definition of WSI makes this point: "Wafer scale integration is a technology that enables the fabrication of monolithic chips as large as the maximum wafer diameter that is in commercial manufacturing in the same time frame."

Probably the first publication to treat WSI issues was by Sack, Lyman, and Chang at Westinghouse in 1964 [2]. These general ideas have now been around for almost a quarter of a century.

Technological features which distinguish WSI from other integrated circuit arts are associated with the management of the special problems of large substrates. There are yield, interconnection, packaging, and thermal management approaches which are usually conceded to be unique to WSI. In most approaches, WSI requires the use of some distinctly different fabrication and design tools. Those tools, which include computer aided design systems, computer aided manufacturing capabilities, form an essential part of the WSI technology base.

On the other hand, WSI is not isolated from technological progress in the general integrated circuit arts. WSI can be based on many different material and/or circuit technologies. For example, both silicon and gallium arsenide WSI circuits are currently under development. Within silicon WSI, both bipolar and MIS devices have been used.

A successful WSI technology should be able to incorporate improvements from the broader area of integrated circuit technology. Stable smaller feature size production capability should be used to further improve WSI density. New device layout schemes which increase packing density, should be usable in WSI circuits.

So as presented here, the WSI concept is quite simple. WSI technology is that collection of integrated circuit arts which deal with the issues of design, manufacture and use of circuits formed in large monolithic substrates.

3 DEVELOPMENT ECONOMICS

For at least the past few years, reduction in feature size has been the primary means for achievement of increased density. Simple economic choices greatly favored that development option. It is now becoming more difficult to continue that trend. Large capital investments are required to build the super fabrication plants capable of volume production at 0.5 micrometer and lower dimensions, and the uncertainties of the market place do not provide any assurance of an adequate return.

In contrast, WSI allows a way of using existing plants to achieve increased density. Defect densities are low enough to support defect avoidance strategies, and the achievable functional content per wafer is adequate to capture market share. The development costs to solve the special problems of large substrates are minor when compared to the alternative of smaller feature size production.

4 WSI WAFER FABRICATION

WSI devices at Westinghouse may be several inches on a side up to the dimensions of the full wafer. A device consists of many independent circuit sites which have been separately tested and then later interconnected by an added top level metal. The manufacturing scheme allows for the elimination of defective circuit sites from the final connecting metal pattern before the metal is formed.

The WSI circuit structure can be thought of as a base CMOS circuit under a WSI metal layer. The base circuit is formed by conventional integrated circuit processing. The base circuit allows for the probe testing of each site. Probe tests may directly evaluate circuit functions, or may rely on built-in test circuits to carry out part or all of the test.

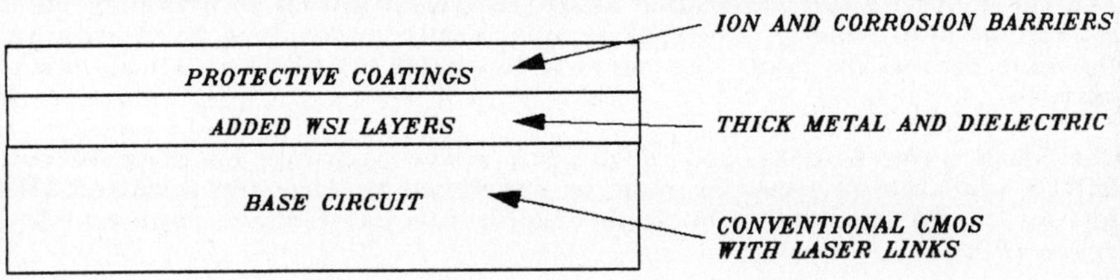

Figure 1 WSI Circuit Cross Section

At this stage, laser operations are performed on the device to remove defective sites. When the top level of metal interconnect is patterned, it connects all working sites into a meaningful functional configuration.

After the WSI device is formed, the wafer is sealed by the application of impervious coatings. This procedure allows the device to be used without need of a hermetic container. The sealing procedure provides multiple levels of protection. First, a layer of silicon nitride is applied using plasma deposition techniques. Then one or more additional coatings of Dow Corning materials are applied using other deposition techniques. The combination of coatings is believed to provide complete assurance of a satisfactory seal.

The baseline fabrication approach builds on well established VLSI technology, and does not depend on a complex major technical break through for success. At the base circuit level an established 1.2 micron CMOS process has been chosen. The tools, materials and equipment for processing the WSI top levels are all conventional.

Even the laser processing equipment is the same as that in common usage for RAM repair. The defect avoidance scheme is essentially that developed by Inova Microelectronics Corporation. The Inova scheme is one of several techniques which will be used by Westinghouse.

Over the past 25 years there have been a number of attempts to develop viable WSI technology. Most of those efforts involved simultaneous bold advances in many facets of the integrated circuit arts. While some of the technical achievements were remarkable, this multiplication of development risk was a key factor in lack of success. The Westinghouse initiative emphasizes evolutionary development to control risk and to insure rapid maturity.

5 INOVA INROUTE SCHEME

In 1984 Inova Microelectronics Corporation [3] was founded with the objective of developing a simple manufacturable approach to WSI. The Inova technique used laser cutting to restructure circuits, and had the advantage of not requiring any nonstandard processing. In 1988, Inova began shipping 1M-bit CMOS static RAM devices, and announced plans for development of other products of greater complexity.

The fabrication approach, called INROUTE (trade mark), is primarily an interconnection scheme. INROUTE is most easily understood by examining the case of a static RAM. The current Inova 1M-bit SRAM is a convenient example.

The SRAM wafer is composed of 40 sites where each site contains 32k-bits. Of the available 40 sites, 32 must be functional to form the required 1M-bit device. To enhance yield, each 32k-bit site has 2 spare rows and 2 spare columns.

Each site contains a number of polysilicon links which can be readily cut by a laser. Some of the links are arranged to allow the row and/or column repair of a site. Other links are associated with each I/O pad. The top level WSI metal layer must connect through these links to gain access to the site. If these links are cut, the site is excluded from the WSI device network.

At wafer probe, a computer is used to control and remember the test results, and to direct the laser operations. If a given site is found to be good, no laser cuts are performed.

If a site is found to be defective, and it is judged to be repairable using the spare rows or columns; the laser is used to accomplish the repair, and the site is retested. If a site is found to be non-repairable, all the I/O links are cut.

Once the initial repair sequence is complete, the computer directs the laser to program an address circuit. Each binary address input is assigned to an available physical site.

This simple laser processing scheme has some very important advantages. At probe test each site is physically and electrically isolated. A power to ground short will not impact adjacent sites. The site test and laser isolate sequence will prevent a local power to ground short from being tied to the WSI level metal.

The entire sequence can be carried out without a great deal of concern as to the details of the base circuit processing. The procedure is flexible enough to allow such yield enhancing operations as row column repairs to be routinely incorporated. Also, the equipment and technology required for the laser and test operations is similar to that already developed for RAM laser processing.

6 WESTINGHOUSE WAFER PACKAGING

The Westinghouse WSI program is oriented toward the implementation of complete signal and data processors. The program is driven by a top down study and understanding of system needs. It includes a new rather general packaging scheme.

Wafer module and subsystem packages are based on composite materials to achieve strong light weight thermally efficient structures. Power and signal distribution are managed by using multilevel flat cables which make contact directly to the top surface of the wafers.

Wafers are mounted unpackaged directly on a composite material which serves as both a structural support and a conductive heat sink. The proprietary Westinghouse material has a thermal coefficient of expansion which closely matches that of silicon, and thermal cycling stress experiments have demonstrated the reliability of the wafer bond.

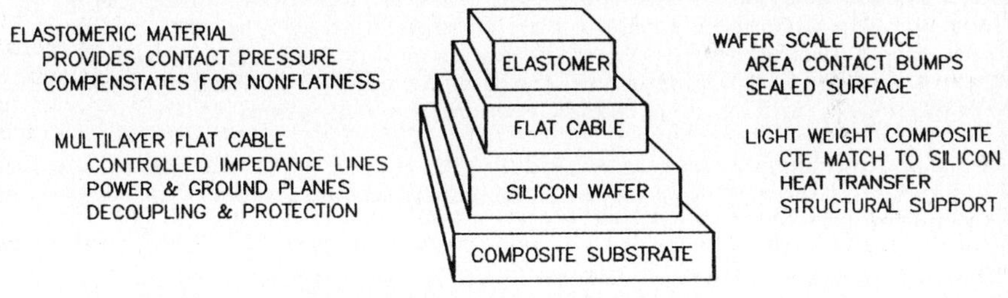

Figure 2 Device Mounting and Connection Concepts

The composite has a very high thermal conductance, and thermal impedances on the order of 1 °C/watt from the wafer to the chassis side wall can readily be achieved. The combination of high physical strength, low weight, and high thermal conductance make the composite an unmatched substrate material.

Interconnection is carried out using flat multilevel multiconductor cables. The novel aspect of the approach is that connections are made directly to the silicon wafer. Because the wafer is sealed against ionic contamination and corrosion using impervious films on the wafer, the entire wafer surface is available for connection.

The controlled impedance cables lie on top of the wafer, and provide near ideal signal transmission and power distribution. Use of such cables allows critical signals to be routed on high quality lines rather than restricting them to the high loss lines on the surface of the wafer.

Compact modules have been devised which house two or more wafers in a low profile completely enclosed unit. These strong low weight modules are compatible with two level maintenance concepts, and can form the building blocks for complete signal and data processors. The packaging scheme is sufficiently flexible to allow the use of discrete VLSI devices or hybrid assemblies to supplement the WSI parts where ever necessary.

Because the overall approach is based on simplification rather than the development of new complex technology, Westinghouse hopes to move it rapidly to production. Memory modules will be the first product to emerge, closely followed by processor functions.

7 WAFER FABRICATION ECONOMICS

WSI offers the promise of cost advantages over alternative subsystem implementation approaches. In order to achieve that potential the product must be prepared in such a manner as to emphasize those attributes which are the source of economic leverage.

One can distinguish several classes of leverage. There are generic savings which result because of the increased level of monolithic integration. There are simplification savings which result because the WSI option allows the accomplishment of a function in a new way. Then there are cases where a special feature of a WSI implementation dominates all other considerations, and thus provides economic justification.

The most appropriate way to consider WSI versus VLSI cost is to compare the yielded "function" per dollar. The essential aspects of this matter can be brought out by a simplified model.

Those operations performed before VLSI die separation, or prior to WSI device packaging, are defined here as the "wafer fabrication level." The intent is to designate logically comparable portions of the two manufacturing sequences.

During wafer level fabrication the item which accrues cost is the wafer. Typically, an average cost will be established for a production lot of wafers. The cost of a yielded wafer will then be taken as the lot cost divided by the average number of yielded wafers.

For WSI approaches currently in use, the wafer level WSI fabrication involves more operations than equivalent VLSI fabrication. Assuming that both processes are comparable in maturity, the average WSI wafer will cost more than the average VLSI wafer. The obvious conclusion is that the WSI process must provide more yielded function per wafer in order to break even.

The cost lever at this stage of production is thus built into the scheme for insuring fabrication yield. A WSI device is composed of a number of interconnected "sites." The sites contain some means for repair in order to manage the impact of defects. Usually, a combination of small area sites and internal repair capability is employed.

In contrast, VLSI die sites are sized to approximately match a projected die yield and packaging cost optimum trade off. The die are as large as the maturity of the line will support with reasonable yield.

For the current state of the art, a 30,000 gate VLSI part 1 cm on aside might achieve about 75% wafer yield and 10% die yield. An 8 cm on a side silicon substrate can accommodate 64 die locations. So the yield would average about 6 die or 180,000 gates per wafer. If the yielded wafer cost is taken as a reference value of 1.0, the bare chip cost would be 0.17 . The cost per gate compared to the normalized wafer cost would be 5.6E-6.

A WSI product made on the same production line could be assumed to have about the same defect density. The wafer yield would be somewhat lower because of additional WSI unique processing. The utilization of the available surface area may be reduced by area required to accomplish site interconnection, and by area devoted to redundancy within a site.

Consider a simple WSI part composed of identical logic blocks of 4000 gate complexity with no internal redundancy. Because of the additional WSI process steps, the number of WSI wafers to reach the end of the line can be expected to be about 92% of the VLSI wafers, or 92% of 75% which is 69%. Assuming the same defect density as the VLSI example, this smaller WSI site will have a yield around 75%.

The additional processing would increase the cost of a lot by about 20%. In terms of the VLSI yielded wafer reference cost of 1.0, the combination of the wafer loss and the additional processes will drive the yielded WSI wafer cost up to about 1.3.

The WSI site would measure 0.365 cm on a side, and a 21 by 21 array of sites would fit into the 8 cm square with a comfortable allowance of space for interconnections. This means that 441 sites of 4000 gates would be available, and 0.75 x 441= 330 sites should yield.

The yielded complexity level would be 0.69 x 330 x 4000 or 910,800 gates. This WSI yield is five times greater than the VLSI complexity of 180.000 gates. The cost per gate compared to normalized wafer cost would be 1.3 divided by 910,800 or 1.43E-6. This is about 25% of the VLSI cost per gate of 5.6E-6.

In practice the design would be targeted to make use of some number of gates lower than the 910,800 gates in order to insure adequate yield margin, but certainly there would be a lot more gates per wafer in the WSI case.

The point of this over simplified example is that the reduced area of the WSI site as compared to a VLSI die can offer a degree of yield enhancement so as to insure a greater amount of yielded function per wafer for the WSI option. It is the reduced site size that can be used as a generic wafer fabrication level cost lever. This advantage can be further enhanced by adding laser (or other) repair techniques to the design.

At the wafer fabrication level it is reasonable to expect that a good WSI design can achieve an advantage over the VLSI alternative. The simplified example showed that the yielded gate cost was reduced by a factor about four. Reduction factors on this order are probably achievable without great effort over a wide range of designs.

The magnitude of these numbers point out the nature of the historical WSI development problem. Cost savings like this have been routinely achieved as part of the learning curve associated with conventional integrated circuit technology. If the WSI development effort takes any appreciable period of time, it is likely that its wafer level cost advantage will vanish. To succeed, the WSI approach must enjoy the use of the same technology as the VLSI option.

The simplified example also points out how this cost lever can be made to be invalid. If the WSI approach fails to make efficient use of the available silicon area, the analysis falls apart. This could happen if the designer calls for an excessive amount of redundancy, or if the interconnect scheme requires a great deal of area.

There are of course other factors which impact the cost of a subsystem rather than just the wafer yield. Nitch markets exist where specific features of a subsystem can only be achieved by the WSI approach. For example, reliability for a critical mission can be so important that it dominates all other considerations. The reliability advantage of WSI would then be the source of economic leverage which makes it the most cost effective alternative.

An interesting case can be made for the weight savings of a WSI subsystem as a special feature lever for space applications. The cost to place a pound of payload in orbit using rocket launches is considerable.

The economic leverage obtained from use of WSI will vary considerably from application to application. Westinghouse studies of cost for military applications at an electronic subsystem level show at least a factor of two advantage for WSI as an alternative to VLSI. Studies of commercial applications conducted by other groups show savings, but they are usually of lesser magnitude.

In state of the art military applications the amount of cost involved in VLSI packaging, test and burn-in is significant. A single WSI part can typically replace 50 to 150 VLSI parts. Thus, on the basis of package count reduction alone, the WSI option gains considerable leverage.

8 WSI RELIABILITY

The most obvious and least controversial promise of a WSI device is that of improved reliability. The claim is based on a dramatic reduction of the number of interconnections required versus those for an equivalent assembly of conventional chips. The implicit assumption is that the WSI approach must not introduce new uncontrolled failure modes.

Reference [4] describes a first order WSI reliability model. The model allows the examination of the reliability impact of water scale integration versus the alternative of VLSI devices. Experiments conducted for a wide variety of hypothetical subsystems show significant advantages for the WSI option. Because of reductions in component count, interconnections, and temperature more than an order of magnitude improvement in serial failure rate is predicted.

Introduction of a WSI technology necessarily involves unknowns related to reliability. In the Westinghouse program, close attention is being given to those aspects of the approach which are new in order to avoid the introduction of new failure modes. For example, the additional top layer processing to accomplish WSI interconnect employs comparatively gross dimensions in order to avoid marginality with respect to yield and/or reliability.

In a WSI approach, the level of circuit integration is increased by increasing the circuit substrate size, not by reducing the feature sizes. This is an important point in that failures normally associated with scaling are avoided.

The excellent heat transfer characteristics of the composite substrates used for wafer mounting are believed to be a reliability asset. Thermal models indicate that WSI thermal stress is likely to be lower than that for discretely packaged VLSI parts.

The most obvious reliability leverage area for Westinghouse WSI is a great reduction in metal to metal bonds. Hybrid circuit approximations of WSI can not provide this magnitude of interconnect simplification.

9 SUMMARY

In 1985 Westinghouse Electric initiated a WSI program for the development of complete signal and data processors. Westinghouse device fabrication makes use of, and extends, the Inova technology base. The development approach is driven top-down by study and understanding of systems needs. It includes a new rather general packaging scheme, and it is not committed to just one device manufacturing strategy. Evolutionary development of WSI technology with a very limited introduction of new processes and materials is a theme of this ongoing program. Wafer module

and subsystem packages are based on composite materials to achieve strong light weight thermally efficient structures. Power and signal distribution are managed by using multilevel flat cables which make contact directly to the top surface of the wafers.

ACKNOWLEDGEMENT

This work was partially sponsored by AFWL/NTC, Kirtland AFB, NM. The technical monitor was Lt. Raymond Ord.

REFERENCES

[1] Ralph K. Calvin and James R. Key, "SRC Wafer Scale Integration Workshop," No. SRC-012, 20-21 September 1984, Durham, NC. (Distribution limited to Semiconductor Research Corporation member companies.)

[2] E. A. Sack, R.C. Lyman and G.Y. Chang, "Evolution of the Concept of a Computer on a Slice," Proceedings of the IEEE, vol. 52, no. 12, December 1964, pp 1713-1720.

[3] Ramesh Varshney, "Wafer Level Integration Technique," U.S. Patent 4703436, filed 1 Feb. 1984, issued 27 October, 1987.

[4] Joe E. Brewer and Dennis W. Beuerle, "WSI Reliability Model," Government Microcircuit Applications Conference Digest of Papers, October 1987, pp 93-96.

Wafer Scale Integration, III
M. Sami and F. Distante (Editors)
Elsevier Science Publishers B.V. (North-Holland)

Automatic Wiring Algorithm for WSI Interconnection Networks

Jeffrey M. Wills Vijay K. Jain

Department of Electrical Engineering
University of South Florida
Tampa, Florida 33620, USA

This paper presents an area efficient algorithm for the omega interconnection network. The algorithm is applicable to multi-level metal designs as well as the traditional two-level metal designs. Such interconnection networks are applicable to wafer scale implementation of many signal processing problems.

I. INTRODUCTION

Intense interest currently exists in massive parallel processing using SIMD and MIMD architectures [6],[7]. The applications include image and speech processing, file processing, database management, knowledge-based systems, and high resolution graphics. The architectures for these and other applications frequently require the use of INterconnection NETworks (INNETs). We examine the geometry of network wires and present an algorithm for minimum area layout of the perfect shuffle for the omega network.

Wiring area is a critical factor in wafer scale integration where the interconnect can take up a significant portion of the total area thus reducing the number of processing elements [4], [5]. This paper presents an area-reducing multilevel-metal wiring algorithm for the omega interconnection network [1]. The objective of multilevel metalization, i.e., more than the traditional two-level metalization, is two-fold: a) to reduce the silicon area used by the wiring, and b) to provide for mask programmability of metal layers three and four for defect tolerance and wafer customization [2], [3]. The emphasis of this paper is on the first of these two aspects.

II. WIRING GEOMETRIES IN MULTILEVEL-METALIZATION

An INNET reorders the data arriving at its input terminals to a desired order at the output terminals. Formally, an interconnection network, or function, is defined to be a set of subfunctions; each interconnection subfunction is a bijection (a one-to-one and onto mapping) on the set of input/output addresses, the integers 0,...,N-1. Throughout we will assume N is a power of two. Thus, the INNET allows a data signal at any input terminal to be routed to any conflictless, but otherwise arbitrary, output terminal. As the definition suggests, one method of providing this capability is to decompose the routing function into several subfunctions that operate on the input data sequentially:

This work was funded in part by DARPA grant No. MDA 972-88-J-1006.

$$y = f(x)$$
$$= f_1(f_2(...(f_n(x))...))$$

where x is the vector of input addresses, and y is the vector of output addresses.

The simplest strategy for such functional mapping of signals is a full crossbar switch, which has the merits of architectural simplicity and minimum propagation delay. However, it requires the use of N^2 elemental switches and is thus prohibitively expensive in terms of area. Multistage INNETS require the use of only on the order of $N\log_2N$ elemental switches. On the other hand, routing of signals from input to output is slightly more difficult than for the crossbar switch, in the case of the omega network.

Fig. 1 is a generic description of a multistage INNET. The network is composed of switching stages which are composed of dynamically controllable switching elements interspersed with exchange stages, which route the outputs of the preceding switching stage to inputs of the next switching stage. The switching stage routes a signal onto one of two, or in some cases three, paths connected to adjacent output terminals. The exchange stage rearrange the order of the switching outputs so that permutations between signals that are not adjacent at the input terminals are possible.

The control lines for the switching stages are routed in the spaces between the switching and exchange stages, which in the generic description are shown as reserved areas of width C and C'. In the remainder of this paper, we will consider that the INNET is arranged as shown in Fig. 1. Generally, this will not alter the assumptions used for the layout of the exchange stages given below, since the height of the exchange stage (long the y axis; see Fig. 2) will be limited primarily by the height of the switching stages, rather than the layout demands of the exchange stages themselves. Additionally, the width of the switching stage (along the x axis), W_S, will be considered to be constant in this paper. We will focus entirely on minizing the area of the exchange stage by reducing W_E - C - C'.

The coordinate system used for wiring layout is shown in Fig. 2. The direction of data flow through the INNET is along the x direction. In the exchange stages, runs along the y direction are required to connect input nodes to different output nodes. Different levels in the z direction correspond to different levels of metal. For example, wires A and B in Fig. 2 have the same projection in the x-y plane; however, since they are implemented in different metal levels, the wires are distinct.

Definition 1: The description vector (DV) of a wire is an N column vector with entries 0 or 1 such that

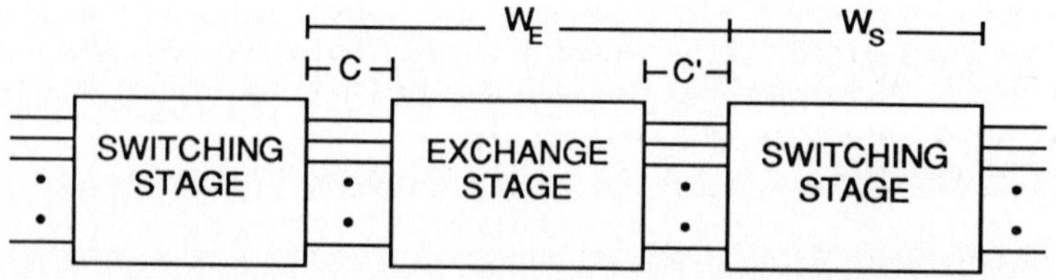

Fig. 1. Generic description of a multistage INNET.

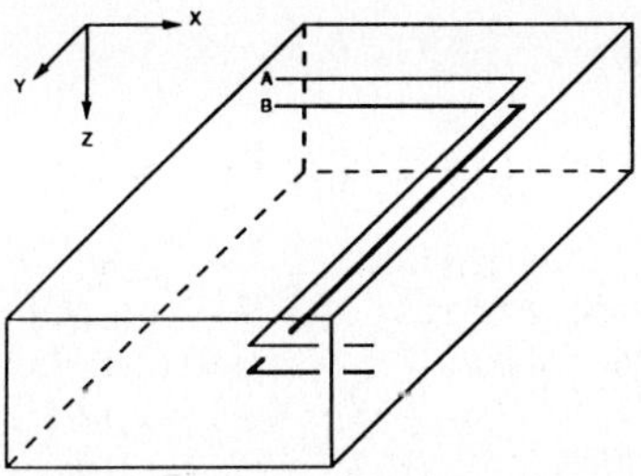

Fig. 2. Coordinate system description.

(a) the 1 entries are contiguous
(b) a string of 1's from entry i to entry j, $0 \le i \le j \le N\text{-}1$. implies that the wire runs in the y direction from the horizontal level of node i to the horizontal level of node j. Thus, the wire runs through each of the coordinates k, $i \le k \le j$.

Example: N = 8. If $\underline{v} = [0\ 0\ 1\ 1\ 1\ 0\ 0\ 0]^T$ is the DV of a wire, then the wire contains a y-segment that runs from the horizontal level of node 2 to the horizontal level of node 4 (see Fig. 3).

The inner product of two DVs can be used to determine if the corresponding wires overlap. Let $\underline{v}_a$ and $\underline{v}_b$ denote the DVs of wires a and b, which we desire to implement in a yz-bundle. Then[1]

(1) the two wires do not overlap vertically if and only if $\underline{v}_a \cdot \underline{v}_b = 0$
(2) wire a completely overlaps wire b if and only if $\underline{v}_a \cdot \underline{v}_b = \underline{v}_b \cdot \underline{v}_b$
(3) wire b completely overlaps wire a if and only if $\underline{v}_a \cdot \underline{v}_b = \underline{v}_a \cdot \underline{v}_a$
(4) Otherwise, wires a and b partially overlap each other.

When several levels of metal are available, some means of describing both the vertical span of the wire and the particular level of metal it is implemented on, is required. For this purpose we define a description matrix for each wire.

Definition 2: The description matrix (DM) W, of a wire is an $N \times m$ matrix whose i-th column is the DV for the wire, while the remaining columns are columns of zeros. Here, N is the number of inputs to the exchange stage, m is the number of levels of metal, and i is the level of metal used to implement the wire.

Using the properties of the DVs and DMs we can make the following assertion. Two wires lying in the same yz-bundle, with DMs W_1 and W_2, are non-overlapping, if and only if[2]

$$\| W_1 + W_2 \|_\infty = 1 \tag{1}$$

---

[1] The notation $\underline{u} \cdot \underline{v}$ means the dot (scalar) product of vectors $\underline{u}$ and $\underline{v}$.

[2] The notation $\| W \|_\infty$ means the L_∞ norm, the largest magnitude of the elements of W.

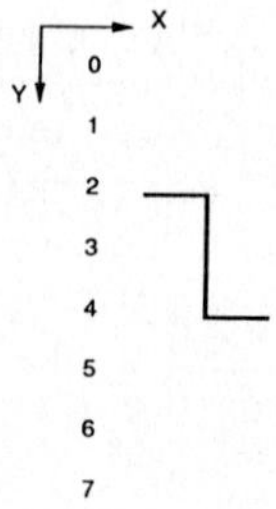

Fig. 3. Layout of a wire with a description vector $[00111000]^T$.

If (1) holds for two wires in the same level of metal, then these wires can be implemented in the same yz-bundle.

The minimum number of metal levels required to layout a set of wire segments with DVs $v^{(i)}$, where i = 1, 2, ..., n, is given by

$$\| \textstyle\sum_i \underline{v}^{(i)} \|_\infty \tag{2}$$

A set of wire segments $\mu_1, \mu_2, ..., \mu_n$ can be implemented successfully in a single yz-bundle if

$$\| \textstyle\sum_i W_i \|_\infty = 1 \tag{3}$$

and

$$\| \textstyle\sum_i \underline{v}^{(i)} \|_\infty \leq m \tag{4}$$

where m is the number of metal levels permitted, and W_i and $v^{(i)}$ are the DM and the DV respectively, of the wire μ_i. If the number of metal levels available is less than the number of levels determined by (2), then the y-wire segments must be spread out over two or more yz-bundles. The algorithm presented in this paper chooses the DMs for wires in the exchange stages from the DVs such that condition (1) is satisfied and the number of yz-bundles is minimized.

III. PERFECT SHUFFLE INTERCONNECTIONS AND THE OMEGA NETWORK

The perfect shuffle interconnection pattern is used in several interconnection networks, including the omega network [1]. The perfect shuffle interconnection pattern intersperses data from inputs 0 through N/2 - 1 with the data from inputs N/2 through N - 1, just as one shuffles a deck of cards.

The height of the perfect shuffle is a function of the size of the switching elements and cannot generally be reduced. However, it is useful to reduce the width of the perfect shuffle layout by reducing the number of yz-bundles.

Remark 1: Due to the symmetry of the perfect, we are focusing attention only on the top half of the input nodes, i.e., nodes 0,...,N/2-1.

When the number of metal levels, m, is even we divide the available levels of metal into two disjoint sets, Γ_0 and Γ_2. As is proven in [8], the perfect shuffle connections for inputs 0 to m-1 will fit in a single yz-bundle using m/2 metal levels. All of the remaining wire segments in the y direction can be placed in yz-bundles containing m/2 wires each. There are N/2 connections for one half of the perfect shuffle. Therefore the minimum of the total number of yz-bundles when m is even, is given by[3]

$$
\begin{aligned}
V_{min} &= 1 + \lceil (N/2 - m)/(m/2) \rceil \\
&= \lceil N/m \rceil - 1 \qquad (5)
\end{aligned}
$$

If the number of levels of interconnect, m, is odd, we choose one level of interconnect γ, to be separate from our sets Γ_0 and Γ_1 , which are chosen as for the even case. The use of the shared level of interconnect γ, depends upon a property of the perfect shuffle: inputs 0 through (N/4-1) do not connect to outputs N/2 through N-1, so that the shared level of interconnect γ, can be used both for input connections 0 through (N/4-1) and 3N/4 through N-1, giving us the (m+1)/2 metal levels to lay out each half-shuffle in these regions. There will be 1 + [((N/4)-(m+1))/((m-1)/2 + 1)] yz-bundles for these connections, since in these regions the layout is the same as if there were m + 1 metal levels. The connections for inputs N/4 through N/2 - 1 will have yz-bundles containing (m-1)/2 wires each, so that there will be [(N/4)/((m-1)/2)] bundles for these connections. Therefore, the total number of yz-bundles when m is odd is given by

$$
\begin{aligned}
V_{min} &= 1 + \lceil [N/4 - (m+1)]/[(m-1)/2 + 1] \rceil + \lceil (N/4)/[(m-1)/2] \rceil \\
&= 1 + \left\lceil \frac{N - 4(m+2)}{2(m+1)} \right\rceil + \lceil N/[2(m-1)] \rceil \qquad (6)
\end{aligned}
$$

Algorithm 1 demonstrates the layout methodology for the case where m is even. The algorithm for the case where m is odd is similar, except that metal level γ is used for connections 0 through N/4-1, and 3N/4 through N-1.

Algorithm 1:
Layout of a Perfect Shuffle Using an Even Number of Levels of Wiring

BEGIN

Step 1: Assign the m levels of interconnect to two equal-sized sets, Γ_0 and Γ_1. Use the interconnect levels in set Γ_0 for inputs 0 through (N/2) - 1, and use set Γ_1 only for the other input-output connections.

Step 2: Create a rectangular grid of height N and width $\lceil ((N/m) - 1) + 2 \rceil$. Number the rows and columns of the grid as shown in Fig. 4. The leftmost column represents input terminals for the shuffle, while the rightmost column represents the outputs. The other columns are used to represent yz-bundles.

---

[3] The notation $\lceil x \rceil$ means the smallest integer greater than or equal to x.

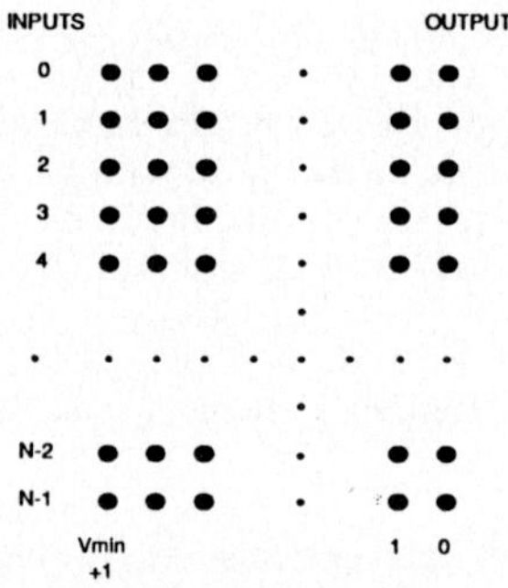

Fig. 4. Grid for layout of perfect shuffle.

Step 3: Let i be an input terminal. Draw the wire segments for each input by the following iterative processes.

$$p(i) = \begin{cases} 0, & \text{if } i = 0 \\ 1, & \text{if } i \le m \\ \lceil 2i/m \rceil - 1, & \text{for } i > m \end{cases}$$

for $0 \le i \le (N/2)-1$
{ connect input i at point $(\lceil N/m \rceil, i)$ to point $(p(i), i)$;
connect point $(p(i), i)$ to point $(p(i), 2i)$;
connect point $(p(i), 2i)$ to point $(0, 2i)$;
Select the level of interconnect for the next input-output connection from set Γ_0;
}

$$q(i) = \begin{cases} 0, & \text{if } i = N-1 \\ 1, & \text{if } N-1-m \le i \le N-2 \\ \lceil 2(N-1-i)/m \rceil - 1, & \text{for } i < N-m \end{cases}$$

FOR $N - 1 \ge i \ge (N/2)$
{ connect input i at point $(\lceil N/m \rceil, i)$ to point $(q(i), i)$;
connect point $(q(i), i)$ to point $(q(i), 2i + 1 - N)$;
connect point $(q(i), 2i + 1 - N)$ to point $(0, 2i + 1 - N)$;
Select the level of interconnect for the next input-output connection from set Γ_1;
}

END

If the preceding algorithm is used, the width of the perfect shuffle interconnection pattern is

$$W = A + B + V_{min} M + (V_{min} - 1)S \tag{7}$$

where A and B are the distances between the shuffle pattern and the preceding and succeeding switching stages, respectively; M is the width of the metal wires, S is the spacing between the wires, and V_{min} is as given in (5) or (6) depending upon whether m is even or odd. Table 1 compares the widths of perfect shuffle layouts for different numbers of metal levels. For this example, M and S are both 4λ, where λ is a scaling factor.

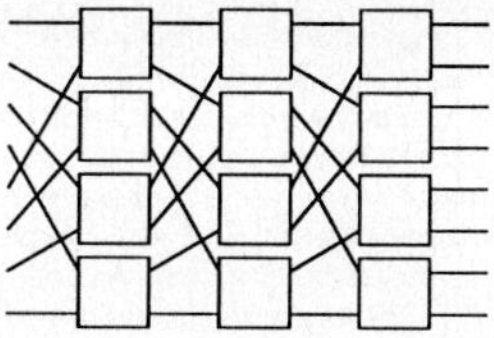

Fig. 5. Omega network for N = 8.

Table 1
Comparison of shuffle layout widths with multilevel metal versus two-level metal

N	m	V_{min}	(W-A-B)
64	2	31	244λ
64	3	22	172λ
64	4	15	116λ
64	5	12	92λ
64	6	10	76λ

Fig. 5 shows an example of the omega network. The omega network consists of $\log_2 N$ stages. Each stage consists of a perfect shuffle connection pattern followed by a set of N/2 switches. A logical schematic for each switch is shown in Fig. 6. The omega network differs from many other INNETs by having switches with four states. Due to this difference, two switch state select lines are required for each switch.

Fig. 7 shows a possible layout for an omega network were N = 8 using four levels of metal for the perfect shuffle patterns. Note that the height of the perfect shuffles is determined by the height of the switches since of the inputs and outputs of the shuffles must come from the switches.

Fig. 8 illustrates the benefits of using more than two levels of metal for layout of the perfect shuffle. Naturally, we note that the width of the perfect shuffle patterns

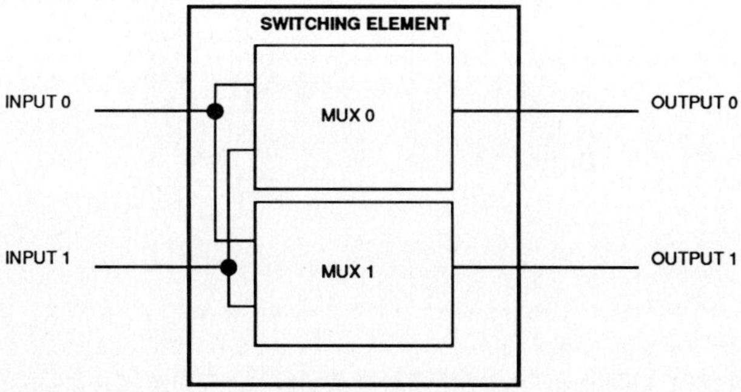

Fig. 6. Schematic for a switching element for the omega network.

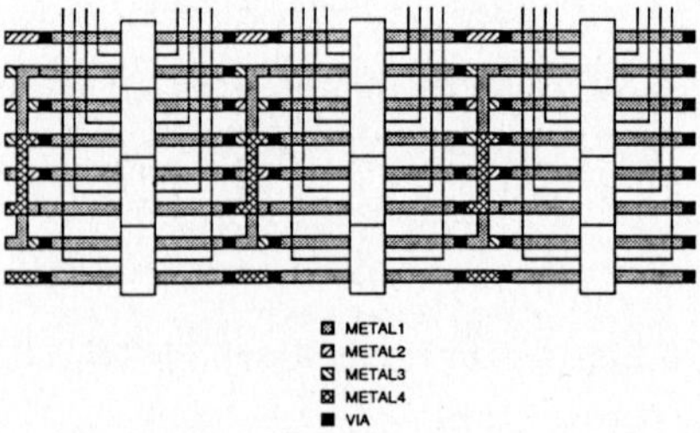

Fig. 7. Layout for an omega network with 4 levels of metal.

decreases as we increase the number of metal levels used. Also, the area required for the control lines is reduced for each increase in metal levels.

IV. OTHER MULTISTAGE INNETS

Other multistage INNETS such as the indirect binary n-cube and the data manipulator can also benefit from the use of multiple metal levels. In [8], we present algorithms for these other networks. The benefits of using multilevel metal are as follows: if three levels of metal are used for layout, the data manipulator connection pattern can be reduced by a factor of 2, if five level of metal are used, the width is reduced by a factor of 2.67. The benefits for the indirect binary n-cube vary with each stage, since the connection pattern is different for each stage. However, in general the benefits are similar to those for the perfect shuffle.

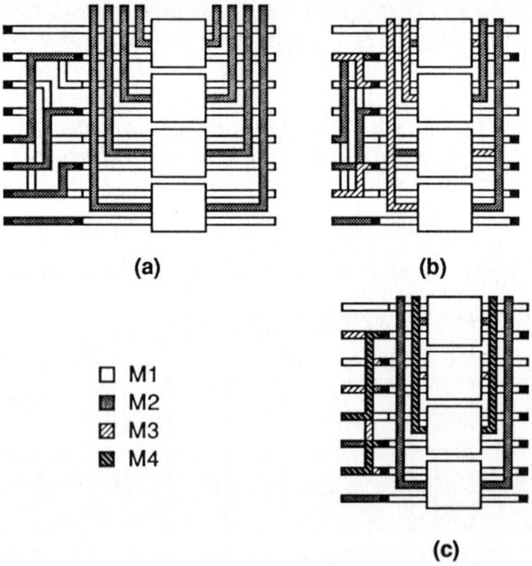

Fig. 8 Examples of layout of a single stage of the omega network where N = 8, and (a) m = 2, (b) m = 3, (c) m = 4.

V. CONCLUSIONS

ULSI/WSI designs are moving toward three dimensional fabrication and hold the potential for almost unlimited computational power, while simultaneously minimizing the usual problems of inter-chip and inter-board connections. We have presented algorithms for area-efficient layout of multistage INNETs using multilevel metal for interconnects. The use of multilevel metal substantially decreases the width of the exchange stages. For example, we find that the width of the perfect shuffle interconnection pattern is reduced by a factor of 1.42 if three levels of metal are used, or by a factor of 2.1 if four levels of metal are available.

Although the emphasis of the paper has been on multiple levels -three or higher- the algorithms described are just as applicable to the traditional two-level metal wiring. Indeed, these algorithms provide systematic procedures for efficient layout for two-level metal INNETs.

REFERENCES

[1] D. H. Lawrie, "Access and Alignment of Data in an Array Processor," IEEE Transactions on Computers, vol. C-24, no. 12, December 1975, pp. 1145-1155.

[2] N. Majid and J. F. McDonald, "Combining Wafer Scale Integration (WSI) and Wafer Scale Hybrid Packaging (WSHP) to Form a Three Dimensionally Stackable Architecture," IEEE International Workshop on Defect and Fault Tolerance in VLSI Systems, 1988.

[3] B. J. Donlan, G. F. Taylor, R. H. Steinvorth, A. S. Bergendahl, and J. F. McDonald, "Wafer Scale Integration Using Discretionary Microtransmission Line Interconnections," in Wafer Scale Integration, ed.s C. Jesshope and W. Moore, 1986.

[4] T. E. Mangir and A. Avizienis, "Fault-Tolerant Design for VLSI: Effect of Interconnect Requirements for Yield Improvement of VLSI Designs," IEEE Transactions on Computers, vol. C-31, no. 7, July 1982.

[5] I. Koren and M. A. Breuer, "On Area and Yield Considerations for Fault-Tolerant VLSI Processor Arrays," IEEE Transactions on Computers, vol. C-33, no. 1, January 1984, pp. 21-27.

[6] A. J. Rushton and C. R. Jesshope, "The Reconfigurable Processor-Array -- An Architecture in Need of WSI," in Wafer Scale Integration, ed.s C. Jesshope and W. Moore, 1986.

[7] J. A. B. Fortes and C. S. Raghavendra, "Dynamically Reconfigurable Fault-Tolerant Array Processors," FTCS-14, pp. 386-392.

[8] J. M. Wills and V. K. Jain, "Multilevel Wiring Algorithms for Interconnection Networks," submitted to IEEE Transactions on Computers.

Wafer Scale Integration, III
M. Sami and F. Distante (Editors)
Elsevier Science Publishers B.V. (North-Holland)

An Architecture-Independent Model for Partitioning and Redundancy in Wafer-Scale Integrated Circuits

Martin F. Beusekamp

University of Twente, Department of Computer Science
P.O. Box 217, 7500 AE Enschede, The Netherlands

1. Introduction.

In Wafer-Scale Integration (WSI), redundancy is used to implement circuit sizes which are significantly larger than state-of-the-art VLSI-techniques permit, and still achieve an acceptable yield. Many models [e.g. 1,2,3] have been published to calculate and optimize the effect of redundancy in more or less specific architectures. This reflects the general policy of implementing redundancy by duplicating certain parts of the micro-architecture of the circuit under design, without paying appropriate attention to the actual sizes of the duplicated sub-circuits and the amount of redundancy which is implemented by this duplication [e.g. 4,5].

This paper, however, will present an architecture-independent model to calculate the optimal partitioning and the required amount of redundancy to achieve a maximum efficiency. It takes several technological parameters into account. At the end, it is indicated how this model can be adapted to any specific architecture. In the appendix, a mathematical substitution is presented, which makes the model completely analytical. This means that by derivation with respect to its parameters, the influence of these parameters can very easily be investigated.

2. List of symbols.

The following symbols will be used in this paper. In the text, the first occurrence of a symbol is printed in boldface.

a	$= A_c/A_{c,50}$
A_c	= area of desired circuitry, if processed in a conventional way
$A_{c,50}$	= area of circuitry which will show a yield of 50% in the given technological process
a_{crit}	= circuit area from which the implementation of partitioning and redundancy becomes favourable
A_r	= redundant area
A_t	= total area
A_w	= area needed for additional wiring and reconfiguration switches
$A_{w,50}$	= area of wiring which will show a yield of 50% in the given technological process
b	$= A_{w,50}/A_{c,50}$

E	= efficiency $Y.(A_c/A_t)$
E_{max}	= maximum efficiency at a given value of k
i	= general integer
k	= number of required basic cells
n	= number of available basic cells
w	= parameter characterizing the overhead of additional wiring and reconfiguration switches
y,y(k)	= single-cell yield
Y	= overall yield
y_c	= yield of circuit part of a single cell
y_w	= yield of wiring part of a single cell

3. The partitioning and redundancy model.

3.a. General assumption.

The main purpose of the model to be presented in this paper, will be to gain insight in the optimal rate of partitioning *any* desired circuit into sub-circuits and the application of such an amount of redundancy that a maximum efficiency is reached. In order to derive a truly architecture-independent model it will, for the time being, be assumed that the desired (large) circuit can be partitioned into any (positive integer) number **k** of identical cells, each having a probability **y** of functioning correctly. In general, this will be an unrealistic assumption, but in section 4 it will be shown that the model can easily be adapted to any specific, realistic architecture by merely adjusting some parameters and limiting the possible solutions to restricted, architecture-dependent values.

3.b. The basics of the model.

In case of a conventional chip without redundancy, consisting of k identical cells with a probability y of functioning correctly and with randomly distributed defects on the wafer, the overall yield **Y** will obviously be

$$Y = y^k \qquad (1)$$

which will be low for k more than a few units, unless y is very close to unity. It is in fact this exponential behaviour which causes the upper limit to the size of conventional VLSI-circuits. In WSI, this problem is solved by implementing redundancy. Let us assume that a "stock" of **n** ($n \geq k$) of the above-mentioned basic cells is available. The probability of finding *at least* k good cells in a stock of n cells is well-known from the basic theory on statistics and is also mentioned in [6]

$$Y = \sum_{i=0}^{n-k} \binom{n}{i} y^{n-i} (1-y)^i \qquad (2.1)$$

$$= n! \sum_{i=0}^{n-k} \frac{y^{n-i} (1-y)^i}{(n-i)!\, i!} \qquad (2.2)$$

A simple example can illustrate the strength of the concept of redundancy. With a single-cell yield $y = 0.9$, a required number of cells $k = 20$ and an available stock of cells $n = 25$, expression (1) leads to a yield of $Y = 0.122$, while (2.2) leads to $Y = 0.967$, a dramatic increase with an investment of only 25% of redundancy.
It is, however, not the yield which is the most important cost factor, but the overall-efficiency **E**. E is defined as the yield divided by the area penalty which has to be paid for implementing redundant circuitry and additional wiring plus reconfiguration switches

$$E = Y.(A_c/A_t) \tag{3}$$

with A_c denoting the circuit area as would be needed in a conventional IC, and A_t denoting the total chip area including redundancy and additional wiring plus reconfiguration switches.
Now, let $\mathbf{A_{c,50}}$ denote that specific area of circuitry which will have a yield of 0.5 or 50% in the given technological process. $A_{c,50}$ is, of course, directly related to the average defect density of the process. Let **a** be the ratio between A_c and $A_{c,50}$

$$a = A_c/A_{c,50} \tag{4}$$

Clearly, the overall yield Y of such a circuit, processed in the conventional way, would then be

$$Y = (0.5)^a = 2^{-a} \tag{5}$$

Let us now partition the desired circuit into k identical cells. Each of these cells will then have a single-cell yield of

$$\mathbf{y_c} = 2^{-a/k} \tag{6}$$

Partitioning a circuit into *two* identical sub-circuits will generally introduce a "lane" between them, required for additional wiring and reconfiguration switches. It will therefore be realistic to assume that partitioning a circuit into k smaller sub-circuits will require an additional area $\mathbf{A_w}$ for wiring and reconfiguration switches which is proportional to $^2\log(k)$

$$A_w = \mathbf{w}.{}^2\log(k).A_c \tag{7}$$

This means that every step of partitioning doubling the total number of required sub-circuits will add an equal amount of re-wiring area, captured in the parameter w. In practice, values for w of 0.1...0.5 may be expected. The inclusion of this re-wiring area will lead to a new total area needed for the circuit

$$A_c + A_w = [1 + w.{}^2\log(k)].A_c \tag{8}$$

The area of the wafer which is only occupied by additional wiring and switches will generally have a higher yield than the same area of circuitry, because fewer critical technological processing steps have been performed on it. Let $\mathbf{A_{w,50}}$ be that area for wiring and switches with a yield of 0.5 or 50% and let

$$\mathbf{b} = A_{w,50}/A_{c,50} \tag{9}$$

In practice, b will be in the order of 1...2. As a consequence, each of the k basic cells will require an additional amount of wiring and switches with a yield of

$$y_w = (0.5)^{w.\,{}^2\log(k).a/bk} \tag{10.1}$$

$$= 2^{-(wa/bk).\,{}^2\log(k)} \tag{10.2}$$

$$= k^{-wa/bk} \tag{10.3}$$

resulting in a total yield for each cell of

$$y(k) = y_c y_w \tag{11.1}$$

$$= 2^{-a/k}.k^{-wa/bk} \tag{11.2}$$

$$= (2.k^{w/b})^{-a/k} \tag{11.3}$$

Introducing redundancy can easily be done by substituting the constant y in (2.2) by the k-dependent y(k) of (11.3). The redundant area A_r of the complete circuit will be n/k - 1 times the area needed for circuitry, additional wiring and reconfiguration switches. Therefore, the total area occupied will now be

$$A_t = A_c+A_w+A_r = (n/k).(A_c+A_w) = [1+w.{}^2\log(k)].A_c.(n/k) \tag{12}$$

This leads to an efficiency according to (3)

$$E = Y.(A_c/A_t) = \frac{n!\sum_{i=0}^{n-k}\dfrac{y(k)^{n-i}\,[1-y(k)]^i}{(n-i)!\ i!}}{[1+w.{}^2\log(k)].(n/k)} \tag{13.1}$$

$$= \frac{k.(n-1)!}{[1+w.{}^2\log(k)]}\sum_{i=0}^{n-k}\frac{y(k)^{n-i}\,[1-y(k)]^i}{(n-i)!\ i!} \tag{13.2}$$

with y(k) as in (11.3).

For every k, there exists an n ($k \leq n < \infty$) for which the efficiency E shows a maximum E_{max}. The problem is to find the maximum of the maxima, for varying k, given the technological parameters w and b and the circuit size a. This can be done by a simple search algorithm. The result will be an optimal rate of partitioning, the required redundancy n/k to achieve the maximum efficiency E_{max}, the yield Y which will be achieved at maximum efficiency and the required total area A_t including the area needed for redundancy and re-wiring.

3.c. Results of the model.

Let us investigate expression (13) by assuming the technological parameters to have constant and realistic values of w = 0.3 and b = 1.5

a	k	n	n/k	Y	A_t/A_c	E_{max}
0.125	1	1	1.00	0.9170	1.00	0.9170
0.25	1	1	1.00	0.8409	1.00	0.8409
0.5	1	1	1.00	0.7071	1.00	0.7071
1.0	1	1	1.00	0.5000	1.00	0.5000
2	26	32	1.23	0.9670	2.97	0.3260
4	62	74	1.19	0.9812	3.33	0.2950
8	135	158	1.17	0.9863	3.66	0.2698
16	312	358	1.15	0.9938	4.00	0.2485
32	673	763	1.13	0.9955	4.33	0.2300

Table 1. Optimal rates of partitioning and redundancy for various amounts of required circuitry (a), at given values for the technological parameters w = 0.3 and b = 1.5.

It can be seen that (in this case) for $a \leq 1$ both partitioning of the circuit as well as application of redundancy are not yet required and in fact undesirable, as the maximum efficiency is achieved for $k = n = 1$. For $a \geq 2$, a rather large and increasing rate of partitioning (k) leads to optimal results, with a gradually decreasing rate of redundancy (n/k). As could be expected, the required total area A_t/A_c increases, and the overall-efficiency E decreases with increasing a. It is surprising, however, that for circuit sizes ranging from a = 2 to a = 32, the maximum efficiency only decreases a factor of 1.42 (0.3260/0.2300).

The fact that for small values of a the application of redundancy is undesirable can be explained by looking at a graph of E as a function of n (or n-k), for various k as depicted in figure 1.

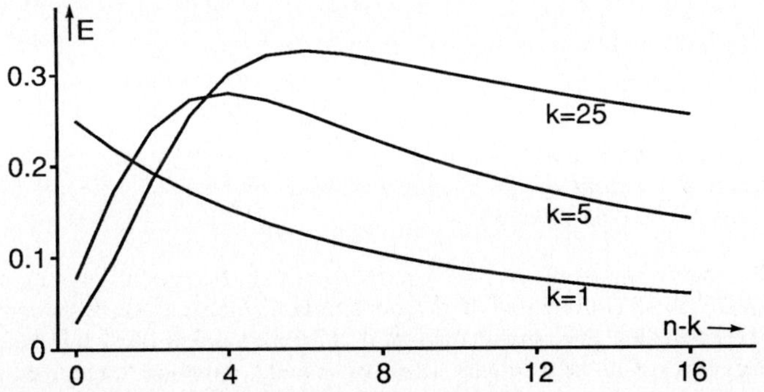

Figure 1. An example of the behaviour of the efficiency E as a function of the available number of cells (n), with the required number of cells (k) as a parameter.

Obviously, for small values of k, meaning large basic cell areas, the area penalty which has to be paid by including redundancy is more severe than the enhancement of the yield which will be achieved. This also explains the behaviour of Y as a function of a. For relatively small circuits ($a \le 1$), no redundancy is implemented and the yield Y will obviously decrease with increasing circuit size a. For larger circuits, the application of partitioning and redundancy causes an important enhancement of the yield, at maximum efficiency, at the cost of not too much area, because a large k means small basic cells and there are not many redundant cells needed to enhance the yield significantly, as was already shown in a numerical example.

Special attention should be given to the fact that there apparently exists a critical circuit-area a_{crit}, below which the partitioning of the circuit is undesirable, but above which a significant degree of partitioning is required to reach the maximum efficiency. This can be explained by looking at a graph of the maximum efficiency E_{max} as a function of the number k of required cells, for various values of a, as depicted in figure 2.

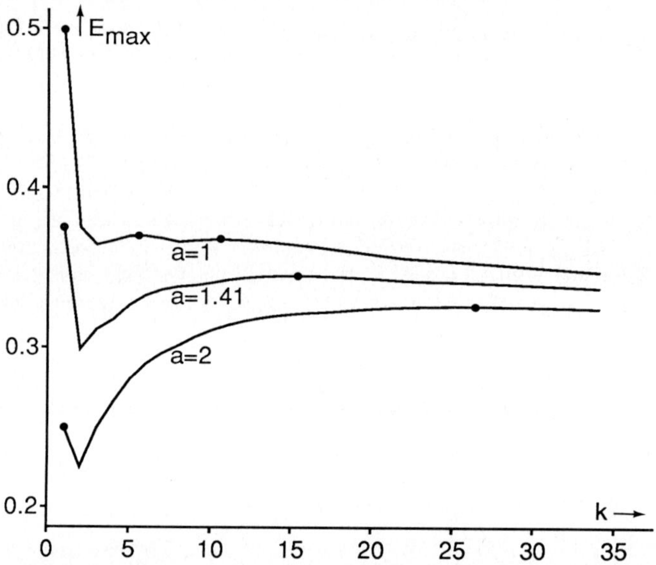

Figure 2. An example of the behaviour of the maximum efficiency E_{max} around $a = a_{crit}$, as a function of the required number of cells (k), with the desired circuit area (a) as a parameter.

As can be seen, the graphs of figure 2 may show more than one (local) maximum, indicated in the figure with dots. The heights of these maxima all depend on the circuit size a, but with different sensitivities. For small values of a, the maximum at $k = 1$ will be the only, or at least the highest one. At a given value of $a = a_{crit}$, another local maximum, at a significantly larger value of k, will be equally high as the maximum at $k = 1$. From this specific a_{crit} upwards, partitioning of the circuit and the application of redundancy will be favourable.

3.d. Influence of the technological parameters w and b.

Of course, all results as given in table 1 depend on the technological parameters w and b. As an illustration, table 2 gives the results calculated for a given a = 8 and for various values of w and b. As can be seen, the required total area as well as the maximum efficiency strongly depend on w, but not so much on b.

w	b	k	n	n/k	Y	A_t/A_c	E_{max}
0.1	1.0	193	211	1.09	0.9912	1.92	0.5154
0.1	1.5	174	190	1.09	0.9920	1.90	0.5208
0.1	2.0	164	179	1.09	0.9924	1.89	0.5238
0.3	1.0	183	213	1.16	0.9902	3.79	0.2614
0.3	1.5	135	158	1.17	0.9863	3.66	0.2698
0.3	2.0	117	137	1.17	0.9863	3.58	0.2752
0.5	1.0	234	277	1.18	0.9911	5.84	0.1697
0.5	1.5	167	199	1.19	0.9908	5.59	0.1772
0.5	2.0	130	156	1.20	0.9870	5.41	0.1823

Table 2. Illustration of the effect of the technological parameters w and b on partitioning, redundancy, area and maximum efficiency for a given desired circuit area a = 8.

Also, the critical circuit area a_{crit}, at which the maxima at k = 1 and at k > 1 in figure 2 have equal heights, depends more on w than on b, as is illustrated in table 3.

w	b	a_{crit}
0.1	1.0	0.71
0.1	1.5	0.70
0.1	2.0	0.69
0.3	1.0	1.62
0.3	1.5	1.57
0.3	2.0	1.52
0.5	1.0	2.27
0.5	1.5	2.19
0.5	2.0	2.13

Table 3. Illustration of the effect of the technological parameters w and b on the critical circuit area a_{crit}.

4. Application to specific architectures.

The model which was presented in the previous section, was put in general terms and was not based on any specific architecture. It was assumed that a circuit could be partitioned into any number of cells, that all these cells were identical and that the circuit could be reconfigurated in such a way that any faulty cell could be replaced by any good cell. Of course, this will generally not be true in practice. Nevertheless, the model can be used for a wide variety of architectures, provided the following adaptations are being made.

1) It is, of course, not always possible to partition a specific circuit into any number of sub-circuits. The micro-architecture will often cause a preference for certain numbers of cells, for instance powers of two. In practice, this means that not all natural numbers should be allowed as possible solutions of k, but that solutions of k should be restricted to a limited set of integer values fitting the micro-architecture of the specific circuit under development. This will not influence the previously mentioned results significantly, because the efficiency of the optimal partitioning will not show a very sharp maximum (see figure 2). This can, for instance, be illustrated by table 4, which shows the calculated results in the case of a = 8, around the maximum efficiency, which will occur at k = 135 (w = 0.3, b = 1.5, see table 1).

k	n	n/k	Y	A_t/A_c	E_{max}
136	159	1.17	0.9862	3.65	0.26982729
135	**158**	**1.17**	**0.9863**	**3.66**	**0.26982730**
134	157	1.17	0.9863	3.66	0.26982559
↓	↓	↓	↓	↓	↓
128	151	1.18	0.9866	3.66	0.26977643

Table 4. The effect of small variations in k around the value of maximum efficiency (a = 8, w = 0.3, b = 1.5).

As can be seen from the above-mentioned and other experiments, a slight variation in k around maximum efficiency will not significantly influence the total area, efficiency and optimal ratio of redundancy.

2) From table 1 it can be seen that a circuit with a size of, for instance, a = 32 has to be partitioned into 673 sub-circuits to give an optimal efficiency. In practical situations, this number may be considered rather large. Table 5 shows the effect of rounding k to the nearest lower and subsequently lower powers of two (w = 0.3, b = 1.5). From this table it can be seen that, in this case, solutions down to, say, k = 128 still show an acceptable efficiency.

3) So far, it was assumed that the desired circuit could be partitioned into any number of identical cells. This may be acceptable for very regular structures, like memories and systolic arrays, but not for all types of circuits. The model of this paper may therefore predominantly be useful for regularly structured architectures. Nevertheless, the architecture of almost any circuit allows a distinction of large functional blocks with a regular structure. Applying the model on each of these blocks separately, will also give a (nearly) optimal solution for the complete circuit.

k	n	n/k	Y	A_t/A_c	E_{max}
673	763	1.13	0.9955	4.33	0.2300
512	601	1.17	0.9951	4.34	0.2291
256	342	1.34	0.9895	4.54	0.2178
128	215	1.68	0.9795	5.21	0.1881
64	160	2.50	0.9618	7.00	0.1374
32	158	4.94	0.9319	12.34	0.0755
16	250	15.63	0.8834	34.37	0.0257

Table 5. The effect of decreasing the rate of partitioning on required redundancy, area, yield and efficiency ($a = 32$, $w = 0.3$, $b = 1.5$).

Moreover, there are three reasons causing the area of a wafer-scale integrated circuit to be significantly larger than that of a VLSI-circuit. Firstly, a WSI-circuit will by nature incorporate a certain amount of redundant circuitry. Secondly, a relatively large area will be occupied by the implementation of the additional wiring and switches needed to (re-)configurate the circuit during the initial test or any power-up routine. Thirdly, the development of WSI-circuits will take some time, during which regular VLSI-circuits will continue to develop as well. In order to produce economically competitive WSI-circuits, one will have to aim at circuit sizes which are at least half an order of magnitude larger than state-of-the-art VLSI-circuits. The size of realistic WSI-circuits will therefore be of such a magnitude that they will inevitably *have* to be more or less regular in order to be designed by presently available CAD-tools.

4) In the model, it was assumed that any faulty cell could be replaced by any spare cell. In a practical situation, this will generally not be true. A faulty cell will have to be replaced by a cell in its close vicinity. Often, the implementation provides for complete rows and/or columns of spare cells, and reconfiguration is performed by implementing the proper "sidesteps". To the model, this means that n can not adopt any integer value $k \leq n < \infty$, but n will be restricted to certain architecture-dependent values. These restricted values of n can, just like the above-mentioned restricted values of k, be incorporated into the algorithm of the model, which will then find the optimal solution under the additional constraints. Moreover, all restrictions considering reconfiguration wiring and switches can be accounted for in the parameters w and b.

5) The model as it is presented in this paper, assumes a random distribution of faults. In practice, however, a clustering of faults is often observed and the fault density is often found to depend on the absolute position on the wafer. It should be emphasized that any mathematical model, describing these phenomena, can easily be incorporated into the partitioning and redundancy model.

Conclusions and summary.

A model was presented which calculates the optimal degree of partitioning, rate of redundancy, total area, yield and efficiency for any desired (large) circuit. It was shown that there exists a critical circuit size, below which the implementation of partitioning and redundancy is undesirable. Above this critical circuit size, partitioning into rather small cells leads to optimal efficiency, in spite of the fact that small cells require a considerable overhead of reconfiguration wiring and switches. This also means that the implementation of very large redundant cells or blocks will cost more area than the enhancement of efficiency justifies. It was shown that the model can be adapted to any specific architecture by adjusting parameters and limiting possible solutions to restricted, architecture-dependent values. By a simple mathematical substitution, the model can be made completely analytic, making it very simple to investigate the sensitivity of the optimal solution to small variations in parameters.

Appendix.

In expression (13), the faculty functions may cause some difficulties, due to the rapid increase of n! with increasing n. In many applications, the numeric value of n! will exceed the range of the integer or even real variables on the used computing instrument. The calculated results, however, are in a reasonable range, as n! is divided by (n-i)!(i)! in all cases, due to the binomial origin of the faculty functions. Out-of-range-problems can therefore be avoided by calculating with the logarithms of the faculty functions.
On the other hand, these faculty functions are the only functions causing (13) to be non-analytic. In order to take the derivative of (13) with respect to its parameters, for sensitivity calculations, it may be found very useful to substitute n! by its well-known Stirling-approximation [e.g. 7,8].

$$n! \approx (n/e)^n . \sqrt{(2\pi n)} . C(n) \tag{A.1}$$

This approximation is based on the theory of gamma-functions. Its derivation is beyond the scope of this paper. C(n) is a correction factor, consisting of a series expansion of n^{-1}

$$C(n) = 1 + (C_1/n) + (C_2/n^2) + (C_3/n^3) + (C_4/n^4) + \ldots \tag{A.2}$$

$C_1 \ldots C_r$ are the results of first order Bernouilli-functions. In most publications, this correction factor is truncated after the fifth term ($C_r = 0$ for $r > 4$). The numeric values of $C_1 \ldots C_4$ are

C_1	=	1/12	=	$8.3333333.10^{-2}$
C_2	=	1/288	=	$3.4722222.10^{-3}$
C_3	=	-139/51840	=	$-2.6813271.10^{-3}$
C_4	=	-571/2488320	=	$-2.2947209.10^{-4}$

Table A.1. Numeric values of the coefficients in the correction factor C(n).

Let SLA ("Stirling Low Accuracy") be the Stirling approximation of n! with C(n)=1 (meaning $C_r = 0$ for $r > 0$), let SMA ("Stirling Medium Accuracy") be the Stirling approximation of n! with C(n) = 1 + (1/12n) (meaning $C_r = 0$ for $r > 1$), and let SHA ("Stirling High Accuracy") be the Stirling approximation of n! with C(n) truncated after the fifth term (meaning $C_r = 0$ for $r > 4$). Then, the accuracies as summarized in table A.2 for

the approximations are obtained. For $n \rightarrow \infty$, n!/SLA, n!/SMA and n!/SHA all approach unity.
Obviously, the SLA-approximation has the disadvantage of an unacceptably low accuracy for low values of n. The SHA-approximation on the other hand, provides an excellent accuracy (for $n > 0$), but will be cumbersome to differentiate, due to the relatively high number of terms in the correction factor C(n). In most applications, the SMA-approximation will show a useful balance between both.

n	n!/SLA	n!/SMA	n!/SHA
0	∞	∞	∞
1	1.08444	1.001019	1.00050078
2	1.04221	1.000519	1.00002102
3	1.02806	1.000279	1.00000300
4	1.02101	1.000171	1.00000073
5	1.01678	1.000115	1.00000024
6	1.01397	1.000083	1.00000010

Table A.2. Accuracies of the Stirling approximation of n! with several numbers of terms in the correction factor C(n) taken into account.

The difficulty of the infinite error arising when calculating 0! can be overcome by recognizing that n! = (n+1)!/(n+1). Combining this with (A.1) and (A.2) where $C_r = 0$ for $r > 1$ yields

$$n! \approx \sqrt{(2\pi/e^3)} \cdot [(n+1)/e]^{n-1/2} \cdot (n+{}^{13}/_{12}) \qquad (A.3)$$

This analytic approximation of n! shows an error ranging from 0.1% for n = 0 to less than 0.01% and approaching zero for n > 5.

References.

[1] Jim C. Harden, Noel R. Strader II. "Architectural Yield Optimization for WSI". IEEE Transactions on Computers, vol. 37, no. 1, January 1988, pp. 88-110.

[2] Hans-Jürgen M. Iden. "On the Optimization of Hierarchical Redundancies Including Configuration Nets and Switches". Earl Swartzlander and Joe Brewer (editors), Proceedings of the First International Conference on Wafer-Scale Integration, IEEE Society Press, ISBN 0-8186-9901-9, January 1989, pp. 173-182.

[3] Tom Leighton, Charles E. Leiserson. "A Survey of Algorithms for Integrating Wafer-Scale Systolic Arrays". Gabrièle Saucier and Jacques Trilhe (editors), Proceedings of the IFIP WG 10.5 Workshop on Wafer Scale Integration, North Holland, ISBN 0-444-70103-6, March 1986, pp. 177-195.

[4] François Blayo, Philippe Hurat. "A Reconfigurable WSI Neural Network". Earl Swartzlander and Joe Brewer (editors), Proceedings of the First International Conference on Wafer-Scale Integration, IEEE Society Press, ISBN 0-8186-9901-9, January 1989, pp. 141-150.

[5] R.M. Lea. "A WSI Image Processing Module". Gabrièle Saucier and Jacques Trilhe (editors), Proceedings of the IFIP WG 10.5 Workshop on Wafer Scale Integration, North Holland, ISBN 0-444-70103-6, March 1986, pp. 43-58.

[6] Ben Warren, Wayne Richardson, Keigi Kanegawa, Cliff Arnell. "A One Megabit SRAM Fabricated with 1.2 μm Technology". Earl Swartzlander and Joe Brewer (editors), Proceedings of the First International Conference on Wafer-Scale Integration, IEEE Society Press, ISBN 0-8186-9901-9, January 1989, pp. 47-53.
[7] I.N. Sneddon (editor). "Encyclopaedic Dictionary of Mathematics for Engineers and Applied Scientists". Pergamon Press, Oxford, ISBN 0-08-016767-5, 1976, First edition, p. 641.
[8] Granino A. Korn, Theresa M. Korn. "Mathematical Handbook for Scientists an Engineers". McGraw-Hill Book Company, New York, 1961, pp. 698-699.

Wafer Scale Integration, III
M. Sami and F. Distante (Editors)
Elsevier Science Publishers B.V. (North-Holland)

WAFER SCALE INTEGRATION FOR SIGNAL PROCESSING

Earl E. Swartzlander, Jr.

TRW Defense Systems Group
Redondo Beach, CA 90278

For a given technology, Wafer Scale Integration is faster and consumes less power. Thus WSI offers the potential to greatly improve the efficiency of signal processors. Package pin constraints represent a potential bottleneck for WSI systems. This paper identifies the problem and shows that systolic arrays require fewer pins than mesh or hypercube architectures. An example signal processor, an FFT, that has been implemented in VLSI and is now being implemented with WSI is described to illustrate the similarities, differences, and advantages of WSI.

INTRODUCTION

Wafer scale integration (WSI) offers the promise of implementing signal and data processors with significantly higher data rates, lower power, and smaller size than conventional VLSI implementations. WSI implementations eliminate most of the signal lines between chips with intra-wafer lines that exhibit significantly less capacitance so they may be driven at higher rates. With the greatly reduced loading, WSI buffer and driver circuits also consume less power than their VLSI counterparts. Thus WSI is much more efficient than VLSI for signal processing.

Packaging Constraints

Currently it is reasonable to fabricate a four inch chip carrier with contacts spaced on 25 mil centers along all four sides which will provide 640 contacts. Alternatively, a four inch square pin grid array with pins on 100 mil centers gives 1600 pins, although there may be difficulty in routing to pins near the center of the package. These considerations suggest an upper limit of about 1000 pins at present.

Another issue that arises with WSI signal processing is the need to be able to connect multiple wafers into regular architectures while using a limited number of package pins. The first part of this paper examines three well known array processor architectures to determine their package pin requirements.

Redundancy

In addition to many issues in fabrication technology, WSI raises a critical issue in processor architecture. WSI needs to employ redundancy to overcome the defects implicit in the fabrication process. There has been much reported on this problem and many potential solutions have been presented in conference proceedings and books [1-6] and journal papers. In the example signal processor that is described in the last section of this paper, functional sparing is employed to produce reasonable predicted yields.

ARCHITECTURE FOR WSI

Regular architectures that appear to be well suited for wafer scale systems are one dimensional systolic arrays, square meshes, hypercubes (i.e., n-dimensional cubes), etc. An important aspect of an architecture is the relationship between the amount of computation and the number of pins required for signals that enter and exit from the wafer. In considering the pin requirements, we assume a system that requires several wafers to achieve the required throughput. In the analysis that follows, the pin requirements for one-dimensional systolic arrays, square meshes, and hypercubes are evaluated. Parallel data ports are assumed as typical for current technology (i.e., CMOS), in the future bit serial data ports may be employed to reduce the pin requirements, but unless extremely high data rates can be achieved serial data ports may severely constrain the system performance.

Systolic Array

Systolic arrays consist of processing elements connected in series, so that a one dimensional systolic array requires only two data ports (i.e., an input port and an output port) irrespective of the number of processors [7, 8] Figure 1 shows a systolic array which consists of five processing elements (denoted by PE in the Figure) connected in series. In addition to this simple one dimensional array illustrated here, two dimensional square, triangular, and hexagonal systolic arrays have been developed for specific applications.

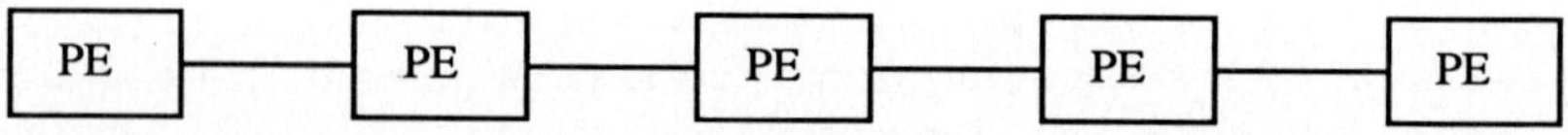

Figure 1. Systolic array.

As is evident from Figure 1, the number of pins required for a systolic array (typically in the range 100 to 200) is independent of the number of processing elements.

Square Mesh

A square mesh array consisting of N processing elements (PE) is shown on Figure 2. Each PE has four data ports to provide connections to upper, right, lower, and left neighbors. A general array consists of N PEs with $4\sqrt{N}$ data ports ($\sqrt{N}$ along each side of the array). Hexagonal meshes are similar except that each processor has six data ports connected to neighboring processing elements resulting in even larger pin requirements than for square arrays with the same number of PEs..

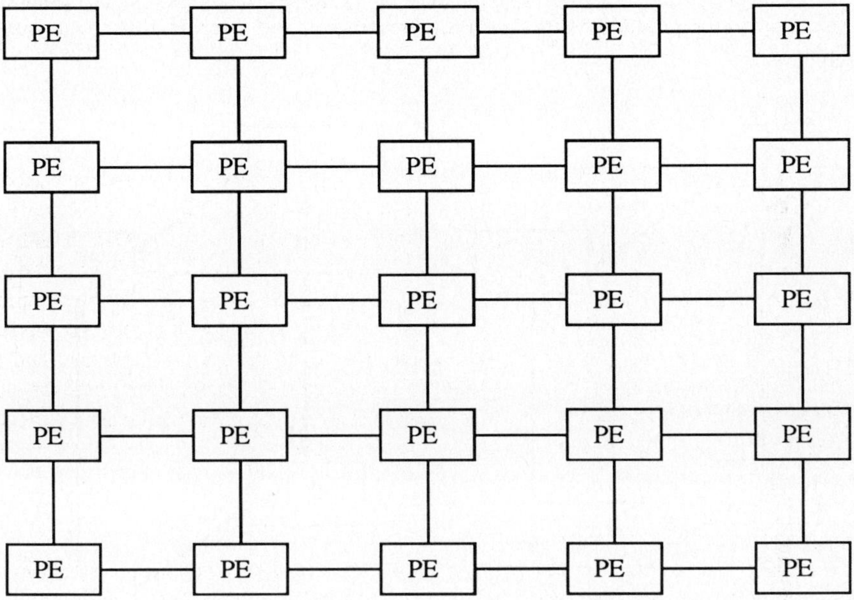

Figure 2. Square mesh of processing elements.

The N processing element square array requires P pins:

$$P = k\sqrt{N}$$

where: k is the number of pins required for a single processing element (typically in the range 100 to 200). Thus for the square array the number of pins increases as the square root of the number of processing elements on a wafer.

Hypercube

In a hypercube array each of the M PEs has connections to the corresponding PEs in Log_2M dimensional hyperspace. An eight processor (three dimensional) hypercube array is shown on Figure 3. A general hypercube system comprises K • N PEs implemented with K chips each containing N PEs.

The N processing element hypercube wafer (part of a system that contains a total of K wafers) requires P pins:

$$P = k \cdot N \cdot \text{Log}_2(K)$$

where: k is the number of pins required for a single processing element (typically in the range 100 to 200). Here the number of pins increases in proportion to the number of processors on a wafer, where the proportionality constant increases with the logarithm of the number of wafers.

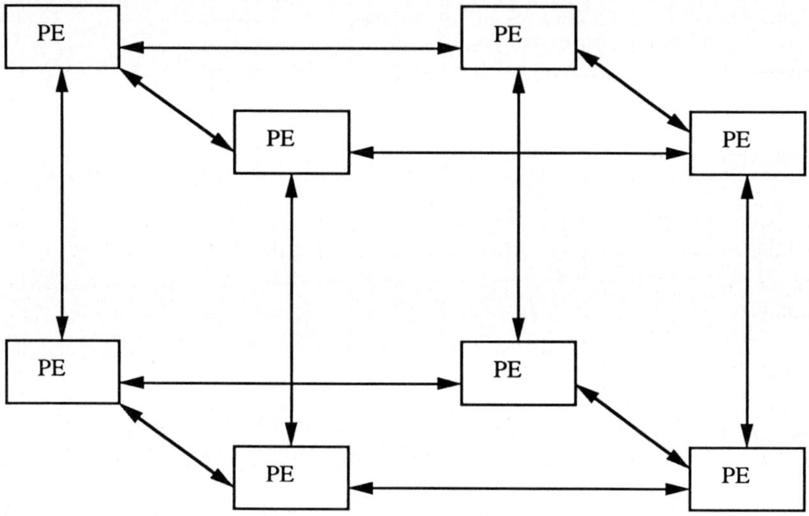

Figure 3. Eight processing element hypercube.

Comparison

Figure 4 compares the pin requirements for systolic arrays, square meshes, and hypercubes. These curves are based on the assumption that the number of pins per processing element is the same for all architectures (i.e., k = 100).

Clearly, if it can implement the desired algorithms, the systolic array is the best choice in view of the packaging considerations. In view of its pin requirements, the hypercube is suitable for WSI only if the required number of processors can be implemented on a single wafer. It is noteworthy that since the systolic array pin requirement is constant, it becomes more efficient with increasing levels of integration.

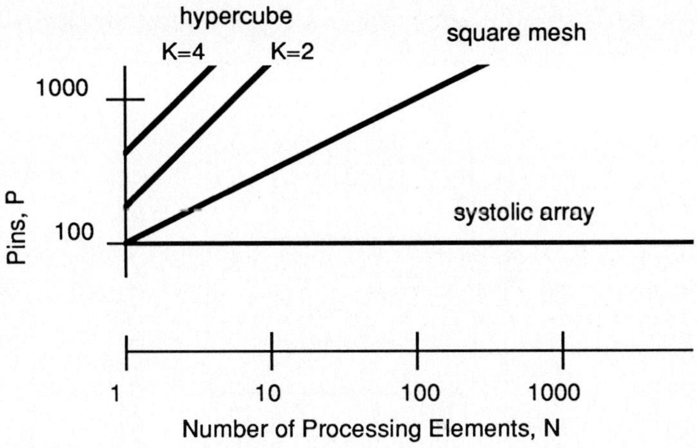

Figure 4. Pin requirements for systolic arrays, meshes, and hypercubes.

WSI SYSTOLIC FFT

The pipeline FFT algorithm [9] was selected as the basis for the wafer scale FFT processor, because it can be realized with a systolic architecture requiring only two processing element types and using nearest neighbor interconnections, thus overcoming many fabrication constraints.

Pipeline FFT Algorithm

The Lincoln Laboratories radix-4 version [10] of the pipeline FFT algorithm is implemented with a four path pipeline, as shown in Figure 5. This is referred to as a radix-4 architecture since each computational element computes a four point FFT. The basic FFT computation is often called a butterfly due to the shape of its signal flow graph. The delay commutator reorders the data between computational stages. The complete processor (for transforms of up to 4096 points) consists of six computational elements interconnected by five delay commutators in a cascade architecture. A system clock speed of 25 MHz, made possible by CMOS technology, yields an overall throughput of 100 MSPS since four data advance in parallel through the pipeline on each clock.

This architecture has been successfully employed in a 40 MSPS FFT processor realized with commercial VLSI technology [11]. It performs both forward and inverse transforms of sequences that are powers of four (i.e., 4, 16, 64, 256, 1024, 4096, etc.) points long. The different computations are effected by varying the number of elements, the counter rate (i.e., the sin-cos sequence) in the computational elements, and the delay length and commutation rate in the delay commutators.

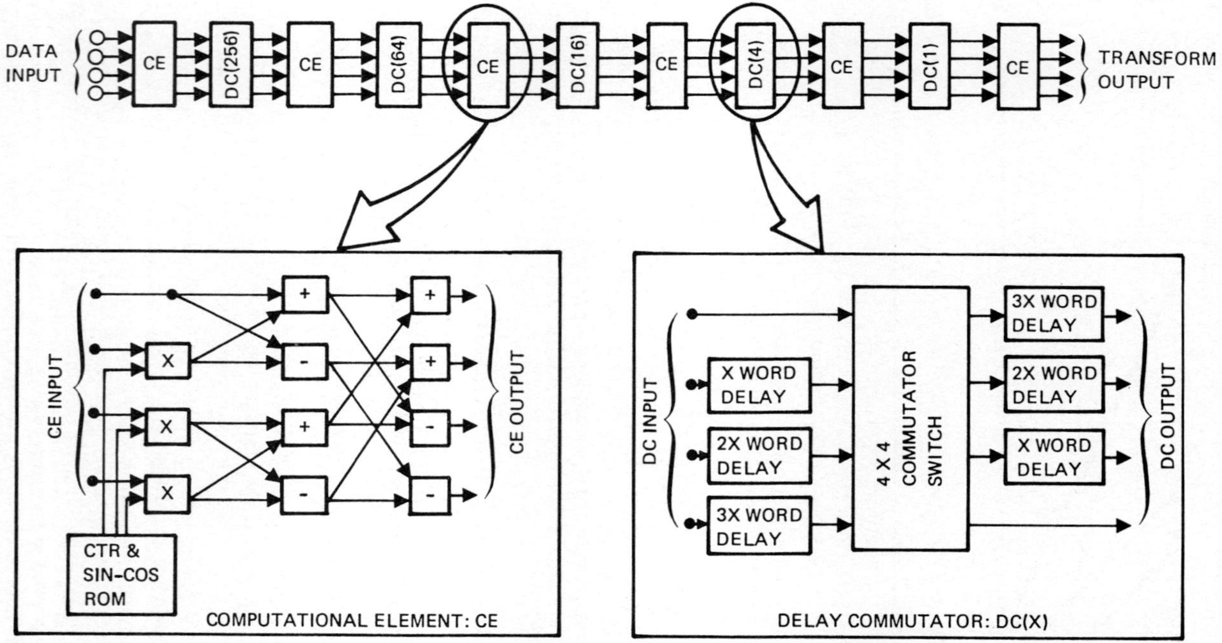

Figure 5. Radix-4 FFT processor architecture.

WSI FFT Implementation

Implementation of a radix-4 FFT processor for 4096 point transforms requires six computational elements and five delay commutators. Each computational element consists of three complex multipliers and eight complex adders, which are realized with twelve real multipliers and 22 real adders. The delay commutators require 12 to 3072 words of memory and a few thousand gates of random logic. Thus the complete wafer must provide a total of 72 multipliers, 132 adders, and 4096 words of memory before sparing. Successful wafer scale fabrication requires: inclusion of sufficient spare parts on the wafer to assure adequate yield and proper organization of the spares to accommodate fault circumvention.

The fault circumvention process is illustrated by examining the trades implicit in realizing a complex multiplier. Current technology exhibits insufficient yield unless redundant elements are used with switching logic (i.e., input demultiplexers and output multiplexers) to improve the yield. The yield with sparing is computed by multiplying the probability that at least one multiplier is good times the yield of the multiplexing logic [12]. As noted previously [13], as the number of computational elements (i.e., multipliers) increases, the probability of finding at least one good element increases, but at the same time the overhead yield decreases.

With the assumed technology, overall yield is maximized with four elements in parallel (quadruple sparing), with a slight penalty for triplicated sparing, and a significant penalty for duplicated sparing. Since there is a 33% area penalty for quadruple sparing relative to the area of triplicated sparing, the latter was selected to implement the real multipliers in the complex multiplier assembly.

To realize 72 multipliers on a wafer, the yield increases by over four orders of magnitude with triplicated sparing, as compared to unspared multipliers. A similar analysis resulted in selection of duplicated sparing for the real adders that comprise the complex multiplier. A block diagram of the complex multiplier with sparing is shown on Figure 6.

Wafer Scale Reordering Logic

Reordering logic between computational elements provides the delay commutator function shown previously on the lower right insert of Figure 5. In contrast to VLSI realizations [13], which use shift registers to provide the delay function, RAM circular buffers are used in the wafer scale implementation. With RAM circular buffers, the contents of only two memory locations change on each clock cycle, instead of all locations changing which occurs with shift registers. As a result the power consumption is greatly reduced. Memory fault circumvention is provided by using an error correcting code. As memory words are stored, error check bits are computed and stored with the data bits. Upon reading data words from the memory, the check bits are used to correct any single errors and detect double errors. Error correction increases the area required to implement the memory, but the increased size is justified since the yield increases dramatically.

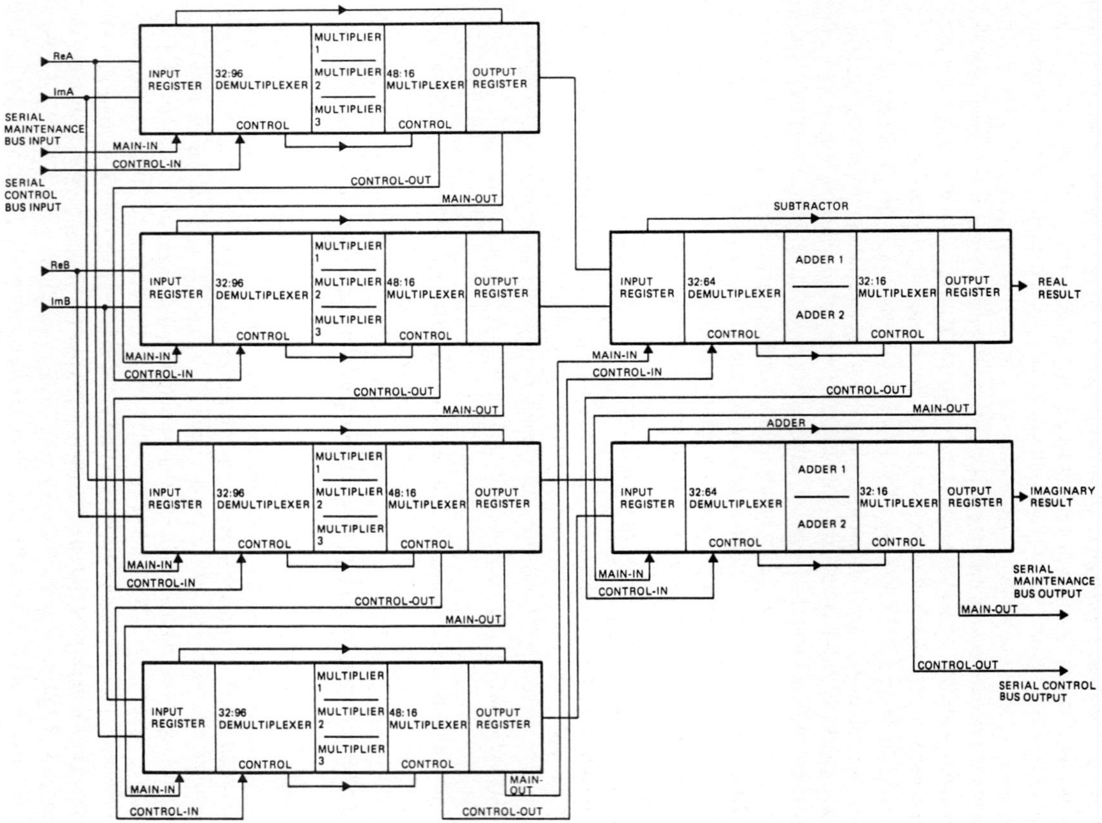

Figure 6. Complex multiplier with internal redundancy.

CONCLUSION

This paper has shown that systolic arrays are attractive for signal processing with wafer scale integration, as they exhibit modest pin requirements. Square mesh and hypercube architectures require significantly more pins and the disparity grows as the WSI integration level increases.

This paper has also summarized design considerations arising in the development of a wafer scale signal processor. This processor is an important building block for a wide variety of signal processing applications, achieving a 100 MSPS data rate in the computation of FFTs of up to 4096 points in length. In contrast, existing special purpose signal processors constructed with conventional VLSI circuits are slower (data rates below 40 MSPS), larger, and consume much more power.

REFERENCES

[1] E. E. Swartzlander, Jr., *et al.*, "A Wafer Scale FFT Processor," Proc. Government Microcircuits Applications Conf. (1987) pp. 45-48, .

[2] C. Jesshope and W. Moore, eds., Wafer Scale Integration (Adam Hilger, Bristol, U.K., 1986)

[3] G. Saucier and J. Trilhe, eds., Wafer Scale Integration (North–Holland, New York, 1986)

[4] R. M. Lea, ed., Wafer Scale Integration II (North–Holland, New York, 1988)

[5] E. E. Swartzlander, Jr., ed., Wafer Scale Integration (Kluwer Academic Publ., Boston, 1989)

[6] E. E. Swartzlander, Jr. and Joe Brewer, eds., Proc. Intl Conf. on Wafer Scale Integration (IEEE Computer Society Press, Washington, 1989)

[7] H. T. Kung, "Why Systolic Architectures?," Computer vol. 15 no. 1 (Jan. 1982) pp. 37-46.

[8] E. E. Swartzlander, Jr., ed., Systolic Signal Processing Systems (Marcel Dekker, Inc., New York, 1987)

[9] H. L. Groginsky and G. A. Works, "A Pipelined Fast Fourier Transform," IEEE Trans Computers vol. C-19 (1970) pp.1015-1019.

[10] B. Gold and T. Bially, "Parallelism in Fast Fourier Transform Hardware," IEEE Trans Audio and Electronics vol. AU-21 (1973) pp. 5-16.

[11] E. E. Swartzlander, Jr., VLSI Signal Processing Systems (Kluwer Academic Publ., Boston, 1989) pp. 117-139.

[12] T. E. Mangir and A. Avizienis, "Fault-Tolerant Design for VLSI: Effect of Interconnect Requirements on Yield Improvement of VLSI Designs," IEEE Trans Computers vol. C-31, (1982) pp.609-616.

[13] E. E. Swartzlander, Jr., *et al.*, "A Radix 4 Delay Commutator for Fast Fourier Transform Processor Implementation," IEEE Jnl Solid-State Circuits vol. SC-19 (1984) pp. 702-709.

Wafer Scale Integration, III
M. Sami and F. Distante (Editors)
Elsevier Science Publishers B.V. (North-Holland)

A MODULAR BIT-SERIAL CONVOLVER

Luigi Dadda, Luca Breveglieri

Department of Electronics
Politecnico di Milano
Milano, Italy

A systolic convolver scheme is illustrated, characterized by serially input samples and serially output convolved: weights are stored in static registers. The interval between adjacent samples can be reduced to zero, thus affording the maximum sampling rate for an assigned number of sample bits and clock rate. The latter is determined by the delay through a full-adder and a flip-flop. The scheme is composed of cascaded modules, each one comprising a serial- parallel multiplier. For samples and weights of equal length, or for weight length multiple of sample length, multipliers can be assembled with bit-slices. Fault tolerance can be easily achieved with additional circuitry. The scheme appears well suited for VLSI implementation.

1. INTRODUCTION

Serial-input serial-output convolvers have raised in the past years a considerable interest for VLSI implementation. McCanny, Evans and McWhirter [4] and Danielsson [2] have proposed new schemes of this type. In comparison to the parallel input-output versions, they require a smaller pin count and a smaller silicon area, affording naturally also a smaller sampling rate.

In [6] a scheme has been proposed offering two main advantages: first, it can be implemented as a bit-slice structure, second it reaches the maximum sampling rate compatible with a serial-input (i.e. with zero interval between adjacent samples). Moreover, the scheme offers the possibility to program a given stack of slices as a convolver with a prescribed operands length and number of terms. A further scheme offering the same properties has been described in [7].

In this paper a serial input-output convolver will be described, capable of a maximum sampling rate. The scheme is modular at the multiplier level and, within each multiplier, at the bit-level. It offers the advantage of a number of components smaller than the preceding schemes.

2. BASIC SCHEMES

For simplicity, the following will consider the case of samples and weights represented by natural integers of equal length n. The case of two's complement numbers and the case of samples and weights of different length will be considered in the following paragraphs. The convolvers introduced in the following belong to a family of circuits

[1] whose general scheme is represented in fig. 1, composed by three main parts: a shift-register storing the previous $N-1$ samples, a linear array of multipliers (including the weight registers) and an accumulator producing the convolved Y.

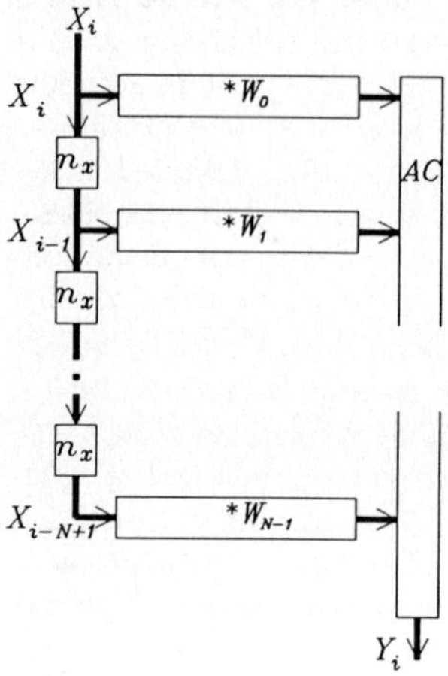

Fig. 1

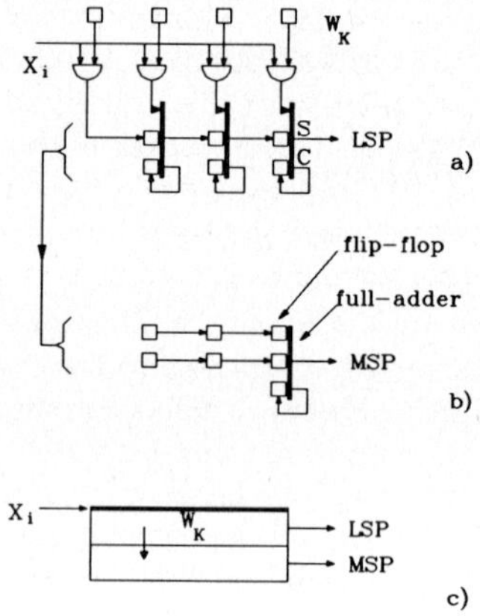

Fig. 2

fig. 1 - *General scheme of the convolver.*

fig. 2.a - *Serial(X)-parallel(W) multiplier.*

fig. 2.b - *Additional registers for a pipelined multiplier.*

fig. 2.c - *A simplified notation.*

In the case of samples in serial form, the simplest multiplier scheme is shown in fig. 2: it is [5] a serial-parallel carry-save scheme, producing at its output port a product bit for each sample bit. In case of samples and weights of equal length n, after the last (most significant) sample bit, the n bits long least significant product part (LSP) has been output, the most significant product part (MSP) being stored in redundant form in the two $n-1$ bit long multiplier registers. If n additional zero input bits are applied, the n bit long MSP will be output. The accumulator is then fed by the N, $2n$ bit long products issued by the multipliers array, and we assume that it must produce their sum at its output as a serial number (the convolved Y_k). The convolved length n_y will be:

$$n_y = n' + 2n \qquad n' = \lceil log_2 N \rceil$$

Consequently, the samples must be separated by an interval of $n_{px} \geq n' + n$ zero bits. This very simple scheme can therefore be adopted provided that the sampling rate f_s is $f_s \leq f_c / n_{tx}$ where f_c is the clock frequency (the bit rate) and $n_{tx} = (n + n_{px}) \geq (n' + 2n)$. Note that during the n_{px} bits of the sample interval, the inputs of the multipliers must be kept zero. This requires $n_{tx} = n_{px} + n$ flip-flops for each sample in the input register. Alternatively n flip-flops will suffice, provided that the clock pulses to the input register are stopped during the sample interval.

In order to increase the sampling rate f_s it is then necessary to increase the clock frequency f_c and/or to reduce the sample period n_{tx}. The latter method is based exclusively on an architectural solution, while the first is based both on IC technology and a suitable circuit choice. The carry-save multiplier scheme is such that the maximum clock rate for a given technology is the inverse of the sum of two delays: the flip- flop delay T_{FF} and the full-adder delay T_{FA}. It is assumed that these components are assigned, so that the multipliers work at a clock rate f_c given by $f_c \leq (T_{FF} + T_{FA})^{-1}$. Also the accumulator circuit can be designed under the above assumptions. An additional condition is nevertheless desirable to obtain an overall circuit with a regular modular structure, useful for VLSI implementation. This means, for instance, having an accumulator structured as a linear array of cells, one for each multiplier.

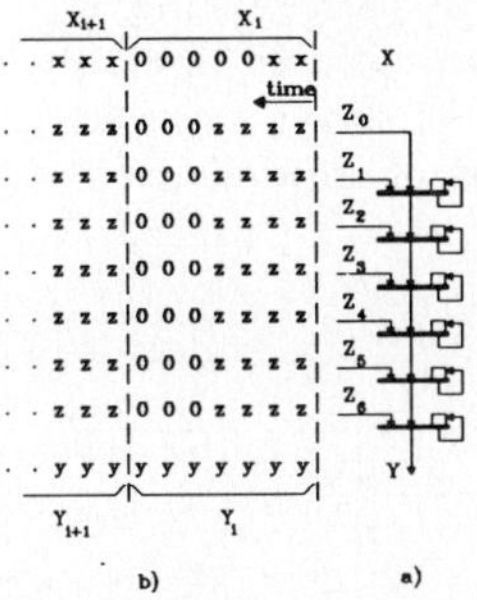

Fig. 3

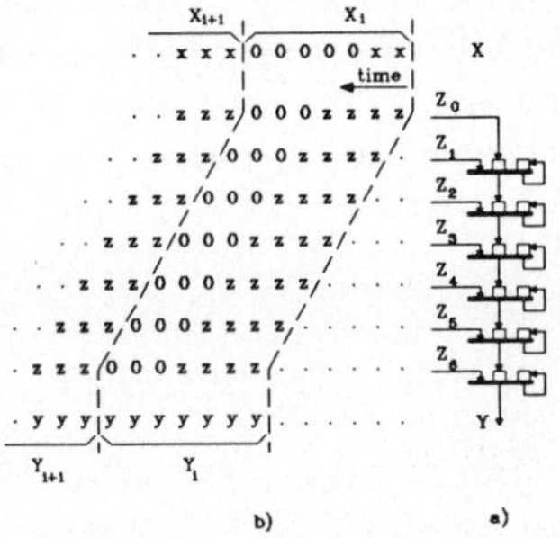

Fig. 4

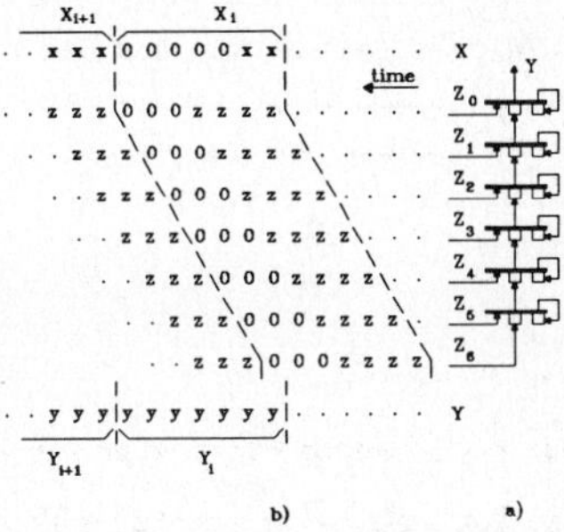

Fig. 5

fig. 3.a - *The accumulator as a multiple input adder of serial numbers ($N = 7$).*

fig. 3.b - *The operation of the accumulator for $n = 2$, $N = 7$, $n_y = 7$.*

fig. 4.a - *A pipelined accumulator for multipliers with input shift-registers of length $n_y + 1$ (fig. 6.b).*

fig. 4.b - *The operation of the accumulator for $n = 2$, $N = 7$, $n_y = 7$.*

fig. 5.a - *A pipelined accumulator for multipliers with input shift-registers of length $n_y - 1$ (fig. 6.c).*

fig. 5.b - *The operation of the accumulator for $n = 2$, $N = 7$, $n_y = 7$.*

Such an accumulator is obtained as follows. Let us consider first the circuit of fig. 3.a, where the accumulator is seen as a multiple input adder of serial numbers, composed by a linear array of full-adders, in which the sum of the two topmost numbers is added to the third one, their sum to the fourth and so on. Fig. 3.b shows the operation of the circuit at subsequent clock periods. Note that the output Y is generated by a network involving $N - 1$ cascaded full-adders. The circuit does not satisfy the condition of a single full-adder delay between inputs and output. This condition can be satisfied by modifying the circuit as shown in fig. 4, i.e. by adopting a pipelined structure.

This requires to delay each input by one clock period with respect to the preceding one. The latter condition can be easily satisfied by adding a stage to each one of the cascaded input shift-registers in fig. 1, as shown in fig. 6.b. A flip-flop is also added at the S (the sum) output of each full-adder in the accumulator, as shown in fig. 4.a. Moreover, a flip-flop (not shown in the figures) must be added at the output of each multiplier. The operation of the new scheme is shown in fig. 4.b. Note that the output of each multiplier is delayed by one bit with respect to the preceding one (while in fig. 3 the outputs of all multipliers are synchronous). Note also that the pipelined accumulator of fig. 4 produces the output with a latency of $N-1$ bits.

A different but similar solution is shown in fig. 5.a and in fig. 6.c. The shift-register length at each multiplier input is $n_y - 1$, and consequently the multiplier outputs are as shown in fig. 5.b. While in fig. 6.b and in fig. 4.a the input and the output of each multiplier are delayed of one bit with respect to the preceding one, in fig. 6.c and in fig. 5.a the one bit delay exists between each multiplier input and output and those of the following one. The accumulator scheme, shown in fig. 5.a, is the same as in the preceding scheme, the outputs of the array of multipliers being connected in a reverse order. As a consequence (fig. 5.b), no latency exists in this circuit.

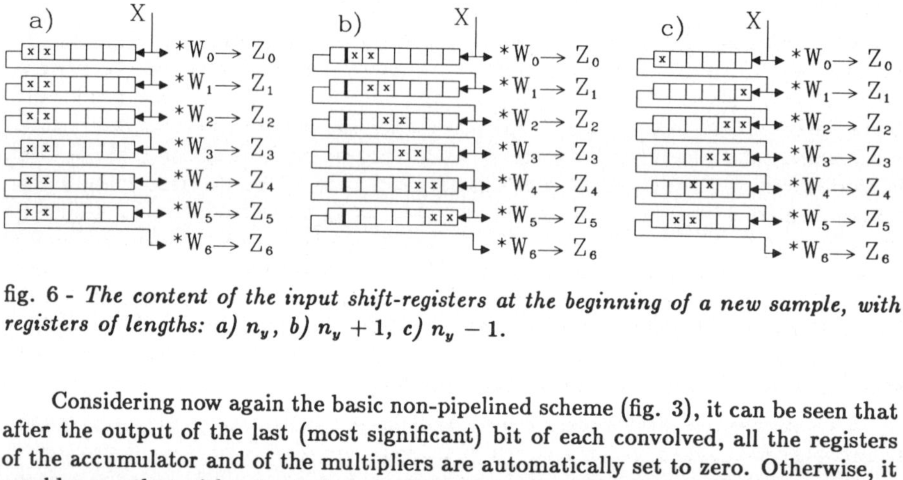

fig. 6 - *The content of the input shift-registers at the beginning of a new sample, with registers of lengths: a)* n_y*, b)* n_y+1*, c)* n_y-1.

Considering now again the basic non-pipelined scheme (fig. 3), it can be seen that after the output of the last (most significant) bit of each convolved, all the registers of the accumulator and of the multipliers are automatically set to zero. Otherwise, it would mean that without entering new inputs to the multipliers, one or more non-zero bits will eventually be output from the accumulator. This cannot happen, since the circuit has been designed in order to obtain an output, the convolved, whose length is determined by n and N as shown before. The whole circuit is therefore in zero state after the output of the last convolved bit without the need of a reset signal. In the pipelined schemes the above property does not apply to the whole circuit, but to each multiplier at different times: precisely, the multipliers registers are automatically set to zero at the end of the n_yth bit after the beginning of the first, least significant bit of each sample (i.e. at the time corresponding to the dotted diagonal lines in fig. 4.b and in fig. 5.b).

At the end of the first bit of a new sample, the flip-flops of the multipliers and of the accumulators will assume values determined by the algorithm and these values will affect the new convolved. The lines (the wavefronts) separating the convolver part generating a convolved from the part generating the following one move downward in fig. 4.a scheme and upward in fig. 5.a scheme. One or more wavefronts (at a distance of n_y stages) can exist at any clock time, depending on N and n_y.

3. REDUCING THE SAMPLE SEPARATION

For simplicity, the following considerations will be referred to the case of a non-pipelined accumulator. The transformation of the new circuits, also not pipelined, into pipelined versions is straigtforward, as shown in the preceding paragraph. The reduction of the samples separation with respect to the value logically necessary for the scheme of fig. 1 can be obtained by modifying its architecture. Note that, whatever is the architecture obtaining such result, the accumulator has to be provided by two or more outputs, due to the constraint given by the number of bits of each convolved, to be output in a time equal to the sampling period. For an assigned $n_{tx} = n + n_{px}$ the number n_o of the accumulator outputs will now be $n_o = \lceil n_y/n_{tx} \rceil$, i.e. the smallest integer not smaller than n_y/n_{tx}. For the case of zero samples separation $n_{px} = 0$, $n_o = 2 + \lceil n'/n \rceil$. This means that with the assumption we made (samples and weights of equal length n) three or more outputs are in general required.

A *first* step in reducing n_{px} can be obtained by observing that all multipliers are in zero state after $2n$ bits have been input (each sample being followed by n zero bits), while the accumulator, having output $2n$ bits, still contains the n' most significant bits of the convolved. Assume that a second accumulator is added to the scheme, with additional circuits producing the transfer of the content of the first accumulator into the second one, under the control of a signal (a "word" clock) representing the $2n$th bit of a sample period. At the end of the $2n$th bit the multipliers and the first accumulator will be in zero state, ready for a new cycle producing the following convolved, while the second accumulator produces the last n' bits of the preceding one, if it is assumed that $n' \leq 2n$ (otherwise a further accumulator is needed). The sample separation of the new circuit is reduced to $n_{px} = n$ and consequently the sample rate to $f'_s = f_c/2n$. The "cost" incurred in this reduction is given by the second accumulator, composed by N stages (each comprising two FF and one FA).

A *second* step in reducing the sample separation n_{px} can be obtained by pipelining each multiplier as shown in fig. 2.b. An additional pair of shift-registers is provided, connected to the multiplier Sum and Carry registers in such a way that at the end of the nth (the last) sample bit the contents of the latter are transferred to the first. In the following n bits a new sample can be applied to the multiplier input (producing the LSP of the new product), while the contents of the shift-registers are added by a serial adder stage producing the MSP of the preceding product [5].

Two accumulators are provided (fig. 7), the first being fed by the LSP multipliers outputs, the second fed by the MSP produced by the shift-registers part of the multipliers. Since the accumulator output is $n' + n$ bits long, i.e. n' bits longer than the samples, these must be separated by at least n' bits. If the sample separation is assumed n' bits long, there will be no separation among the successive Y_L and Y_M accumulator outputs. Y_M and Y_L must be added in order to obtain the convolved (fig. 7.b).

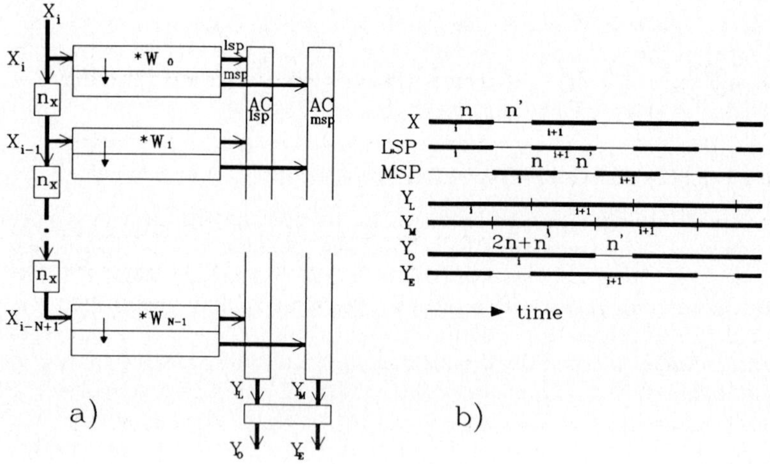

fig. 7.a - *A convolver with two outputs using pipelined multipliers (fig. 2.a, 2.b).*

fig. 7.b - *The operation of the pipelined convolver with two outputs.*

Note that Y_M is delayed, with respect to Y_L produced by the same set of samples, by n bits. The adder must therefore have two outputs, and it can be designed so that, for instance, the odd Y_O and the even Y_E indexed convolved are output respectively by the two ports. In another solution the convolved are split in two parts Y_{MSP} and Y_{LSP}, composed by the most significant and the least significant bits, respectively, of the convolved Y and such that their length is smaller than $n + n'$. All Y_{MSP} and all Y_{LSP} are output from separated ports. Note (fig. 7.b), that the convolved issued by each port are separated by n' bits, as the samples. This separation results in a sampling rate $f_s'' = f_c/(n + n')$. The corresponding increase in "cost" with respect to the basic scheme is given by: two shift-registers of $n - 1$ bits each one, a serial adder in each multiplier and an additional accumulator in the final adder.

Finally, a *third* scheme with a zero sample separation can be obtained, leading to a maximum sampling rate for a given clock rate and sample length n. The scheme is represented in fig. 8 and, assuming $n' \leq n$, it differs from the scheme in fig. 7 for the addition of a third accumulator AC_3. Its operation, represented in fig. 8.b, is as follows. From the first to the last bit of each sample X_i, the n least significant product bits (LSP) are fed to the first accumulator AC_1, producing at its output the n least significant convolved bits Y_{Li} of the convolved Y_i. At the end of the last, most significant sample bit, the S and C multiplier registers are transferred to the two auxiliary shift-registers in each multiplier (fig. 2.a, 2.b). Simultaneously, the contents of the registers in accumulator AC_1 are transferred into the corresponding registers of accumulator AC_2 and likewise a transfer occurs from AC_2 to AC_3.

During the following sample X_{i+1}, AC_2 will produce the following more significant bits Y_{Mi} of Y_i, while AC_1 will output Y_{Li+1}. At the end of the sample X_{i+1}, the previously described transfer between registers will occur. During the sample X_{i+2}, AC_3 will produce the $n' \leq n$ most significant bits Y_{MMi} of Y_i. Simultaneously, AC_2

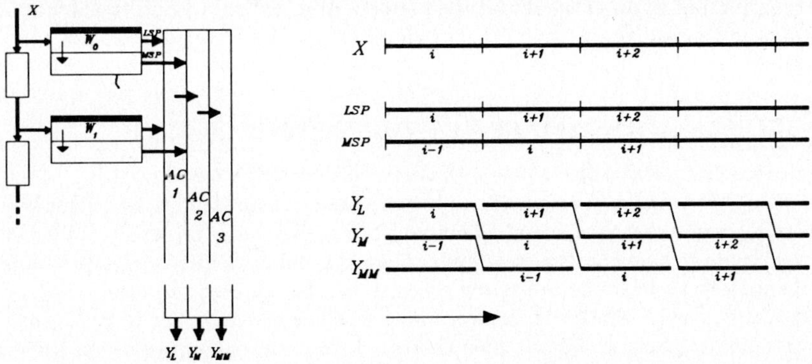

fig. 8 - *A convolver with zero interval between samples for* $n_w = n_x = n$, $N \leq 2^n$.

will output $Y_{M\,i+1}$ and AC_1 will output $Y_{L\,i+2}$. Fig. 8 scheme can handle convolutions with $n' \leq n$, i.e. with a number of terms $N \leq 2^n$. For a larger N, and up to 2^{2n}, a fourth accumulator and a further couple of shift-registers in each multiplier will be necessary.

4. CONVOLVERS FOR TWO'S COMPLEMENT NUMBERS

In the preceding schemes it has been assumed for simplicity that both weights and samples are natural integers. In case where weights and samples can assume both positive or negative values, it is convenient to represent them in two's complement form. The preceding schemes must then be modified accordingly. Among the many ways of treating the multiplication of numbers in two's complement form, we have shown [3] that a most convenient method for serial-parallel multipliers (as those used in this paper) is based on the following transformation. It is well known that in a two's complement number the most significant bit can be assumed as sign bit with a negative weight -2^{n-1}, n being the number of bits. Consequently, in the array of a multiplication involving two's complement factors the array terms containing one (and only one) sign bit have negative weight. It can be shown that such an array can be evaluated (to obtain the product) with the following algorithm:

1. replace all negative weight terms in the multiplication array with their complement;
2. evaluate the transformed array (composed now of positive weighed terms only);
3. add a negative constant: $C_1 = -2^{2n-1} + 2^n$.

If such an algorithm is adopted, the only transformation necessary in the previously illustrated schemes (and also in the following ones) are those implementing steps 1 and 3. For step 1 the transformation consists in replacing in each multiplier (fig. 2) the "and" gate, fed by the most significant weight bit and by the sample bits, with a gate producing the true array term at the time corresponding to the sample sign bits and the complement of that term otherwise; viceversa for the remaining gates. Note that the signal corresponding to the sample sign bits is also used for the transfer between registers, as illustrated before. For step 2, since all multipliers outputs are added into the accumulators in order to obtain the convolved, it suffices to add to the convolver output a single constant: $C = NC_1$. For simplicity, such corrections are not shown in the figures.

5. WEIGHTS AND SAMPLES OF DIFFERENT LENGTH

In the preceding paragraphs it has been assumed that weights and samples have equal length n. It is common in applications to have cases where weights and samples have different n_w and n_x lengths, usually being $n_w > n_x$. Whenever n_w and n_x differ by few bits, a simple solution consists in adding an equal number of bits as sign prolongation to the smallest, so that the preceding schemes can be adopted. In other cases, n_w is multiple of n_x (or it appears to be possible to add few bits to n_w or to n_x so that this condition holds). Assume for example that $n_w = 2n_x$. In such a case a simple solution consists in splitting each weight W_i into two parts W_i' and W_i''. W_i' is composed by the $n_w/2$ least significant bits and W_i'' is composed by the same number of most significant bits, so that:

$$W_i = 2^{(n_w/2)} W_i'' + W_i'$$

It is therefore easy to obtain a convolver as a combination of two convolvers using weights W_i'' and W_i', respectively.

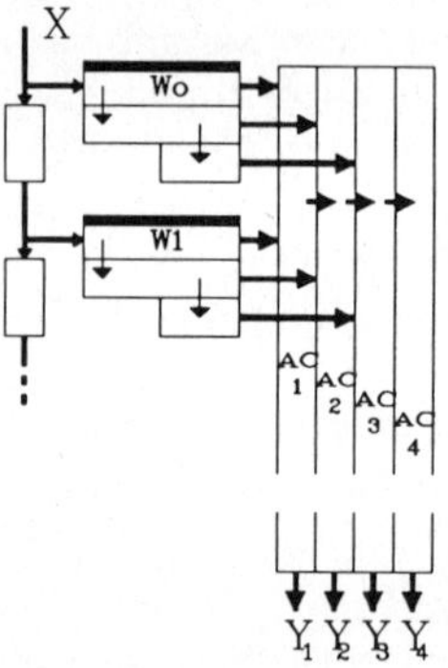

fig. 9 - *A convolver with zero interval between samples for* $n_w = 2n_x$, $N \leq 2^{n_x}$.

A better solution is represented by fig. 9 scheme. The multipliers have a length n_w and include two pair of additional shift-registers: one having the same length, the second having a length $n_w/2$. An additional accumulator AC_4 is necessary, assuming $N \leq 2^{n_x}$. Its operation is similar to the one of fig. 8 scheme. Schemes for $n_w/n_x > 2$ can be easily drawn.

6. FAULT TOLERANCE

The proposed convolver scheme is particularly suited for WSI integration. In fact, the structure is completely modular, it forms a simple linear array and the signals connecting adiacent cells are few. A WSI solution seems to become attractive for convolvers of hundreds or thousands of terms.

Two main solutions for WSI integration are possible. First, one can directly realize individual multipliers (with their accumulator sections) on the wafer and then can structure a convolver by connecting them. This solution requires to realize many connections on the wafer (fig. 10.a).

Second, one can implement a convolver of a certain length on a single chip and then can connect a number of chips on a wafer to form a longer convolver. Its advantage consists in a limited number of connections to realize on the wafer (fig. 10.b).

Convolvers of such lengths, however, require the introduction of fault tolerance, to obtain acceptable production yelds. The two solutions for WSI integration lead to substantially different ways to implement fault tolerance [9].

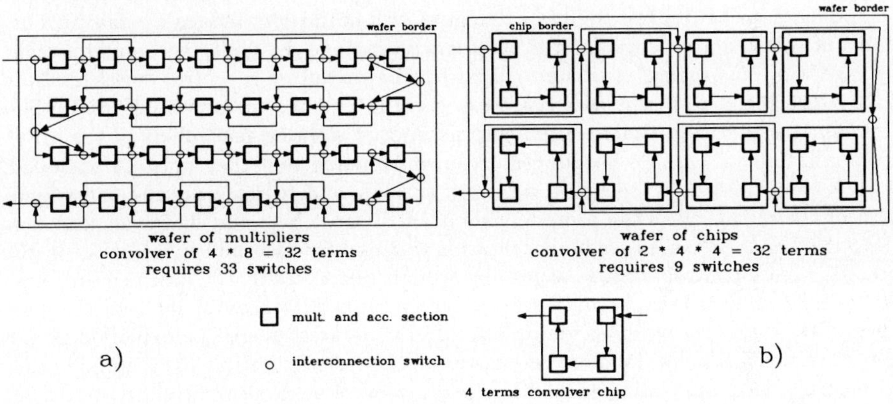

fig. 10.a - *A wafer implementing* $N = 32$ *multpliers.*

fig. 10.b - *A wafer implementing* $N = 8$ *chips, each one containing* 4 *multipliers.*

In the second case (fig. 10.b), we suppose that a whole chip can be extracted from the array, if it is faulty. To obtain this it is sufficient to introduce a variable (i.e. a flip-flop) to control suitable multiplexers that switch the chip to or from the array. The configuration of these control variables can, for instance, be programmed at start-up or after executing a test. All control variables could be connected to form a shift-register, easily programmable as it requires only one input and few control signals. Thus, the increment of input and output pins to the whole structure is irrelevant. Clearly, the control variables and the multiplexers are critical as they cannot be faulty. This condition is statistically matched if they are designed in a "robust" way and if their area occupation is small in comparison to the area of the controlled chips. Thus, the balancing between production yelds, the areas of the chips and the control logic leads to the definition of the optimum size of the chips. A similar case has been studied in [8]. However, this solution always leads to a certain waste of working multipliers, as a whole chip is discarded if a single multiplier is faulty. Alternatively, chips can be extracted from the array by restructuring the connections at production time. Of course, this works only for a single and irreversible configuration of the wafer at the end of production.

In the first solution (fig. 10.a), it is necessary to extract from the array a single multiplier or its accumulator section, if they are faulty. Faults in the multiplier require different handling from faults in the accumulator section. Faults in the multiplier require its exclusion from the array. This can be done by forcing its outputs to the accumulator to zero by means of suitable switches. In this case, the residual effect of the accumulator section is merely to introduce a delay. Such effect is taken into account by modifying the length of the shift-register leading the samples to the multiplier. This can also be obtained by means of switches (fig. 6.b, 4.a, 6.c, 5.a). This last modification is simple and details are not discussed.

Faults in the accumulators could be accepted as fatal, due to the small number of components, to be implemented in a "robust" way in order to reduce the fault probability. If this is not acceptable, it is necessary to exclude the faulty accumulator stage by means of "bypassing" switches. These can be driven by a control variable. Note that bypassing the accumulator stage can exclude also the corresponding multiplier. If it is necessary to mantain a minimum number of working multipliers in the array, spare multipliers must be introduced. It must be noted that they must be excluded, if they are not used. Once again, a control variable and switches are necessary, handled in the same way of the preceding solution. It must be noted, however, that now their number is increased, as fault control is required for each multiplier, while in the preceding case control is made at groups of multipliers. Thus, the area requirements of the fault control logic become a more critical drawback than in the preceding solution. However, the working multipliers can in the present case be exploited at full extent. Attention must therefore be focused on the reconfiguration algorithm, in order to obtain a high harvesting (i.e. a low percentage of working multipliers that must anyway be discarded as they cannot be connected). In the preceding solution, the reconfiguration algorithm was less important, as it had to restructure the connections among few large chips. The linear structure of the array simplifies this problem. Cases similar to the present one have also been treated in [8] and [9]. An algorithm to restructure the connections at production time can be used to reconfigure the array, but it also requires a high harvesting performance to exploit fully the working multipliers. However, a longer time to restructure the wafer may be required with respect to the preceding solution.

7. CONCLUSION

A family of serial-input serial-output convolver schemes has been discussed, based on the use of serial-parallel multipliers. The convolver family includes schemes working with no interval between successive samples. The convolvers are composed by identical modules, each one comprising a pipelined multiplier and circuitry necessary for cascade connection. Moreover, each module can be obtained by assembling bit-slices, permitting to design convolvers with prescribed weight and sample length and assigned number of convolution terms. Fault tolerance can be achieved by means of simple circuitry. The above properties make the proposed schemes good candidates for VLSI and WSI implementation.

8. REFERENCES

[1] H. T. Kung, "Why systolic Architectures", Computer, vol. 15, n. 1, pp. 37-46, 1982

[2] P. E. Danielsson, "Serial-parallel Convolvers", IEEE Trans. Comput., C-33, 7, pp. 652-667, July 1984.

[3] L. Dadda, "Fast Multipliers for two's Complement Numbers in serial Form", Proc. of IEEE, Symp. on Computer Arithmetic, pp. 57-63, 1985

[4] J. V. McCanny, R. A. Evans, J. G. McWhirter, "Use of unidirectional Data Flow in Bit-Level systolic Array Chips", Electronic Letters, 22, 10, pp. 540-541, 8 May 1986.

[5] L. Dadda, "A serial-Input pipelined Multiplier", Dept. of Electronics, Politecnico di Milano, internal report n. 87-030, 1987

[6] L. Dadda, L. Breveglieri, "A serial-Input serial-Output Bit-sliced Convolver", Proc. of ICCD 88, pp. 490-495, New York, 1988

[7] L. Dadda, "Serial-Input Polyphase Convolvers", Systolic Array Processors, pp. 185-194, Prentice Hall, 1989

[8] S. Lasserre, "A single Wafer Bit-sliced Convolver: an optimum statistical Solution for the Implementation of Redundancy", Systolic Array Processors, pp. 514-524, Prentice Hall, 1989

[9] R. Negrini, M. G. Sami, R. Stefanelli, "Fault Tolerance through Reconfiguration in VLSI and WSI Arrays", Computer Systems Series, the MIT Press, Cambridge, Massachusetts, 1989

Wafer Scale Integration, III
M. Sami and F. Distante (Editors)
Elsevier Science Publishers B.V. (North-Holland)

WSI Implementation of a Video Rate 2 Dimensional FFT

R. Philhower and J.F. McDonald
Center for Integrated Electronics
Rensselaer Polytechnic Institute
Troy, New York 12181
(518) 276-2919

1 Introduction and Choice of Architecture

The potential benefits of wafer scale integration are linked to its ability to realize architectures beyond the scope of a single VLSI chip. [1, 2] WSI must offer compelling advantages in either net cost, physical size, weight, reliability, power consumption, or speed to justify the extra complexity in processing. In addition, architectures too large for integration on a single wafer must be able to be partitioned between multiple wafers cleanly. There should not be a large number of pins for wafer to wafer communications.

We have chosen the two-dimensional Fast Fourier Transform as one of the demonstrator architectures in our wafer scale integration project. It is a flexible, well-known algorithm that meets the above criteria. Even with projected technology improvements, two-dimensional image processing algorithms that operate at video rate on data sets of any reasonable size will remain outside the realm of single-chip VLSI implementation. Also, the 2D FFT lends itself well to a parallel implementation.

The design of the 2D FFT is made to accommodate the two major forms of WSI under development at Rensselaer. The first is a traditional WSI with discretionary wiring formed by electron-beam lithography. The second scheme under consideration is a form of wafer-scale hybrid packaging (WSHP) called the wafer transmission module (WTM).

1.1 Mathematical Considerations

For square data sets, the computation of a 2D FFT simplifies to performing a 1D FFT on each row of the image data followed by performing a 1D FFT on each column of the output of the first set of FFTs. This can be expressed as:

$$\begin{aligned} X(k_1, k_2) &= \sum_{n_1=0}^{N-1} \left[\sum_{n_2=0}^{N-1} x(n_1, n_2)\omega^{n_2 k_2} \right] \omega^{n_1 k_1} \qquad (1) \\ \text{where } \omega &= e^{-\jmath\left(\frac{2\pi}{N}\right)} \end{aligned}$$

Thus, to perform an N by N 2D FFT, $2N$ 1D FFTs must be performed. The choice of N was the object of much simulation and design work. From the start, we wanted to use mature VLSI technology to increase yields for this demonstration system. We found that a 256 by 256 2D FFT can be performed at video rate by a

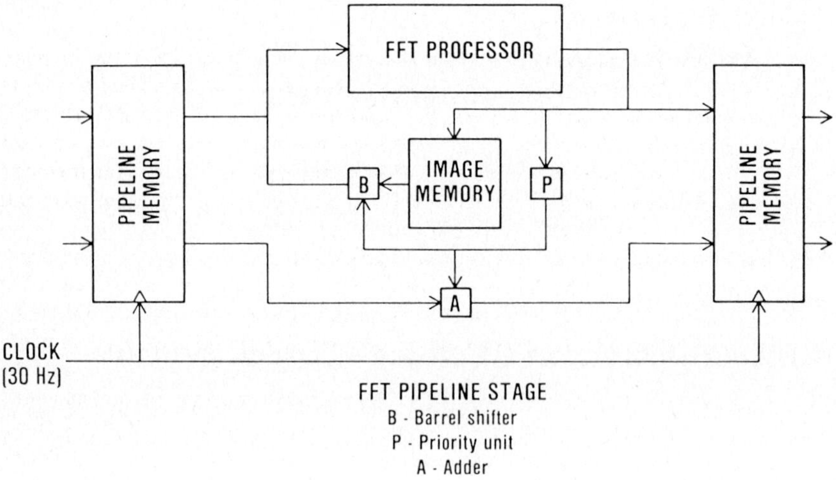

Figure 1: Two-Dimensional FFT Processor Design

collection of 2K gate building blocks. These building blocks have been implemented as two different gate-arrays using $3\mu m$ technology. Obviously, by taking advantage of recent lithography shrinkages or full custom design we could perform an FFT at the new high-definition video rate (HDTV) or enlarge the data set to provide a larger image.

Extensive simulation by Taylor [3, 4] demonstrated that sufficient resolution could be obtained by storing each pixel as a 16-bit real and a 16-bit imaginary part and by using scaled fixed-point arithmetic. The single scale factor is common to all pixels in the image. By using a common scale factor (or exponent), the internal operations are essentially fixed-point, and thus the hardware requirements are reduced greatly.

2 2-D FFT Processor Design

The overall design of the 2D FFT processor is shown in Figure 1. The image to be transformed is loaded into the pipeline memory which feeds the rows of data into the 1D FFT processor. The 1D FFT processor is designed with dual input and output memories. Thus, while one set of rows is being transformed, the results from the previous set are being read out and the data for the next set is being written. As the transform results are filling the image memory, the priority encoder determines the minimum number of leading zeroes in the data. When the pixels are read back out by columns, they are shifted by this amount. This shift factor is then added to the overall scaling for that image. The shifting reduces the loss of precision inherent in the repeated row and column FFT operations.

The design of the image memory, barrel shifter, priority encoder and exponent adder are relatively straightforward and will not be discussed here.[3] We will focus, instead, on the internal design of the 1D FFT processor. The 1D FFT processor is made up of four processor clusters (PC). Each PC can perform a complete 1D FFT on 256 elements. Thus, four rows or four columns can be transformed simultaneously.

2.1 WSI Considerations

It is our aim that a PC should fit on a four-inch wafer. Larger wafers would increase production costs and produce an awkwardly sized package. Along with the PCs, there must be a coefficient memory sending the Fourier coefficients to each PC. Because the PCs all operate in lock step, they all could share one coefficient memory. In view of this situation, there are two choices for the placement of this coefficient memory. A separate one could be provided for each processor cluster, on its wafer; or a central one could be provided elsewhere and used in common for all clusters.

Deciding whether to put put a coefficient memory on each wafer involves a wafer area for wafer pinout tradeoff. If there is one common coefficient memory, each PC wafer has a pinout of about 120 pins—not counting power and ground. Of these, approximately 32 are used to transmit coefficients to the wafer. All thirty-two could be saved if each wafer had a coefficient memory. The coefficient memory could be loaded (at initialization time) through one of the working memory data ports.

Adding the coefficient memories increases the number of chips in a PC from thirty-two to thirty-six. Our proposed wafer transmission module (WTM) design can accommodate approximately thirty-seven sites for these large dies.[5] Adding the coefficient memories will cut the opportunity to repair or make use of partially working WTM substrates. Using discretionary wiring techniques, a four-inch wafer should have approximately sixty-four die sites.

The proposed form of WSI with discretionary wiring also creates a requirement on the artwork of the individual chips. The EL-2 electron beam system at Rensselaer requires registration marks be placed at 4 mm intervals across the entire wafer.[6] These $200\mu m$ by $200\mu m$ marks are necessary for the EL-2 to navigate across the wafer and cannot contain any other structures. Because our chips measure approximately 7 mm on a side, there must be space for a registration mark near the center of each chip. Figure 2 shows a routed wafer for a processor cluster containing a coefficient memory.

2.2 Processor Cluster Design

Figure 3 shows the design of a PC. A 256-point 1D FFT calculation involves eight sets of 128 butterfly operations each. Each PC has four butterfly processors (BP). Thus, each BP must perform 32 butterfly operations per 1D FFT. For increased efficiency, while pairs of pixels are sent to the butterfly processors from the transmitting memory, the results from previous operations are read simultaneously into the receiving memory. For each successive set of butterfly operations, the roles of transmitting and receiving memories are reversed. In the "constant geometry" model used (see Figure 4 for a diagram of this algorithm), every set of butterfly operations requires the same memory access patterns. This consistency simplifies the design of the memories. At the end of the set of butterflies, the results are in the same memory that the original pixels were in. A final dummy "pass" cycle is required to move the results into the output memory.

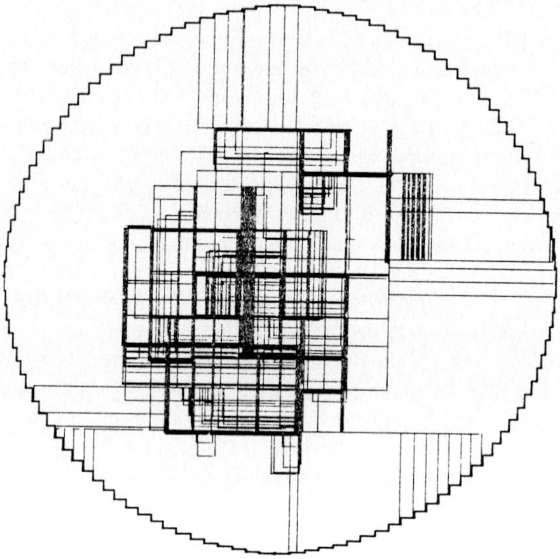

Figure 2: Wafer Routing for a Processor Cluster

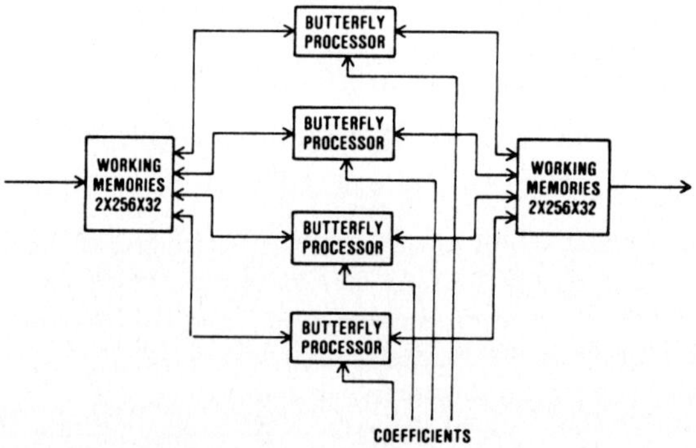

Figure 3: Design of a Processor Cluster

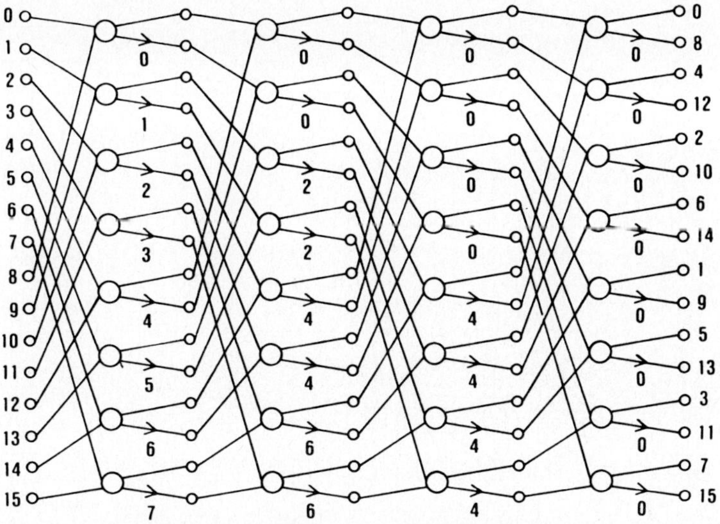

Figure 4: Constant Geometry FFT Algorithm

3 Working Memory

Each processor cluster (PC) must have two pixel memories (PM). These memories alternately supply and receive data during successive sets of butterfly operations. (Their design is in Figure 5.) Gate arrays are notoriously poor at implementing RAM memory. Thus, a portion of the gate array cells were removed, and a custom RAM designed at RPI was substituted. The functionality of the pixel memory has been combined with that of the coefficient memory to reduce the number of different chips to two. This versatile memory has been called a working memory (WM).[7]

Each PM must communicate with four BPs—sending two pixels to each (or receiving two pixels from each) every butterfly cycle. To provide four pixels simultaneously, the memory was divided into four blocks. If the butterfly operations are divided among the four processors as shown in Figure 4, then the memory can be partitioned as also shown on that figure. Butterfly processors 0 and 1 receive all of their data from memory modules 0 and 2; while butterfly processors 2 and 3 receive all of their data from memory modules 1 and 3. Each butterfly processor sends data to its respective numbered memory module. Thus, in the four by four crossbar switch in Figure 5, only the indicated connections need be supported. The "B" and "C" crossbar patterns indicated are used to send the pixels to the BPs. The "B" pattern sends the first pixel; the "C" pattern sends the second. The "A" crossbar pattern is used when receiving data from the BPs. A side effect of this system is that the odd numbered BPs receive their 'B' pixel first while the even numbered BPs receive their 'A' pixel first.

Separate address and data ports are provided to allow the memory to be written to and read from by outside circuitry. This enables the memory to be loaded with image data.

Control logic and address generation internal to the WM provide proper sequencing of addresses to match the constant geometry FFT algorithm. Also, the WM

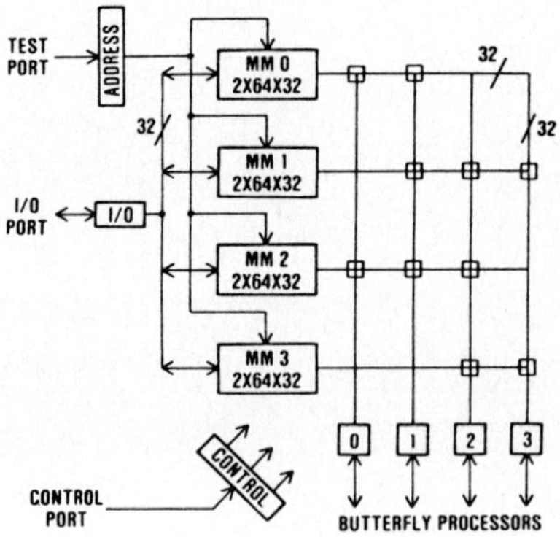

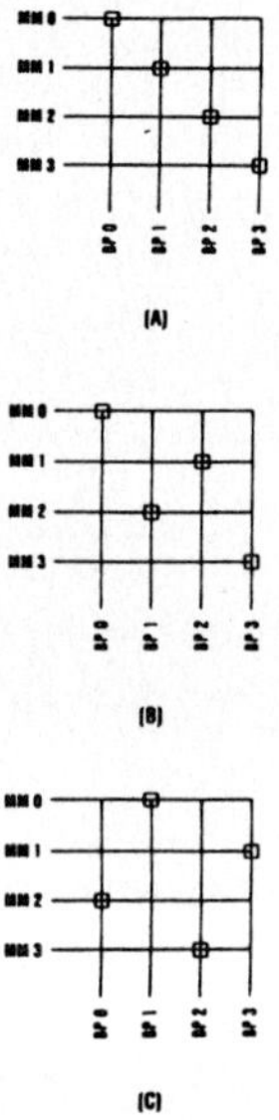

Figure 5: Working Memory Block Diagram

control logic contains a second address generation system so that it can function as a coefficient memory. An advantage of this is that the coefficients are stored in RAM, allowing them to be changed to produce inverse FFTs or other algorithms.

3.1 Implementation

Both the working memory chip and the butterfly processor chip were implemented hierarchically. Thus, verification of small sub-circuits increased our faith in the larger circuits.

3.1.1 Memory

The major feature of the WM chip is the memory itself.[8] It is a 2K bit static RAM which fits into approximately half of the gate-array space. A static RAM was chosen to eliminate the need for refresh which could complicate memory accesses. The memory is divided into four 512-bit sections (8 bits by 64 words) to provide the partitioning mentioned above. (See floor plan in Figure 6.) The memory cell (schematic in Figure 7) is a six-transistor fully static CMOS cell. Because it is CMOS, it has a small static current, but a large footprint.

The combination sense-amplifier/data output buffer is shown in Figure 8. This circuit differs from the standard implementation in two ways. First, the current source device normally has its gate voltage at a generated reference voltage to control current flow. Second, the top portion of the circuit is typically a current mirror configuration rather than a cross-coupled p-channel pair. Through SPICE simulation, we found that using a cross-coupled pair in conjunction with the pre-charge signal (PCN) controlling the current source provided adequate performance while not drawing a static current. Also, this scheme improves the rising-edge performance.

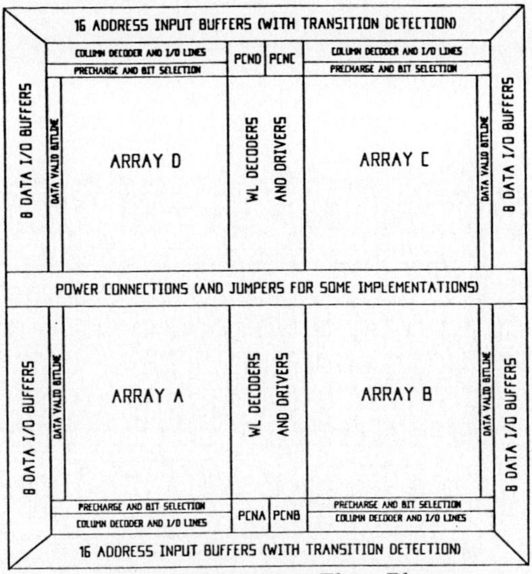

Figure 6: Memory Floor Plan

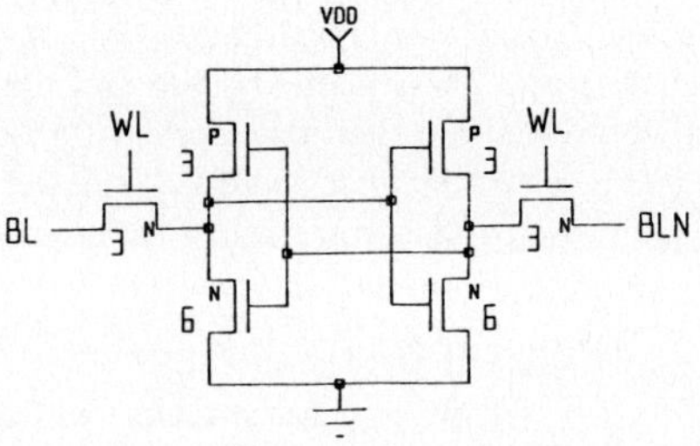

Figure 7: Memory Cell

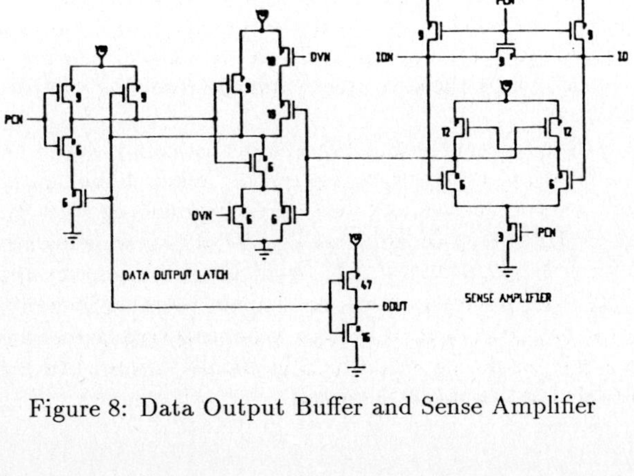

Figure 8: Data Output Buffer and Sense Amplifier

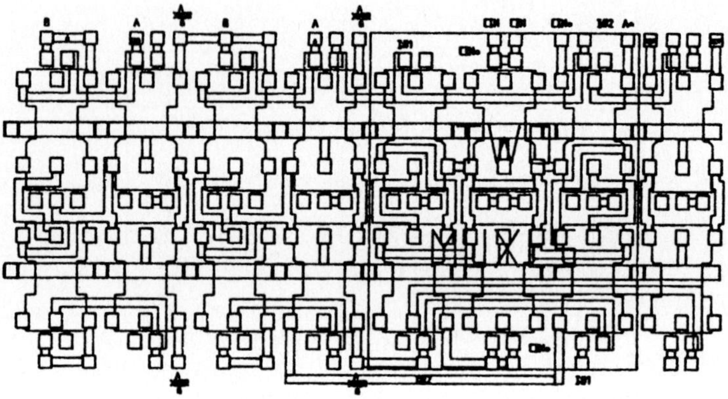

Figure 9: Sample Crossbar Switch Circuit

3.1.2 Crossbar Switch

The design of the crossbar switch was simplified greatly by the assumption that the mapping between memories and ports is one-to-one and onto. Thus there is no need to allow for a memory to drive more than one port, or for a port to drive more than one memory input. The crossbar becomes merely a series of and-or-invert gates used to select the proper input-to-output connection. (See Figure 9.)

3.1.3 Control

The working memories each contain a state machine which provide the distributed intelligence which controls the processor cluster. At any time, each working memory is acting in one of three functions: coefficient memory, transmitting memory, receiving memory. The coefficient memory keeps track of what set of butterflies is being performed, and which butterfly within that set is being calculated. It sends the proper coefficients to each BP. On the ninth cycle, it generates the PASS signal which tells the BPs to pass the data through unchanged. During the PASS cycle, butterfly processors ignore the coefficients they are receiving and just pass the pixels through unchanged.

The transmitting memory begins sending pixels one-by-one to the BPs. It holds its DATA VALID output high as each one is sent. When it has sent all 64, it switches state to become a receiving memory. The receiving memory waits for DATA VALID to come from the BPs (they pipeline DATA VALID for three cycles to mimic their internal pipeline.) Once DATA VALID from the BP rises, the receiving memory reads the pixels from its input ports and stores them in memory. When it has received all 64, it switches state to become a transmitting memory and immediately begins transmitting pixels. By this time, the memory which had been transmitting is already awaiting the receipt of the pixels.

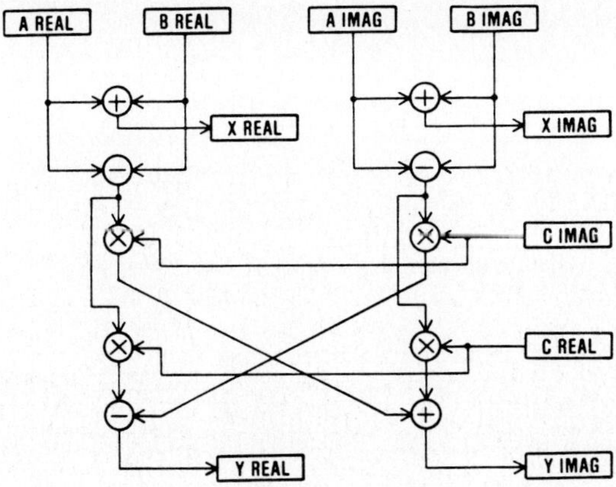

Figure 10: Butterfly Processor Function

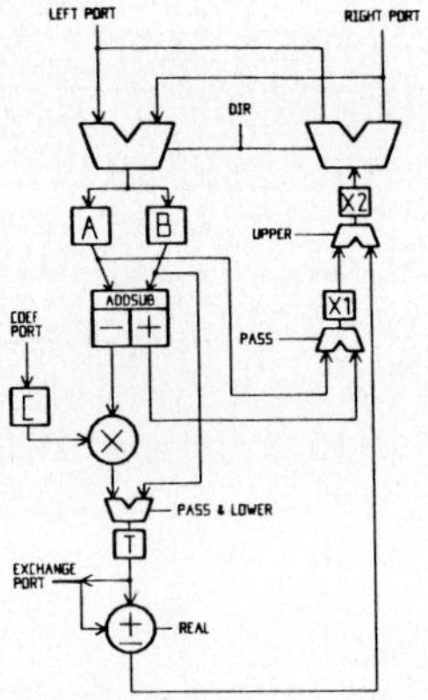

Figure 11: Gate Array Chip Schematic

4 Butterfly Processor

Each butterfly processor must perform the function diagramed in Figure 10. A minimal implementation would consist of three adders, three subtractors and four multipliers. However, preliminary calculations showed that the 16 by 16 multiplier would take one-third to one-half of the desired die size. Thus, the design was split into sections, each containing one multiplier. The partitioning chosen allowed each partition to be made from the same chip. This chip is shown in Figure 11.[9] The chip used requires two "personality signals" to tell it which of the four positions it occupies. A third "personality signal" tells the chip whether it receives its 'A' pixel or its 'B' pixel first from the WM.

4.1 Implementation

Various algorithms exist for the realization of the adders, subtractors, multipliers, and registers required in the BP. We chose algorithms to meet the overall timing

Adder	*Relative Delay*	*Relative Size*
RCA	1.0	1.00
CLA	0.4	2.50
CSA	0.5	1.25

Table 1: Comparison of Adder Designs

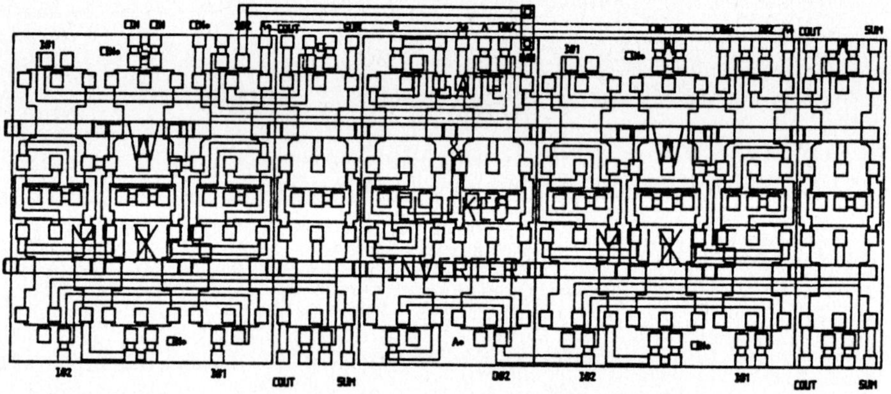

Figure 12: Artwork for Combined Adder-Subtractor

goals of the project while taking as little area as possible.

4.1.1 Adders and Subtractors

Three possible adders were considered for this chip, the ripple carry adder (RCA), the carry look-ahead adder (CLA), and the carry skip adder (CSA). These three adder implementations were simulated using SPICE and compared. The results are summarized in table 1. The CSA was chosen because it is far smaller than the CLA, but not much slower.

As shown in Figure 11, the BP requires two adder-like circuits. The first computes the sum and difference of the A and B values simultaneously. In twos complement arithmetic, the customary way to perform subtraction is to invert the bits of one of the inputs to the adder and set the carry in to 1. The first stage of such a carry-skip subtractor would compute $(A \oplus \overline{B})$. However, $(A \oplus \overline{B})$ is equivalent to $(\overline{A \oplus B})$ which is an intermediate value in the CSA. Thus, a savings of two gate-array cells per bit can be achieved if the adder and subtractor are combined into one module. Artwork for one bit of the adder-subtractor is in Figure 12.

The second "adder-like" circuit can be configured to calculate either the sum or the difference of two inputs. (The selection of addition *vs.* subtraction is determined by one of the "personality signals" controlling the BP.) Here also, a great savings can be achieved by combining the adder and subtractor into one module. Figure 13 has

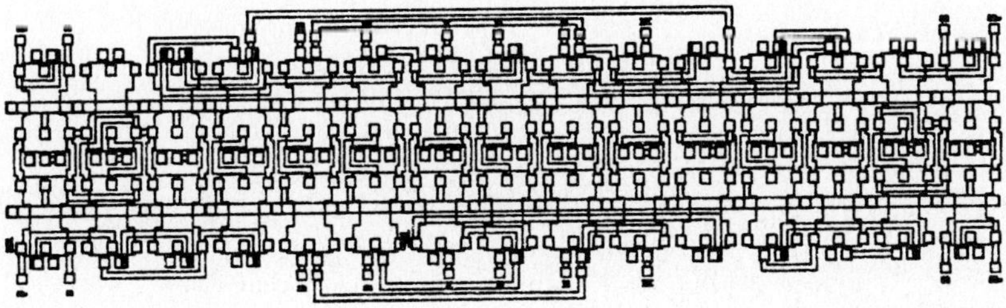

Figure 13: Artwork for Selectable Adder/Subtractor

artwork for one bit of this module.

4.1.2 Multipliers

The multiplier design must be capable of multiplying two 16-bit signed numbers (in twos complement form) to produce a signed 16-bit product. A Booth's recoded multiplier was chosen for implementation. In a Booth's multiplier, the multiplication is accomplished by successive stages of pipelined addition. Successive stages each look at three bits of the multiplier; there is a one-bit overlap between stages. Thus, to multiply two 16-bit numbers, eight stages are necessary. The three components of each stage are:

- *Mapping Logic* which looks at three bits of the multiplier to determine which multiple of the multiplicand should be added at that stage.
- *Multiplicand Converter* which generates the proper signed multiple of the multiplicand as requested by the mapping logic.
- *Adder* which shifts the previous partial product right by two bits and adds in the output of the multiplicand converter. A modified partial carry-save adder is used to speed up the addition.

4.1.3 Multiplexors and Registers

The BP chip requires seven registers (one is internal to the multiplier). The design of the standard registers is a master/slave combination with separate clocks. The separate clocking allows both control over when a value is latched and control over when it is available on the output lines.

Two of the registers, *X1* and *X2*, are preceded by multiplexors. The multiplexors and the registers are combined into one circuit because layout considerations preclude

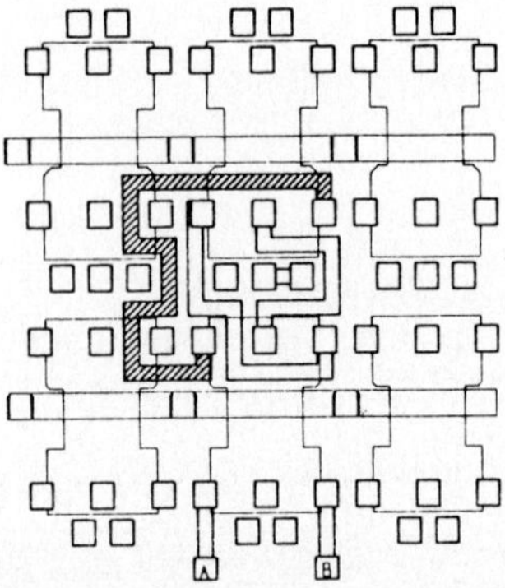

Figure 14: A One-Bit Multiplexor and Adjacent Cells

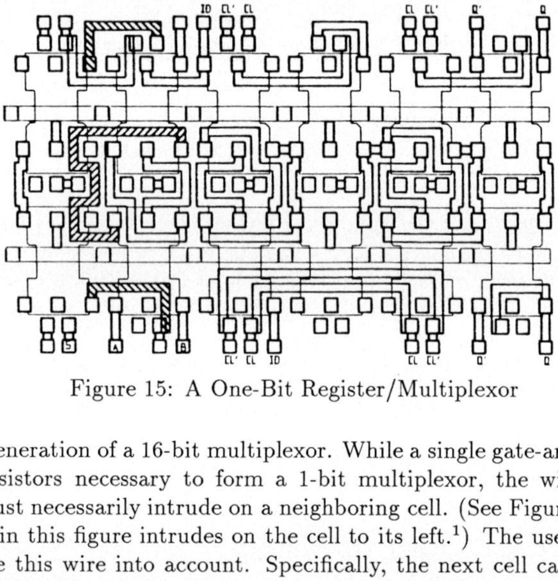

Figure 15: A One-Bit Register/Multiplexor

the efficient generation of a 16-bit multiplexor. While a single gate-array cell contains the four transistors necessary to form a 1-bit multiplexor, the wiring among the transistors must necessarily intrude on a neighboring cell. (See Figure 14. The cross-hatched wire in this figure intrudes on the cell to its left.[1]) The use of the adjacent cell must take this wire into account. Specifically, the next cell cannot be another 1-bit multiplexor. Thus, if a 16-bit multiplexor were implemented, it would include fifteen unused cells separating the 1-bit multiplexors. However, a 1-bit multiplexor followed by a 1-bit register can be developed that wastes no gate-array space. It is shown as Figure 15. The same wire is cross-hatched (///) in this figure as in Figure 14. The wires cross-hatched in the other direction (\\\) show how the blocked area is worked around. These wires both make the same connection. These cells can be placed side-by-side without wasting any space. Sixteen of these cells make the multiplexor/register sets that hold the output values.

[1]Some mask layers are omitted from this figure for clarity.

References

[1] J. F. McDonald, E. H. Rogers, K. Rose, A. J. Steckl, "The Trials of Wafer-Scale Integration", IEEE Spectrum, October 1984, pp 32–39.

[2] B. Donlan, J. McDonald, G. Taylor, R. Steinvorth, A. Bergendahl, "Computer-Aided Design and Fabrication for Wafer Scale Integration", VLSI Design, April 1985.

[3] G. F. Taylor, "A Two Dimensional Fast Fourier Transform Processor Suitable for Wafer Scale Integration", Rensselaer Polytechnic Institute Technical Report CIE 85-02, October 1985.

[4] G. F. Taylor, R. H. Steinvorth, J. F. McDonald, "An Architecture for a Video Rate Two-Dimensional Fast Fourier Transform Processor", IEEE Trans. Comput., September 1988, pp 1145–1148.

[5] B. Donlan, J. McDonald, R. Steinvorth, M. Dhodhi, G. Taylor, A. Bergendahl, "The Wafer Transmission Module", VLSI Design, January 1986, pp 54–56, 88–90.

[6] J. F. McDonald, M. Stanton, R. Rajapakse, H. Lin, R. Selvaraj, N. King, D. King, M. Haslam, "Adaptive Discretionary Wiring for Wafer Scale Integration Using Electron-Beam Lithography", Proc. SPIE, Vol 773, 5-6 March 1987, pp 140–149.

[7] O. W. Freitas, "Design and Implementation of a Working Memory for a Wafer Scale Integration Two Dimensional Fast Fourier Transform Processor", Rensselaer Polytechnic Institute Technical Report WSI 84-2, December 1984.

[8] G. T. Koker, "Design and Implementation of a Configurable 2K Static RAM for a 3 Micron CMOS Gate Array", Rensselaer Polytechnic Institute Technical Report WSI 85-1, July 1985.

[9] R. Philhower, "Design and Verification of a Butterfly Processor for a Two Dimensional Fast Fourier Transform Processor Suitable for Wafer Scale Integration", Rensselaer Polytechnic Institute Master's Thesis, December 1988.

Wafer Scale Integration, III
M. Sami and F. Distante (Editors)
Elsevier Science Publishers B.V. (North-Holland)

The Reliability of Large Arrays for Matrix Multiplication with Algorithm-Based Fault-Tolerance

William G. Bliss
Department of Electrical Engineering
Texas A&M University
College Station, Tx.

Michael R. Lightner
Department of Electrical Engineering
University of Colorado
Boulder, Co.

Algorithm-Based Fault-Tolerance (ABFT) applies linear block error codes to array implementations of linear-algebraic operations to detect, locate, or correct fault-induced errors. We consider run-time error detection and location, which would serve as input to reconfiguration and/or recalculation algorithms. The probability of undetected run-time errors, called the *infidelity rate*, is derived for a large class of matrix products and mesh array architectures under a high level multiplier fault model. The pure matrix algebraic form of ABFT is shown to have infidelity q^d times worse (larger) than new implementations which decode all implicit vector-scalar products, where q is the conformal dimension and d is the Hamming distance of the code. Low redundant overhead can be achieved with the $d = 2$ checksum code. More powerful codes with low overhead are formed as multi-dimensional products of the checksum code. The two-dimensional code is advantageous for VLSI/WSI because of ease of layout. Vector outer-products allow low overhead and low infidelity with more powerful ($d > 2$) linear basis codes. A new decoding strategy achieves minimum infidelity from a Single Error Correcting (SEC) code while only decoding the SEC code every $q'th$ iteration, but decoding the checksums each iteration.

1. Introduction

This paper concerns the general problem of increasing the reliability of computations for signal processing and linear algebraic operations, and specifically addresses the problem class characterized as being both 'number crunching' and highly amenable to parallel processor implementations. We will consider in detail the enhanced reliability derived from applying Algorithm-Based Fault-Tolerance (ABFT) to various arrays computing various matrix products. These results apply directly to large custom WSI or VLSI arrays, and may be simply extended to various general multi-processor systems.

1.1 Algorithm-Based Fault-Tolerance

The term Algorithm-Based Fault-Tolerance [1][4][9][10] was used to describe the general technique of detecting and/or correcting fault-induced errors using special properties of a given algorithm. In practice, ABFT has meant the application of linear block error codes to array implementations of certain linear algebraic operations. There are three main distinctions between the methods of ABFT considered here and the general fault tolerant literature. First, ABFT applies redundancy in the hardware domain instead of in the time domain. Second, ABFT strives to keep the redundancy (overhead) to a small fraction of that required by the unprotected algorithm. And third, ABFT is applied at a high level which makes it (hopefully) independent of many low level details, such as multiplier architecture.

The simplest ABFT code is the *checksum code* which appends one redundant number (symbol) to n data (message) numbers. The appended number, called the *checksum*, a_n, is just the negative sum of the data, denoted:

$$a_n = -\sum_{i=0}^{n-1} a_i. \tag{1.1}$$

The use of the checksum code on a linear array is illustrated in figure 1. The top row of adders calculates the checksum, which is called *encoding*. The encoded vector (of length $n+1$) is multiplied by the scalar x in the row of $n+1$ multipliers, one multiplier for each element of the encoded vector. The elements of the resultant vector are denoted $y_i = a_i x$. Finally, the bottom row of adders sum the $n+1$ elements of the vector $\vec{y}$, which is called *decoding*. The

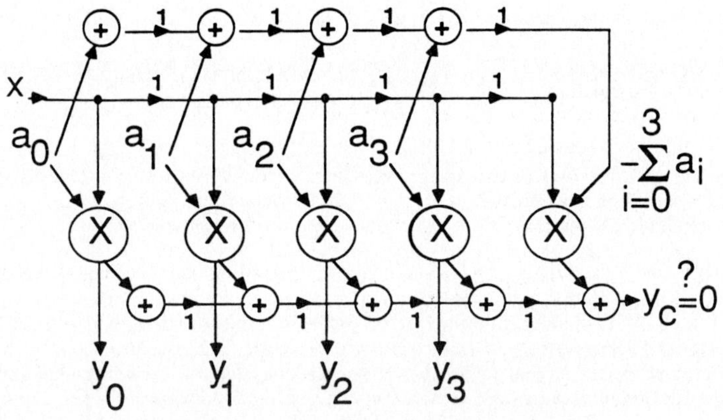

Figure 1: Linear array for vector-scalar product with checksum code. Encoding in top row, decoding in bottom row.

result of decoding is called the *syndrome*, which for the checksum code is the single number:

$$S_0 = -y_n + \hat{y}_n = -x\sum_{i=0}^{n-1} a_i + \sum_{i=0}^{n-1} x a_i. \tag{1.2}$$

It is clear that in the absence of faults, if multiplication distributes over addition, i.e. in a distributive algebra, then S_0 is identically zero. This is the usual condition for error control codes, which are normally implemented in finite (Galois) fields. In that case the checksum code has minimum distance $d_{min} = 2$, which means only two or more errors can produce a false zero syndrome. However, for the typical application where calculations use approximations to the real or complex number fields, then certain small magnitude errors will be indistinguisable from normal roundoff errors. This degradation in theoretical coding power is considered extensively in [2][3], where it is shown that additional hardware is needed to mitigate or overcome these numerical problems. In this paper these numerical effects will be simply ignored, i.e., the arithmetic is considered to be perfect.

More powerful codes can be applied to ABFT arrays. For example, the exponentially weighted *single-error-correcting* (SEC) code introduced in [4] adds an additional redundant symbol, denoted a_{n+1}, besides the checksum,

$$a_{n+1} = -\sum_{i=0}^{n-1} 2^i a_i. \tag{1.3}$$

The weight multiplications by powers of two are generally simple to implement, but have inferior numerical performance to a minimum dynamic range code introduced in [2][3].

The application of the linear basis codes above to matrix algebraic formulations was described in [1][4]. Consider the matrix-vector product:

$$\vec{y}^{n\times 1} = A^{n\times q}\vec{x}^{q\times 1}, \tag{1.4}$$

where the superscripts indicate dimensions. The column-encoded ABFT form of this equation is derived by encoding each column of A as a vector in equation 1.1. The resultant vector is then also column encoded:

$$\vec{y}_c^{(n+1)\times 1} = A_c^{(n+1)\times q}\vec{x}^{q\times 1}, \tag{1.5}$$

where the subscript c indicates column encoding. While technically correct, we will show that this formulation ignores the q vector-scalar products inherent in the matrix-vector product, and thus is not optimal.

Now consider the matrix-matrix product:

$$Y^{n\times m} = A^{n\times q}X^{q\times m}. \tag{1.6}$$

Again, encode each column of the A matrix according to the linear basis code. In a like manner, encode each row of the Q matrix. The resultant matrix is clearly both row and column encoded:

$$Y_f^{(n+1)\times(m+1)} = A_c^{(n+1)\times q}X_r^{q\times(m+1)}, \tag{1.7}$$

where the subscript r indicates row encoding, and f indicates simultaneous row and column encoding. Again, this formulation ignores the constituent vector-scalar products and is thus not optimal. The matrix form of ABFT for linear basis codes other than the checksum code used above is straightforward.

In this paper we will introduce other coding methods and examine their effect on reliability.

1.2 A Fault Model and Infidelity

An advantage of ABFT is that the fault model can be defined at a very high cell level. Here we take a 'cell' to be a dedicated simple collection of a multiplier and a few adders. In general, we consider arrays composed of many simple (small) cells. We consider a run-time fault model where faults are spatially independent, i.e. a faulty cell doesn't influence its neighbors to fail. This work also applies to transient (i.e., non-permanent) errors, as long as the errors remain spatially independent. Notice that spatially correlated errors could be dealt with by using traditional error codes for bursty channels, such as interleaved codes.

The reliability of a system at any time t is the probability that the system has performed adequately at all times up to t, given that it performed adequately at time $t = 0$. Typically, no external repair or replacement, nor any down time is allowed. The availability of a system at time t is conventionally defined as the unconditional probability of adequate operation at time t, regardless of previous status. These measures are inadequate for our purposes because they do not consider applications where no result (error flagging) is preferred to erroneous results, and their calculation requires specific knowledge about the reconfiguration algorithm and hardware. Therefore, we are motivated to define the time independent probabilities of unavailability and infidelity.

Define the unavailability rate, $\tilde{U}$, as the probability that the system reports an error is detected, and define the infidelity rate, $\tilde{F}$, as the probability that data in error is not detected by the system. Notice that it is trivial to obtain a system which is never unavailable by simply not detecting errors. Similarly, it is trivial to obtain a system which is never unfaithful by always flagging (detecting) every result. However, it is clearly not trivial to simultaneously achieve low probability of unavailability and infidelity. To simplify our analysis and discussion, we will limit our consideration to adding redundancy to completely available systems so as to decrease the infidelity rate, which must increase the unavailability. Such systems are sometimes called *concurrent error detecting* (CED).

Define n as the number of simple cells in an original unprotected system, N as the number of cells in the redundant system, and ρ as the probability of fault in any cell during one word-level operation (i.e., time of one multiplication). The difference, $N - n \equiv p$, is the redundant overhead measured in cells. The infidelity rate of the unprotected original array is then:

$$\tilde{F}_{orig} = 1 - (1 - \rho)^n \approx n\rho, \tag{1.8}$$

where we assume $n\rho \ll 1$, which means the original array is fairly reliable. We are thus considering an incremental approach to improving the reliability (e.g. decreasing infidelity here) of reasonable original arrays by the use of small amounts of redundancy.

To further illustrate our notation and assumptions, consider the common error detection system built by duplicating each of the n cells and comparing results. Optimistically ignoring the overhead and the unreliability of the comparison circuits, the infidelity rate is dominated by faults in an original and duplicate cell;

$$\tilde{F}_{Dup} \approx n\rho^2. \tag{1.9}$$

The unavailability rate increases from 0 for the unprotected original array to $\tilde{U}_{Dup} > 2n\rho$, which is dominated by a single cell failure.

Notice that while we have assumed the time of occurence and the location of fault-induced errors are spatially independent, the infidelity rates above have implicitly assumed that error values are correlated in a worst-case sense, e.g. equal and opposite errors in the duplication code. For most fault models and transient error models however, these errors are also independent, so the true infidelity rate is lower. In a perfect arithmetic system using B-bit number representations, e.g. the Galois field $GF(2^B)$, the infidelity rate for the duplication code is decreased to;

$$\bar{F}_{Dup} \approx n\rho^2 2^{-B}, \tag{1.10}$$

where the overbar denotes the effect of the independent and uniform error distribution.

Unless otherwise noted, we will assume that all problem instances have all inputs as variables, i.e., there are no 'constants'. Therefore, each problem instance must be encoded as well as decoded. The essence of the design problem considered in the rest of the paper is to distribute computations over an array of cells so that any faulty cell can only introduce a single error in any syndrome component.

We of course want designs with low redundant overhead that still yield low infidelity.

2. Vector-Scalar Products

In this section we consider various fault-tolerant array implemtations of vector-scalar products and their infidelity. Vector-scalar products are especially important, since we will show that matrix-vector products and matrix-matrix products are best understood in terms of their implicit vector-scalar products.

2.1 Checksum Code

Consider the checksum code implemented on the array of figure 1, where the overhead is $p = 1$ cell. However, each cell now includes two adders, one for encoding and one for decoding. We will uniformly assume that the hardware complexity of an adder is much less than that of a multiplier, otherwise a larger effective ρ must be used.

Theorem 2.1 The checksum code (on the linear array) has infidelity rate;

$$\tilde{F}_{cs} < 1 - [(1-\rho)^N + N\rho(1-\rho)^{N-1}] \approx \frac{(N\rho)^2}{2} = \frac{((n+1)\rho)^2}{2}. \tag{2.1}$$

Proof: The proof is direct from the assumptions and definitions. The inequality comes since we ignore the case where errors in the redundant checksum go undetected. •

Notice that the unavailability is increased from 0 to $\tilde{U}_{cs} \approx N\rho$.

In perfect arithmetic, assuming independent uniformly distributed errors, the infidelity is decreased to $\bar{F}_{cs} = \tilde{F}_{cs}2^{-B}$. For systems using approximate arithmetic, fault-free (and transient error free) operation doesn't produce indentically zero syndromes because of roundoff errors. In that case, some small magnitude errors are indistinguishable from roundoff errors, and the syndrome must be compared against some threshold, $\Delta \equiv 2^{-Q} > 2^{-B}$. The infidelity rate is then only decreased to $\bar{F}_{cs} = \tilde{F}_{cs}2^{-Q}$.

2.2 Minimal Redundancy Codes

The exponentially weighted checksum code with p=2 as described by equations 1.1 and 1.3 is a minimally redundant SEC code (with mininum distance $d = 3$).

Theorem 2.2: The minimally redundant ($p = 2$) SEC codes have infidelity rate;

$$\tilde{F}_{mrsec} \approx \binom{N}{3}(3\rho)^3 \approx \frac{9}{2}N^3\rho^3, \tag{2.2}$$

Proof: A minimum of three errors are needed to produce an undetected error, and since $n\rho \ll 1$ the three error case dominates. Each data multiplier cell has an encoder multiplier and a decoder multiplier, so the effective failure rate of each supercell is $\rho' = 3\rho$. •

It is well known that minimally redundant codes with distance $d = p+1$ can be constructed using p redundant symbols.

Theorem 2.3: The minimal redundancy $d = p + 1$ codes have infidelity rate:

$$\tilde{F}_{mrd} \approx \binom{N}{d}((2d-3)\rho)^d. \tag{2.3}$$

Proof: Such codes have p-1 'weighted checksums' which each take one encode and one decode weight multiplier for each data element. Thus $2(p-1)$ redundant multipliers are used for each data element, so $\rho' = (2d-3)\rho$. The failure rate is dominated by d faults. •

The high overhead, $\mu > p$, makes these codes expensive and also lowers the availability for large p.

2.3 Generalized Hamming and Locator Codes

Powerful codes without expensive encoding and decoding multiplications can be constructed for ABFT in a manner analagous to the generalized Hamming codes. The ternary generalized Hamming codes [5] can be considered as an ensemble of checksum codes, where each checksum is over a specific subset of the data symbols. For the SEC case, the number of redundant symbols required is $p > log_3 2n \approx 0.631(1 + log_2 n)$. Similar codes which are simpler to decode in the ABFT setting are described in [3].

Theorem 2.4: The generalized Hamming and explicit locator codes have infidelity rate:

$$\tilde{F}_{gh} \approx \binom{N}{3}\rho^3 \approx \frac{(n + log_3 2n)^3 \rho^3}{6} \tag{2.4}$$

Proof: Because there are no weight multipliers, $\rho' \approx \rho$. The infidelity rate is then dominated by 3 faults. •

These codes can be similarly designed for greater detection/correction power. However, the number of encode and decode additions grows with order $O(np)$, and there is no known good method to layout and route these adders.

2.4 Multi-Dimensional Checksum Codes

A similar method of generating greater power codes without encoding and decoding weight multiplications is to form the multi-dimensional product of the checksum code. General product codes are due to [6], and are also known as iterated or N-dimensional codes. The geometric interpretation is to arrange the n cells uniformly inside a ν-dimensional hyper-cube of unit volume n, with edges of length $s = n^{1/\nu}$. The hyper-cube has ν pairs of parallel faces, and each face has $s^{\nu-1}$ elements. The multi-dimensional product code extends each dimension (edge) by p, where p=1 for the checksum code. Each 'line' of length s perpendicular to a face is encoded. Note that redundant symbols are also encoded in a self-consistent manner. The number of redundant symbols in a multi-dimensional product code is

$$p' \equiv N - n = [(s+p)^\nu - s^\nu] > \nu p s^{\nu-1} \tag{2.5}$$

The minimum distance of the product code is known to be the product of the distances of the underlying codes, so $d' = d^\nu$.

For the two-dimensional ($\nu = 2$) product of the checksum code, $p' = 1 + 2\sqrt{n}$, and $d' = 4$, as in figure 2. Direct application of equation 2.3 bounds the infidelity rate as:

$$\tilde{F}_{2dcs} < \binom{N}{4}\rho^4 \tag{2.6}$$

But the bound above isn't tight because it ignores spatial dependence.

Theorem 2.5: The two-dimensional checksum code has infidelity rate:

$$\tilde{F}_{2dcs} \approx \left[\binom{\sqrt{N}}{2}\right]^2 \rho^4 \approx \frac{N^2\rho^4}{4}, \tag{2.7}$$

Proof: All weight four faults which produce all zero syndrome components must be arranged as rectangles. There are less than $N^2/4$ such rectangles, because there are only $\sqrt{N}$ choose two possibilities in each of the two degrees of freedom. •

Theorem 2.6: The ν-dimensional checksum code has infidelity rate:

$$\tilde{F}_{\nu dcs} \approx \frac{N^2\rho^{2^\nu}}{2^\nu} \tag{2.8}$$

Proof: The product code has minimum distance $d' = 2^\nu$. The only weight 2^ν undetected error patterns are ν-dimensional boxes in the hypercube. There are less than $N^2/2^\nu$ of these boxes possible, since there are only $\binom{N^{1/\nu}}{2}$ combinations in each of the ν dimensions. •

Notice that the assumption, $\rho' \approx \rho$, is less valid for high dimension codes. The number of adders used to implement the code increases linearly with ν, i.e., a ν dimensional code requires νN adders. Also, more difficult than the number of adders is the interconnection of them for large dimensions. For example, the three dimensional code in a VLSI model of computation requires greater than $N^{4/3}$ unit area, even when the processor cells are of unit area.

The two-dimensional checksum codes have been extended to distance 6 and 8 by adding checksums on the extended diagonals and anti-diagonals respectively [3]. While the coding gain is not as large as with higher dimension codes, inter-cell communication is still only with nearest neighbors, which is an important advantage for VLSI/WSI.

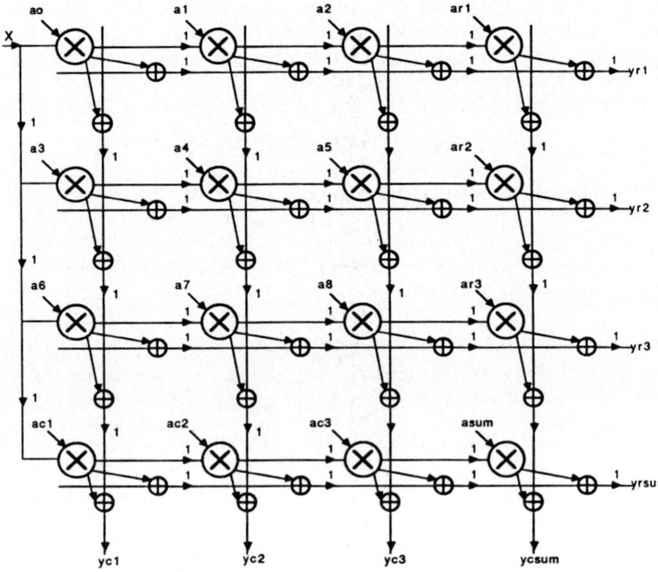

Figure 2: Mesh array for vector-scalar product with two-dimensional checksum code. Encoding not shown.

2.5 Error Distribution

Finally, we comment that all of the codes for vector-scalar products provide additional error detection based on the uniform distribution of errors. For example, the minimally redundant SEC codes may not detect certain three error failures, but of the three error 'words', only two word degrees of freedom produce this condition. The average infidelity is then $\bar{F}_{mrsec} = \tilde{F}_{mrsec}2^{-2B}$ in perfect arithmetic and $\bar{F}_{mrsec} = \tilde{F}_{mrsec}2^{-2Q}$ in approximate arithmetic. Similarly, consider the two-dimensional checksum code. Only certain geometric arrangements of four errors can produce an undetected error, but of those four words, only one word degree of freedom exists. The average infidelity is then $\bar{F}_{2dcs} = \tilde{F}_{2dcs}2^{-3B}$ in perfect arithmetic and $\bar{F}_{2dcs} = \tilde{F}_{2dcs}2^{-3Q}$ in approximate arithmetic.

3. Matrix-Vector Products

Consider the matrix-vector product, also known as a linear transformation:

$$\vec{y}^{n\times 1} = A^{n\times q}\vec{x}^{q\times 1}, \tag{3.1}$$

where the superscripts indicate the dimensions. The problem in general requires nq multiplications (we don't consider special structured or sparse matrices).

Theorem 3.1: The inherent infidelity of an unprotected solution of the matrix-vector product is:

$$\tilde{F}_{original} \approx nqp. \tag{3.2}$$

Proof: The proof is direct from the assumptions and definitions. •

By encoding each column of the A matrix with a p-redundant code we get:

$$\vec{y}_c^{(n+p)\times 1} = A_c^{(n+p)\times q}\vec{x}^{q\times 1}, \tag{3.3}$$

where $\vec{x}$ is not encoded, and the subscript, c, indicates column encoding. This is the matrix algebra form of ABFT. It checks qn multiplications with qp extra multiplications, but reduces that information down to p checks on $\vec{y}$, which throws away most of the error information.

We view the result as a sum of the products of columns of A with components of $\vec{x}$:

$$\vec{y}^{n\times 1} = \sum_{i=1}^{q} \vec{A}_i^{n\times 1} x_i^{1\times 1}, \tag{3.4}$$

where $\vec{A}_i$ is the i'th column of A. Thus, we again encode the columns of A, but we check all the column vector times scalar products, which requires qp extra multipliers as above, but detects errors out of groups of n multipliers. Thus, we view a matrix-vector product not as the inner-products that form elements of the output vector, but rather as a group of implicit vector-scalar products. We will look at several strategies for error detection on arrays, and compare the fidelity they achieve.

3.1 Linear Array

We first consider the common vector or linear array type of solution. If we treat the problem as computing inner products,

$$y_i = A_{row\ i}\vec{x}, \tag{3.5}$$

then ABFT is not applicable, as there are no vector-scalar products. But, if we view the problem as the summation of vector-scalar products as in equation 3.4, then we can apply ABFT. First consider the simple checksum code on each column of A.

Theorem 3.2: The infidelity rate for a matrix-vector product executed recursively on a linear array with checksum code is:

$$\tilde{F} \approx q(N\rho)^2/2. \tag{3.6}$$

Proof: The infidelity rate for each column of A (each vector-scalar product) is $(n\rho)^2/2$, so the infidelity rate for the full matrix of q columns implemented recursively column by column is approximately q times worse. •

Consider a minimal redundancy code with $p = d - 1$ redundant symbols of section 2.2.

Theorem 3.3: The infidelity rate for a matrix-vector product executed recursively on a linear array with a minimal redundancy code of minimum distance d is:

$$\tilde{F} \approx q\binom{N}{d}((2d-3)\rho)^d. \tag{3.7}$$

Proof: The proof follows theorem 2.3. •

These codes generally have too much overhead to be practical.

3.2 Scalar-Vector Mesh Array

Now consider a rectangular mesh array of q by n data elements, which is essentially q stacked copies of the linear array of figure 1. We refer to this array (not illustrated here) as a scalar-vector mesh array because it is like q copies of the linear array for scalar-vector products, but where only the *i'th* components of each vector are summed. The *i'th* element of $\vec{x}$ is broadcast to the *i'th* row of the array, where the *i'th* column of A is stored (or input). In the matrix algerbra form of ABFT, only a single overall check of $\vec{y}$ is implemented.

Theorem 3.4: The infidelity rate of a matrix-vector product on the scalar-vector mesh with a single overall checksum is:

$$\tilde{F} \approx \binom{N}{2}(q\rho)^2 \approx (qN\rho)^2/2, \tag{3.8}$$

Proof: One fault in a column (with q cells per column) will introduce one fault in the checksum code, so the effective $\rho' = q\rho$. •

Notice that this is inferior to the linear array, even though the same percentage of redundant multipliers are used (and the same amount of encoding is performed), because the checking (decoding) is not as frequent.

Consider adding decoding hardware to each column of the partial sum of $\vec{y}$, which are rows of the array. This is q copies of figure 1, with an additional overall check on the sum of the columns of the array.

Theorem 3.5: The infidelity rate of a matrix-vector product on the scalar-vector mesh with each vector-scalar product decoded with checksum is identical to that of the linear array in equation 3.6.

Proof: The scalar-vector mesh array is now used like q independent linear arrays. •

Now consider encoding each column of A with a $d = 3$ minimally redundant (SEC) code. Consider the infidelity rate when only the result $\vec{y}$ is decoded, which is the matrix algebraic form of ABFT.

Theorem 3.6: The infidelity rate of a matrix-vector product on the scalar-vector mesh with one overall decoding of a minimally redundant SEC code is:

$$\tilde{F} \approx \binom{N}{3}((2q+1)\rho)^3 \approx \frac{4}{3}(qN\rho)^3, \tag{3.9}$$

Proof: Any fault in a column of the mesh contributes an error to the code. Each column has q data multipliers, q encoding weight multipliers, and one decoding weight multiplier, so $\rho' = (2q+1)\rho$. •

Note that if contrary to assumption, the A matrix is a constant, then the encoding need only be done once, and the effective $\rho' = (q+1)\rho$.

Consider decoding the SEC for each vector (row in array)

Theorem 3.7: The infidelity rate of a matrix-vector product on the scalar-vector mesh where each vector-scalar product is decoded with a minimally redundant SEC code is:

$$\tilde{F} \approx q\binom{N}{3}(3\rho)^3 \approx \frac{9}{2}q(N\rho)^3, \tag{3.10}$$

Proof: The proof follows theorem 2.2, and the results are the same. •

An interesting alternative to decoding each vector-scalar product (row of array) for the SEC at each iteration is to instead decode each row for the checksum code and to only decode the single overall result $\vec{y}$ for the SEC code. This greatly reduces the decoding overhead, and we now prove that it achieves approximately the same reliability as decoding each vector for the SEC code.

Theorem 3.8: The infidelity rate of a matrix-vector product on the scalar-vector mesh where each row is decoded for a checksum code and where a single overall decoding check is for the minimally redundant SEC code is:

$$\tilde{F} \approx (q+1)\binom{N}{3}(2\rho)^3 \approx \frac{4}{3}(q+1)(N\rho)^3, \tag{3.11}$$

Proof: Any undetectable (fooling) error pattern must now be made up of columns (rows of array) with two or more errors, and there must be three or more rows (columns of array) affected. This condition is of course met by three errors in any single column, which has a probability the same as SEC checking each vector. The most likely alternative is that two rows of the array have two errors in each row, which has probability

$$\binom{q}{2}\left[\binom{N}{2}\rho^2\right]^2 \approx \frac{1}{8}q^2(N\rho)^4, \tag{3.12}$$

But $N\rho \ll 1$ by assumption, so $qN\rho < 1$, and therefore this fault pattern is less likely than three errors in a single vector. There are q+1 rows that can fail (counting the decoding row), and $\rho' \approx 2rho$, since there is an encoding weight multiplication and not a decoding weight multiplication. •

We therefore achieve performance comparable to checking every vector with the full SEC code, while actually only checking every vector for the checksum code. This technique is especially useful if A is a constant. In that case the encoding is performed only once, and the overhead is almost halved to approximately $2(n+q)/nq$.

This technique also applies to more powerful codes. Consider using a $d = 5$, double error correcting (DEC) code.

Theorem 3.9: The infidelity rate of a matrix-vector product on the scalar-vector mesh where each row is decoded for a $d = 3$ SEC code and where a single overall decoding check is for the $d = 5$ (DEC) code is:

$$\tilde{F} \approx (q+1)\binom{N}{5}(4\rho)^5 \approx 8.53(q+1)(N\rho)^5, \tag{3.13}$$

Proof: Any fooling (undetectable) error pattern must now be made up of columns (rows of array) with three or more errors, and there must be five or more rows (columns of array) affected. This condition is of course met by five errors in any single column, which has a probability the same as checking each vector with the $d = 5$ code. The most likely alternative is that two rows of the array have three errors in each row, which has probability

$$\binom{q}{2}\left[\binom{N}{3}(3\rho)^3\right]^2 \approx 10.125q^2(N\rho)^6, \tag{3.14}$$

But $N\rho \ll 1$ by assumption, so $qN\rho < 1$, and therefore this fault pattern is less likely than five errors in a single vector. There are $q+1$ rows that can fail (counting the decoding row), and $\rho' \approx 4\rho$, since there are three encoding weight multiplications. •

Again, if A is constant the overhead is significantly lowered by this technique.. However, if the A matrix is constant for several vectors $\vec{x}(t)$, then we can formulate the problem as a matrix-matrix product, which allows more effective coding yet.

4. Matrix-Matrix Products

Now consider multiplying full and dense matrices:

$$Y^{n\times m} = A^{n\times q}X^{q\times m}. \tag{4.1}$$

Theorem 4.1: The infidelity rate for an unprotected implementation of matrix-matrix multiplication is:

$$\tilde{F} \approx nqm\rho. \tag{4.2}$$

Proof: The proof is direct from the assumptions. •

The matrix algebraic form of ABFT checks the columns of A and the rows of X to give a result Y that is both row and column encoded. Assume p redundant symbols are used to encode the columns of A and the rows of X.

$$Y_f^{(n+p)\times(m+p)} = A_c^{(n+p)\times q}X_r^{q\times(m+p)}, \tag{4.3}$$

The result of $(n+p)q(m+p)$ multiplies is thus compressed to $p(n+m+p)$ checks. Using our method, we can view the problem as both $q(m+p)$ vector-scalar products of length $n+p$, and simultaneously as $q(n+p)$ vector-scalar products of length $m+p$.

4.1 Linear Array

First consider the performance possible from a linear (vector) array.

Theorem 4.2: The infidelity rate for a matrix-matrix product calculated recursively on a linear array with checksum code is either:

$$\tilde{F} \approx mq((n+1)\rho)^2/2 \qquad \text{column checksum A and Y}, \tag{4.4}$$

$$\tilde{F} \approx nq((m+1)\rho)^2/2 \qquad \text{row checksum X and Y}, \tag{4.5}$$

Proof: First consider column encoding of A for equation 4.4. Compute Y one column at a time, where each column is a sum of q vector-scalar products of length $n+p$. We now have m matrix-vector problems of the same form as section 3. Notice that row encoding of X can't add any guaranteed coding powers in this case, so we just column encode A, and hence Y. Doing so, we have mq uses of the length n linear array, and hence equation 4.4. Now consider row encoding of X for equation 4.5. Calculate Y one row at a time, where each row is the sum of q vector-scalar products of length $m+p$. Notice that column encoding A doesn't add to the guaranteed coding power, so we just row encode X, and hence Y. We now have nq uses of a linear array of length m, and hence equation 4.5. •

The selection between these two methods might be based on selecting the longer of n or m, so as to reduce the redundancy overhead. Of course, if n and m are too long, the problem can be simply partitioned, or more powerful codes can be used. If more powerful codes such as the minimal redundancy basis codes are used, then the infidelity rate for column encoding of Y is m times that in theorem 3.3, etc.

4.2 Scalar-Vector Mesh Array

The matrix-matrix product on this mesh is simply multiple uses of the mesh as in multiple calculations of matrix-vector products. Using column encoding the infidelity rates for the whole problem are just m times worse than the matrix vector problem discussed previously. The technique for checking all the vectors with a less powerful decoder than the overall check still applies. For the row encoding approach, the values for n and m simply switch places, as above for the linear array.

4.3 Outer-Product Mesh Array

The simplest method to take advantage of simultaneous row encoding of X and column encoding of A, which produces the fully encoded Y, is to distribute the calculations of the outer-products of the $i'th$ column of A with the $i'th$ row of X over a mesh of mn multipliers. Such an outer-product mesh is illustrated in figure 3 for the checksum code, where both the elements of A and X are broadcast, and Y accumulates in place. The calculation takes q iterations, where each iteration forms an outer-product and adds it into the sum of previous outer-products:

$$Y_f^{(n+p)\times(m+p)} = \sum_{i=1}^{q} A_{c\ col\ i}^{(n+p)\times q} X_{r\ row\ i}^{q\times(m+p)}, \tag{4.6}$$

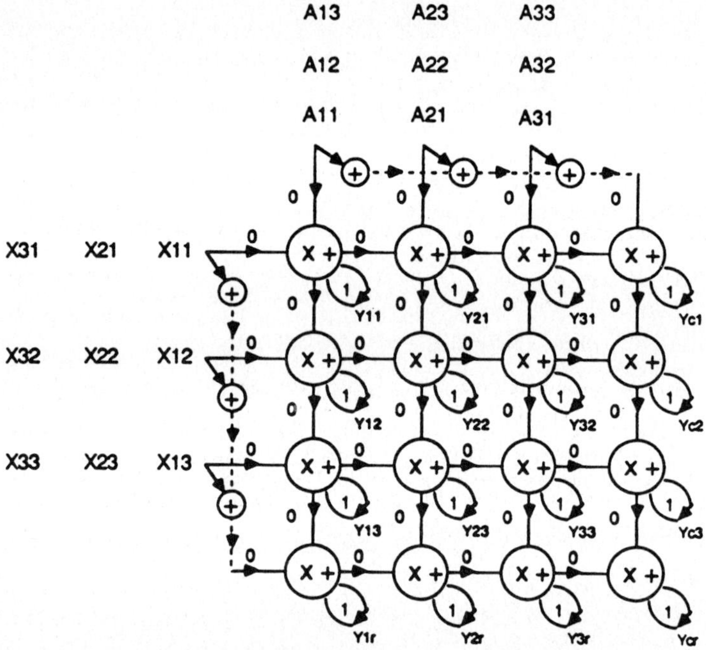

Figure 3: Outer-product array for matrix-matrix products, non-systolic, row and column encoding shown.

Notice that each outer-product is also a full checksum matrix. Checking for errors (decoding) just the final result, Y_f, is the matrix form of ABFT. Because the array must go through q recursions, we denote the decoding of the result (and not the intermediate products) as a 'final check.'

Theorem 4.3: The infidelity rate of matrix-matrix multiplication on the outer-product array with both row and column checksum codes and only decoding of the final result Y_f is:

$$\tilde{F} \approx [(m+1)(n+1)]^2(q\rho)^4/4 \approx (mn)^2(q\rho)^4/4. \tag{4.7}$$

Proof: The array is $(m+1)$ by $(n+1)$ and like the two-dimensional checksum code in theorem 2.5, the only weight four undetected errors form a rectangle on the mesh. •

Now consider decoding (checking for errors) after each of the q iterations on the array.
Theorem 4.4: The infidelity rate of matrix-matrix multiplication on the outer-product array with both row and column checksum codes and decoding at each of the q iterations is:

$$\tilde{F} \approx q\binom{m+1}{2}\binom{n+1}{2}\rho^4 \approx q(mn)^2\rho^4/4. \tag{4.8}$$

Proof: Like theorem 4.3, the weight four errors must form a rectangle. But since decoding is at each iteration, the factor q multiplies the probability of error in a single iteration, which is weighted by ρ^4. •
Notice that this is a factor of q^3 better than the single overall matrix check.

If a minimal redundancy SEC basis code is used for both row and column encoding, then the full check code has distance $d = 9$.
Theorem 4.5: The infidelity rate of matrix-matrix multiplication on the outer-product array with both row and column SEC code and only a final decoding of the result Y_f is:

$$\tilde{F} \approx \binom{m+1}{3}\binom{n+1}{3}(q\rho)^9 \approx \frac{1}{36}(mn)^3(q\rho)^9. \tag{4.9}$$

Proof: All weight nine undetected errors must line up as 3 errors in 3 rows and 3 columns. There are qm weight multiplications for row encoding and qn weight multiplications for column encoding. For the overall decoding there are $2mn$ weight multiplications. This number of weight multiplications is small compared to the mnq data multiplications (assuming moderate q), so the effective $\rho' \approx q\rho$. •
The multiplier overhead is quite small as long as n, q, and m are all moderate in size (not small). Notice that we have not specified how data leaves the outer-product array, and how the overall decoding is performed.

Consider decoding each iteration of the outer-product array with SEC row and column coding.
Theorem 4.6: The infidelity rate of matrix-matrix multiplication on the outer-product array with both row and column SEC code and decoding of each of the q iterations is:

$$\tilde{F} \approx q\binom{m+1}{3}\binom{n+1}{3}(3\rho)^9. \tag{4.10}$$

Proof: As in theorem 4.5, the weight nine undetected errors are in 3 rows and 3 columns. The effective $\rho' \approx 3\rho$ since each data multiplier has two decode multipliers, one for the row code and one for the column code. The $m+n$ encoding multipliers are negligible in comparison. •

Consider a technique similar to that used in theorem 3.8 Encode with a SEC code, decode each iteration for the simple checksum code, and provide an overall SEC decoding.
Theorem 4.7: The infidelity rate of matrix-matrix multiplication on the outer-product array with both row and column SEC code, checksum decoding at each of the q iterations, and a final SEC decoding of the result, is approximately the same as in equation 4.10 from decoding the SEC at each iteration.
Proof: The proof is based on showing that all undetected overall SEC errors that are also not detected by the checksum at each iteration are less likely than weight 9 errors at a single iteration. The most likely overall error that comes as errors over more than one iteration is two groups of 6 errors that cover 4 rows (3 errors per row) and 3 columns (4 errors per column) or vice versa. The probability of this type of event, assuming $m \approx n \approx q$, is:

$$\binom{q}{2}2\binom{n}{3}\binom{n}{2}\binom{n}{2}\rho^{12} \approx \frac{1}{24}n^9\rho^{12} \ll n^7\rho^9, \tag{4.11}$$

Other multi-iteration error patterns, such as 3 iterations of four errors on rectangles, have probabilities with the exponent of ρ four greater than the exponent of n. •
This is a strong result because the performance of a powerful distance nine code is achieved with very little multiplier overhead for reasonable values of n, q, and m. Notice that for the outer-product array only the input vectors need to be encoded, and there are less of them than encoded partial products.

Thus, we see that calculating matrix-matrix products as sums of outer-products has the strong performance advantage of two-dimensional coding, while the geometry of the data structures exactly fits with no penalty.

5. Banded Matrices

The material on banded matrices [3] is deleted due to space limitations. Similar results are achieved, despite iterative linear arrays and hexagonal arrays.

6. Conclusions

We have shown that all of the 'matrix' multiplication problems can best be understood from an ABFT point of view by considering the implicit vector-scalar products and vector outer-products. This formulation allows simple application of any desired linear basis code. From this vantage, matrix-vector products are simply a convenient method to denote many independent vector-scalar products. Vector outer-products allow the two-dimensional product of the linear basis code, which has strong coding advantages, while still conforming with the data structure for the problem itself. The vector outer-product formulation, known as the full check matrix form of ABFT is shown to be a special case of multidimensional product codes, which we have applied to vector-scalar products. These results also apply to banded matrices.

We have shown that for the greatest reliability enhancement from ABFT, all the vector-scalar products in a problem generally need to be checked. Checking all these vector-scalar products decreases the infidelity rate by a factor q^{d-1}, where q is the common dimension of the multiplier matrices, and d is the minimum distance of the overall product code. A novel decoding technique that decodes infrequently with a powerful code and more frequently with a less powerful code is shown to achieve the maximum reliability possible for various applications. These results have been extended to matrix triangularization array implementations for dense and banded matrices in [3].

References

[1] K.H. Huang, and J.A. Abraham, "Algorithm-Based Fault Tolerance for Matrix Operations," IEEE Trans. on Computers, Vol. C-33, No. 6, June, 1984, pp.518-528.

[2] W.G. Bliss, M.R. Lightner, and B. Friedlander "Numerical Properties of Algorithm-Based Fault-Tolerance for High Reliability Array Processors," Proc. of the 22nd Asilomar Conf. on Signals, Systems, and Computers, Nov. 1988.

[3] W.G. Bliss, *Area-Time Efficient and Fault-Tolerant VLSI Arrays for Digital Signal Processing*, P.h.D dissertation, Univ. of Colorado, 1988.

[4] J.Y. Jou, and J.A. Abraham, "Fault-Tolerant Matrix Operations on Multiple Processor Systems using Weighted Checksums," SPIE Vol. 495 on Real Time Signal Processing VII, 1984, pp.94-101.

[5] W.W. Peterson, and E.J. Weldon, *Error-Correcting Codes*, MIT Press, Cambridge, Mass., 1972.

[6] P. Elias, "Error-Free Coding," IRE Trans. Professional Group on Information Theory., Vol. PGIT-4, 1954, pp. 29-37.

[7] H.T. Kung, and C.E. Leiserson, "Systolic Array for VLSI," in *Sparse Matrix Proceedings*, 1978, SIAM, pp.256-282.

[8] U. Weiser, and A. Davis, "A Wavefront Notational Tool for VLSI Array Design," in *VLSI Systems and Computations*, ed. H.T. Kung, et. all., Computer Science Press, Rockville, Md., 1981, pp.226-234.

[9] J.Y. Jou, *Fault-Tolerant Matrix Arithmetic and Signal Processing on Highly Concurrent VLSI Systems,* Ph.D. dissertation, Univ. Illinois at Urbana, 1985.

[10] J.Y. Jou, and J.A. Abraham, "Fault-Tolerant Matrix Arithmetic and Signal Processing on Highly Concurrent Computing Structures," Proc. of the IEEE, Vol. 74, No. 5, May 1986, pp.732-741.

Wafer Scale Integration, III
M. Sami and F. Distante (Editors)
Elsevier Science Publishers B.V. (North-Holland)

A PRACTICAL EXPERIMENT IN DEFECT TOLERANCE : THE HYETI 16-BIT MICROPROCESSOR

R. Leveugle *, M. Soueidan *, N. Wehn **

* Institut National Polytechnique de Grenoble / CSI
46, avenue Félix Viallet
38031 GRENOBLE CEDEX
FRANCE

** Technische Hochschule Darmstadt
Schloßgartenstrasse, 8
6100 DARMSTADT
F.R.G.

This paper investigates microprocessor design techniques using redundancy in order to increase the yield by means of end-of-manufacturing defect tolerance. A lot of effort was done to obtain a regular design allowing the introduction of standby elements at an adequate level. In the HYETI microprocessor chip, the area of the redundant elements was limited to 25% and the goal was to achieve a good compromise between area overhead and yield enhancement. The HYETI chip is being manufactured within the ESPRIT 824 project and this paper reports mainly on practical results on silicon.

INTRODUCTION

The goal of the HYETI demonstrator of the ESPRIT project (824) was to prove the feasibility of a high yield defect tolerant 16-bit microprocessor. Such a device is supposed to become the core of an application specific microprocessor-based system integrated on a single chip.

In the past, defect-tolerant techniques have been mainly applied to regular arrays. The reconfiguration of Random Access Memories (RAMs) was addressed, as well at the chip level as at the wafer level ([1], [2], ...). A lot of work has also been done on the reconfiguration of regular arrays of Processing Elements (PEs) ([3], [4], ...). In that case, spare PEs are provided to replace the defective ones. The reconfiguration is performed after manufacturing by interconnecting the good PEs. This reconfiguration may involve one or several levels of hierarchy but the lowest one never involves the use of standby elements in the PE itself, except for eventual RAM parts. In other respects, the microprocessor studies reported up to now explored three types of crucial related problems: how to implement Built-In Self-Test (BIST) devices for off-line test purposes ([5], [6], ...), how to concurrently check some elements of the processor ([7], [8], ...) and how to improve the yield by defining a good design methodology [9]. Very few papers deal with end-of-manufacturing defect tolerance. The implementation of a correlator with redundancy is reported in [10] but only the reconfiguration of accumulators is considered and the circuit was realized in a bipolar technology. The introduction of redundancy to non regular CMOS circuits is considered in [11]; some figures are provided about area and yield but no practical information is given about the reconfiguration methodology. In fact, no paper reported the possibility to successfully use redundancy in a complex processing element in order to cope with some manufacturing defects.

This paper investigates microprocessor design techniques using redundancy. A lot of effort was done to obtain a regular design allowing the introduction of standby elements at an adequate level. The reconfiguration strategy must allow the tolerance of a large number of defects in order to increase the yield and to be able to use the processor in a WSI system. Nevertheless, all defects are not necessarily tolerated. The interconnection lines between the controller and the datapath are for example not reconfigurable. Some elements in the datapath are also not reconfigured because this would require a large area overhead. The area of the redundant elements was limited to 25% and the goal was to achieve a good compromise between area overhead and yield enhancement. The expected yield increase was theoretically found to be about 15 to 20%.

1. GENERAL PRESENTATION OF HYETI

The HYETI microprocessor is dedicated to highly dependable real time control systems and its characteristics have been determined in cooperation with potential industrial customers.

Special emphasis has been given to on-line test. A compaction device allows to compact any information on the bus through a multiple-input Linear Feedback Shift Register with a programmable characteristic polynomial. Dedicated on-line test instructions perform the verification of the instruction code sequences according to a "adjusted signature" method [12].

1.1. Software characteristics

The instruction set has been defined according to the application area (masked operations, boolean matrix processing, ...).

Eight addressing modes are provided : relative, inherent, immediate, direct, register indirect, register indirect with post-modification and register indirect with pre-modification. For the two last modes, the indirection register is automatically modified by adding a programmable arithmetical offset. These modes were introduced for efficient matrix processing.

The instruction set is divided into 10 classes :

- Arithmetic instructions,
- Logical instructions, which can be masked or not,
- Shift and rotate instructions,
- Comparison instructions with or without mask,
- Transfer instructions with or without mask,
- Instructions with implicit operand,
- Branch and jump instructions,
- Stack instructions,
- Interrupt instructions,
- Test instructions (for on-line test).

The masked instructions allow the programmer to process selected bits in a 16-bit word.

1.2. Hardware characteristics

HYETI is a 16-bit microprocessor with an addressing capability of 16 pages of 64 K words. The memory interface signals are compatible with those of Motorola family. Two hardware interrupt levels are provided, as well as a halt control signal.

The circuit is implemented in the CMOS double metal 1.2μ technology of SGS-Thomson Microelectronics. Its general architecture is shown in Figure 1. The elements of the datapath include registers, the ALU and two special devices : a masking device for masked operations and a signature device for on-line test purposes.

2. DESIGN METHODOLOGY

The underlying strategy consists of dividing the microprocessor into blocks where one defect can be corrected, except some fatal ones on critical lines (for example, control signal lines, which are not duplicated). This partitioning takes into account the general architecture of the chip : the datapath and the controller are considered separately.

Furthermore, the definition of the blocks has to be closely associated to the elements of the circuit. For example, it is impossible to consider that two PLAs are in the same block. They have obviously to be reconfigured separately. So, architectural and topological aspects of the design have to be considered during the partitioning.

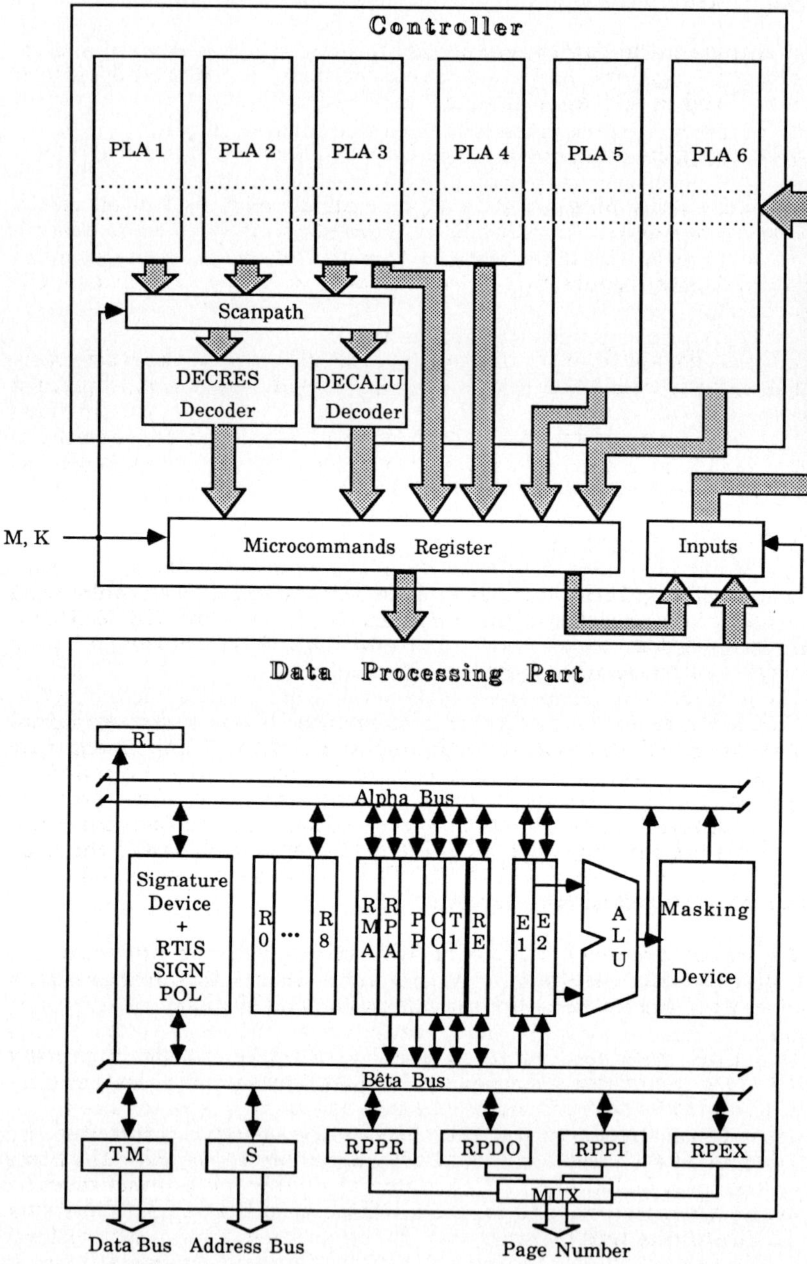

Figure 1 : General architecture of the HYETI microprocessor

2.1. Datapath reconfiguration

Generally speaking, the datapath of a microprocessor is made up of several functional units (registers, ALU, bus precharge logic, specialized operators) which are realized in a classical bit slice architecture.

The partitioning of the datapath can be either functional or structural [13].

The functional partitioning consists in a separate reconfiguration of the different functional units. This partitioning may be necessary when there is no regular structure. The main drawback of this strategy is the complex connection network required between blocks in order to achieve the reconfiguration.

In the HYETI datapath, we take advantage of the bit slice structure (structural partitioning). One spare slice is provided and can replace a defective one. Providing more spare slices would have led to an area increase superior to the limit fixed by the ESPRIT project. Furthermore, the area of the datapath is about 7 mm^2 and this part of the circuit can be considered as only one block.

2.2. Controller reconfiguration

A global lack of regularity can be noticed in the most commonly used controller structures. Several types of elements are necessary (ROM, PLAs, decoders and random logic) and the reconfiguration requires a lot of redundancy and several reconfiguration strategies.

In the HYETI controller, only one type of element (PLAs) is used. An initial virtual PLA is partitioned into several ones, some of them implemented as decoders. This organization in two stages allows to reduce the silicon area (a single PLA would be relatively sparse). Furthermore, the first stage is composed of several PLAs for topological reasons. A method has been defined in order to minimize the global size of the controller and the area needed for the interconnections [14]. The final controller is composed of six PLAs in the first stage and two decoders.

The PLAs which are used in the HYETI microprocessor have clocked dynamic planes with switched grounded array transistors for the AND plane and directly grounded array transistors for the OR plane.

The initial PLA matrices, obtained after a complex synthesis process (optimized state assignment, minimization ...) with the ASYL system [15], have been modified for test purpose in order to be able to localize a defect. The PLA testing strategy is based on the testing scheme presented by Bozorgui-Nesbat and McCluskey [16] and improved later [17], [18]. The basic idea of this scheme is the ability to activate an individual product term by adding some additional inputs to the AND-plane of the PLA. With this scheme, the test procedure requires only the application of some test vectors; no register is needed to control the activation of product or input lines. A more detailed presentation of the method can be found in [19].

Each PLA is considered as a block from the reconfiguration point of view.

Test inputs and redundant programmable product terms are provided. When taking into account the external interconnection overhead needed for redundant inputs and outputs, it is obvious that redundant product terms are a good compromise between area overhead and fault occurence.

Input and output lines are designed on two layers with multiple connections. This technique ensures fault-tolerance with regards to open defects. On the other hand, decreased speed due to the enhanced capacitance must be accepted.

3. RECONFIGURATION DEVICES

3.1. Different types of switching devices

Reconfiguration involves disconnection of defective elements and connection of spare ones. Several methods, software or hardware, can be used. Three types of hardware switches were used in a first version : laser fuses, antifuses and floating gate FETs.

Laser fuses and antifuses can be used when no reversibility is needed. Floating gate FETs can be used when several programming attempts are necessary.

3.2. Datapath modifications

A bit slice partitioning is chosen. Two problems have to be overcome : the bus connections on the good slices and the bypass of the defective slice for the signals propagated between adjacent slices.

The first problem arises because some devices are not reconfigurable without a large area overhead. For example, in the case of the status register, each bit has a specific meaning and is associated with some random logic. The datapath is therefore composed of three parts, according to Figure 2. Depending on the slices used in the reconfigurable parts, the buses have to be reconfigured in order to correctly connect the non reconfigurable block. Furthermore, the bus structure is a complemented precharged and amplified one. Since two different buses are available in the datapath, four bus lines have to be reconfigured on each slice (two bus lines and the complementary lines).

The initial connections of the buses are made with laser fuses. After test, the reconfiguration consists in cutting some fuses and connecting new lines. The connections may use antifuses. Nevertheless, it is useful from an economical point of view to minimize the number of different switches, namely the number of machines necessary during reconfiguration. Therefore, the connections are made by means of transfer gates. A special reconfiguration signal Crec is generated for each slice. It indicates if the

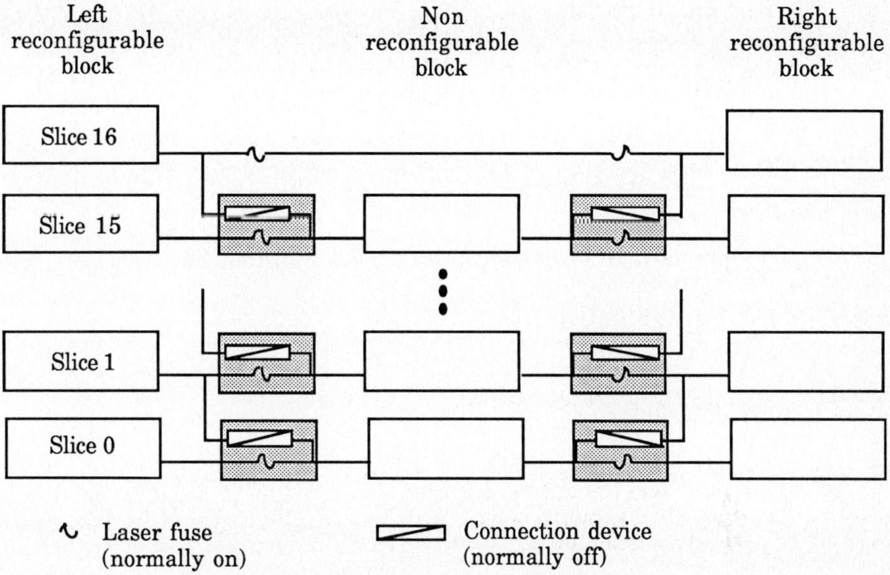

Figure 2 : General floorplan of the datapath

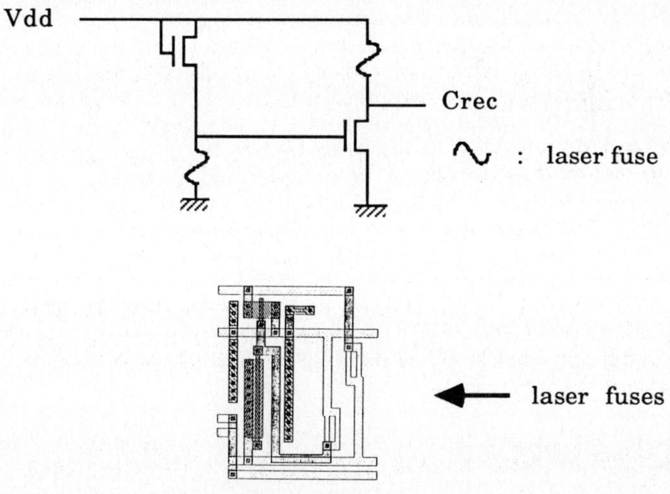

Figure 3 : The circuit used for the generation of the Crec signal and its layout

slice must be bypassed and its logic level is determined by laser fuses. The Crec signal is generated on each slice by the circuit of Figure 3. On slice 0, the transfer gates are directly controlled by the Crec signal; on the other slices (1 to 15), the transfer gates are controlled by a logic combination of the Crec signals:

$$C_i = Crec_i \,.\, C_{i-1},\ 1 \leq i \leq 15 \qquad \text{with} \quad C_0 = Crec_0$$

The connection circuit for these slices is shown in Figure 4. The Crec signals are also used to propagate correctly information between the slices according to Figure 5.

The area increase for reconfiguration is about 20 to 25%.

3.3. PLA modifications

As mentioned earlier, the reconfiguration of the PLAs is made by means of spare programmable product terms which can replace the defective ones. The programmation uses either laser fuses or N channel depleted floating gate FETs. In our case, fuses were used because reversibility is not essential and fuses are already present in the datapath. In that way, only one type of fuse is used on the chip and the reconfiguration process is shorted.

A normal product term is implemented as shown in Figure 6. A laser fuse allows to disconnect it from the OR array when it is proved to be defective. The programmable product terms are implemented as shown in Figure 7. These product terms have, at the end of manufacturing, no influence since transistors are connected in the AND array on all the input and complemented input lines. Some of these transistors are disconnected from the AND array when the redundant product term has to be used. An example of reconfiguration is given in Figure 8. The programmable product terms can also be disconnected from the OR array when necessary. The layout of a complete PLA is given in Figure 9. The programmable product terms can be found at the top of the AND and OR planes.

The main characteristics of the reconfigurable PLAs used in the HYETI controller are summarized in Table 1. The number of test inputs and programmable product terms, as well as the area overhead for reconfiguration, are listed in this table. Taking into account the routing overhead due to the test inputs of the PLAs, and even the scanpath register area, the complete area overhead in the controller is about 21%.

4. THE COMPLETE CHIP

The chip sent to the foundry is shown in Figure 10. It incorporates about 50,000 transistors in a total area of 35 mm^2 (including the laser positioning patterns). The microcommands register can be loaded or read serially in a

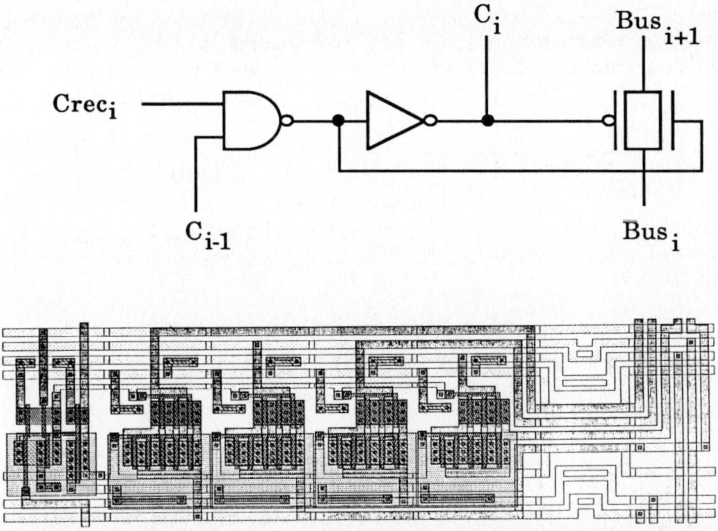

Figure 4 : The connection circuit for slice $1 \leq i \leq 15$ and its layout

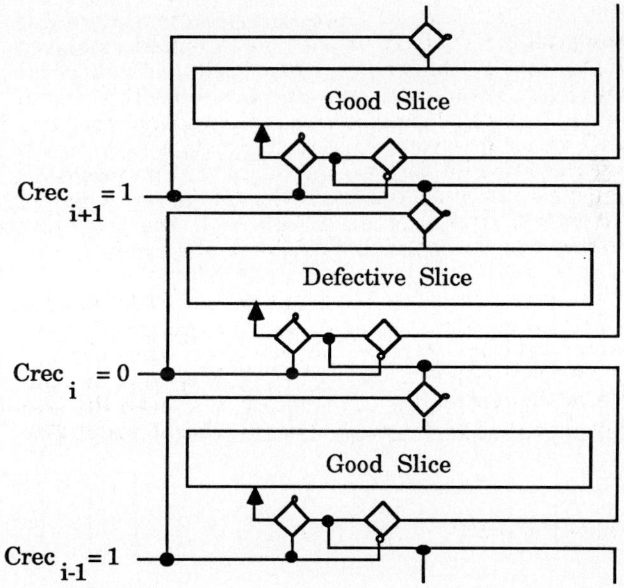

Figure 5 : Reconfiguration of signals propagated between slices

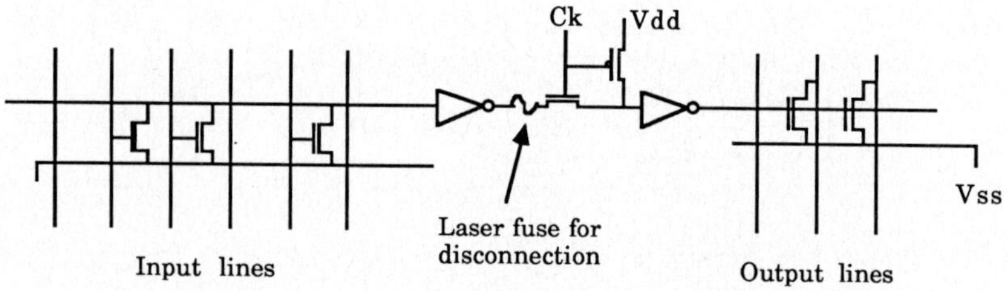

Figure 6 : Normal product term of a reconfigurable PLA

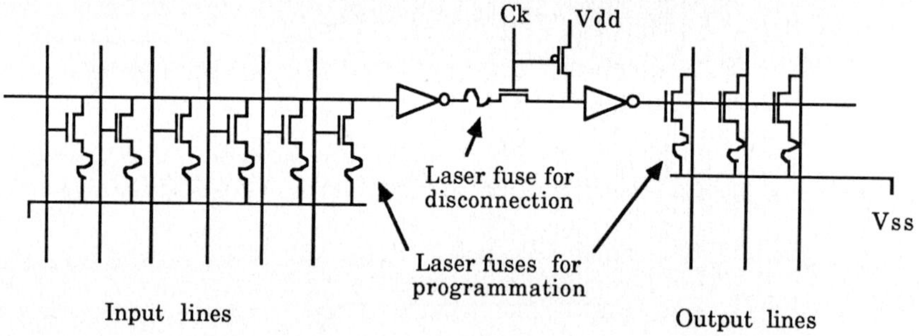

Figure 7 : Programmable product term of a reconfigurable PLA

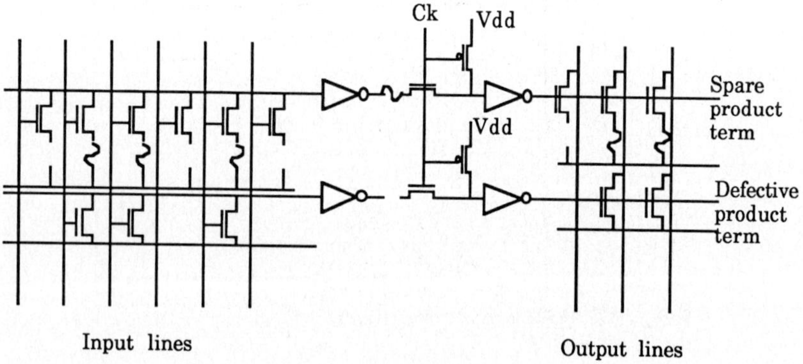

Figure 8 : Reconfiguration in a PLA

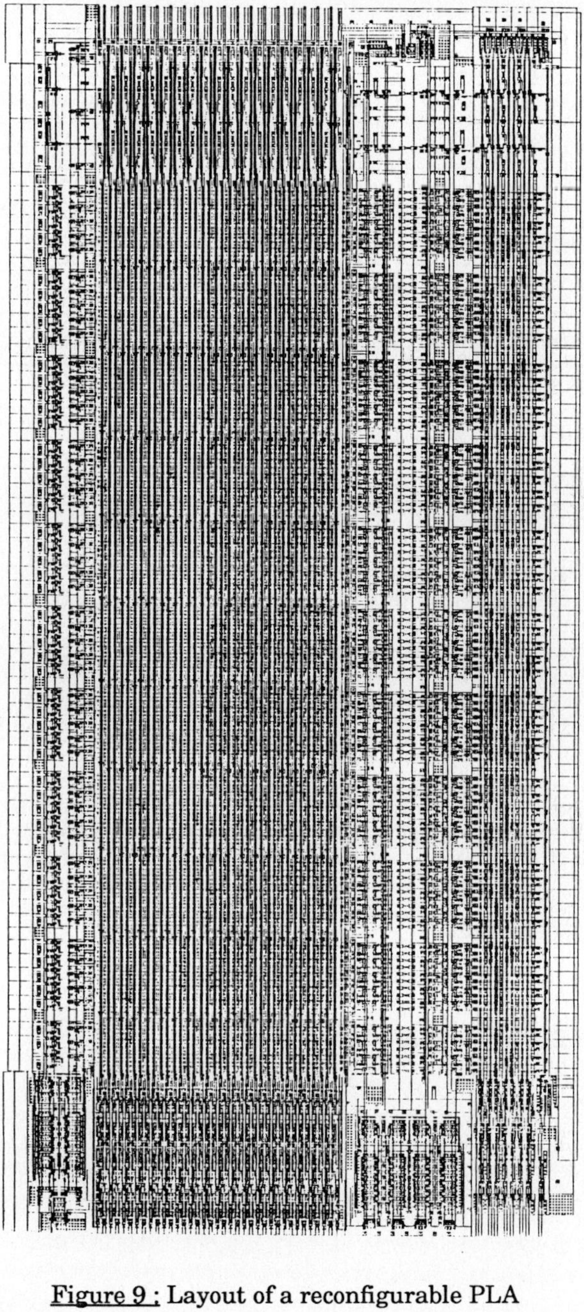

Figure 9 : Layout of a reconfigurable PLA

Table 1 :

Characteristics of the reconfigurable PLAs used in the HYETI controller

PLA name	P1	P2	P3	P4	P5	P6	DECALU
No. inputs	15	15	12	26	26	26	4
No. outputs	7	5	29	23	3	3	10
No. product terms	96	94	101	108	166	178	13
No. test inputs	3	2	3	3	3	3	1
No. test product terms	1	1	2	1	2	2	1
No. prog. product terms	2	2	2	2	4	4	1
Complete size (μm^2)	703881	662088	976080	1205988	1379530	1501650	225853
Area overhead (%)	19	15	15	13	18	17	43

test mode. Furthermore, two scanpath registers are added onto the inputs of the controller and between the two stages of PLAs in the controller. Two signals M and K are provided to control the different scanpaths. Among the 60 pads, 3 pads are redundant (one per external bus: adresses, pages and data). Six pads are used only for test (two control signals for the scanpaths, the serial input of these registers and three serial outputs). The area needed for the two scanpaths between the controller and the datapath is about 2 mm^2. A master/slave register is mandatory from a functional point of view (microcommand synchronization) between the datapath and the controller. The buffers implemented on the inputs of the controller are also unrelated to the test. Furthermore, the size of the chip is essentially defined by the core and not by the ring of pads (in consequence, the removal of the pads used for test would not lead to significant area savings). Taking this into account, the total silicon overhead, including both test and reconfiguration devices, can be evaluated about 8 mm^2.

5. OFF-LINE TEST AND RECONFIGURATION

At the end of manufacturing, a very efficient off-line test can be performed. This is due to the modifications introduced in the PLAs, to a special design of the ALU which has an Iterative Logic Array (ILA) C-testable structure and to the implementation of the scanpath registers.

Figure 10 : Layout of the HYETI chip

The test after manufacturing involves the following steps :

- Parametric test (general validation of the chip).
- Test of the three scanpath registers.
- Test of each PLA in the controller, using the scanpath registers.
- Test of the datapath elements, using the scanpath registers.
- Test of the whole microprocessor in working mode if all the other tests are successful.

When defects are detected in the reconfigurable parts of the processor, a reconfiguration phase takes place after this first test. In the different PLAs, the defective product terms are disconnected and replaced by spare ones which are programmed by laser cuts to implement the right function. In the datapath, the Crec signal of the defective slice D is set to 0 and all the fuses on the buses are cut on slices D ($0 \leq D \leq 15$) to 16. When no defect is detected or when the defective slice is the 17th one, only the fuses on this slice have to be cut.

When the reconfiguration is achieved, a second test is performed to verify the effectiveness of the corrections.

Notice that multiple consecutive test and reconfiguration runs are not allowed. A defect can be repaired only if it can be localized during the first test with the general test patterns. This restriction is economically motivated, since test and reconfiguration of a chip is a time-consuming process.

CONCLUSION

The reconfiguration of a non-regular circuit has been described. The general method is given. The different circuits dedicated to reconfiguration are shown with their layout. The overall cost on silicon of the approach is also discussed.

The method presented here may be applied to any type of circuit with a bit slice datapath and a controller. Nevertheless, the necessary modifications of the circuitry have to be considered at an early stage of the design, especially for complex dedicated operators. These techniques can be used to improve the productivity of VLSI circuits and may lead in the near future to the possible integration of whole systems on a single chip. As far as yield increase is concerned, the prototype which was developped will allow to validate the theoretical results.

ACKNOWLEDGMENT

This project has been developped under the EEC ESPRIT 824 grant, under the leadership of Dr. J. Trilhe from SGS - Thomson Microelectronics. This work has been done under the responsability of Professor Saucier (I.N.P.G.) and Professor Glesner (T.H.D.).

REFERENCES

[1] W. K. Fuchs, M.-F. Chang, Diagnosis and repair of large memories : a critical review and recent results, International Workshop on Defect and Fault Tolerance in VLSI Systems, Springfield, Massachusetts, USA, October 6-7, 1988

[2] B. Nasreddine, E. F. Kouka, Y. Wang, D. Marron, J. Trilhe, A reconfigurable SRAM 4.5 Mbits WSI memory, International Workshop on Defect and Fault Tolerance in VLSI Systems, Springfield, Massachusetts, USA, October 6-7, 1988

[3] M. G. Sami, R. Stefanelli, Reconfigurable architectures for VLSI processing arrays, Proc. of the IEEE, vol. 74, no. 5, May 1986

[4] T. Leighton, C. E. Leiserson, A survey of algorithms for integrating wafer scale systolic arrays, in : "Wafer Scale Integration", G. Saucier and J. Trilhe, ed., Elsevier Science Publishers, Amsterdam, 1986

[5] P. P. Gelsinger, Design and test of the 80386, IEEE Design & Test of Computers, vol. 4, no. 3, June 1987

[6] J. Kuban, J. Salick, Testability features of the MC68020, International Test Conference, 1984

[7] S. Chau, D. Rennels, Design techniques for a self-checking self-exercising processor, International Workshop on Defect and Fault Tolerance in VLSI Systems, Springfield, Massachusetts, USA, October 6-7, 1988

[8] T. Nanya, T. Kawamura, Error secure / propagating concept and its application to the design of strongly fault-secure processors, IEEE trans. on Computers, vol. 37, no. 1, January 1988

[9] B. Maytal, A. Danor, V. Karpati, R. Nassrallah, Y. Sidi, E. Shihadeh, Designing for high yield : the NS32532 microprocessor, International Workshop on Defect and Fault Tolerance in VLSI Systems, Springfield, Massachusetts, USA, October 6-7, 1988

[10] G. H. Teepe, W. L. Engl, A bipolar correlator with redundancy, IEEE Journal of Solid-State Circuits, vol. SC-22, no. 6, December 1987

[11] G. W. Sumerling, G. E. Dixon, A. K. J. Stewart, An assessment of non-regular cell based architecture for ULSI and WSI, in : "Wafer Scale Integration", G. Saucier and J. Trilhe, ed., Elsevier Science Publishers, Amsterdam, 1986

[12] R. Leveugle, M. Soueidan, X. Delord, HSURF : un microprocesseur à test intégré pour les applications à haute sûreté de fonctionnement, Sixième Colloque International de Fiabilité et de Maintenabilité, Strasbourg, France, October 1988

[13] P. Genestier, C. Jay, G. Saucier, A reconfigurable microprocessor for Wafer Scale Integration, in : "Wafer Scale Integration", G. Saucier and J. Trilhe, ed., Elsevier Science Publishers, Amsterdam, 1986

[14] R. Leveugle, M. Soueidan, Design of an application specific microprocessor, International Workshop on Logic and Architecture Synthesis for Silicon Compilers, Grenoble, France, May 1988

[15] G. Saucier, M. Crastes de Paulet, P. Sicard, ASYL: a rule-based system for controller synthesis, IEEE trans. on Computer-Aided Design, vol. CAD-6, no. 6, Novembre 1987

[16] S. Bozorgui-Nesbat, E. J. McCluskey, Lower overhead design for testability of PLAs, International Test Conference, 1984

[17] S. Bozorgui-Nesbat, J. Khakbaz, Minimizing extra hardware for fully testable PLA design, International Conference on Computer-Aided Design, 1985

[18] E. H. L. Aarts, F. P. M. Beenker, M. M. Ligthart, Design for testability of PLAs using statistical cooling, 23th Design Automation Conference, 1986

[19] N. Wehn, M. Glesner, K. Caesar, P. Mann, A. Roth, A defect-tolerant and fully testable PLA, 25th Design Automation Conference, 1988

[20] S.-Y. Kuo, W. K. Fuchs, Fault diagnosis and spare allocation for yield enhancement in large reconfigurable PLAs, International Test Conference, 1987

[21] T. E. Mangir, A. Avizienis, Fault-tolerant design for VLSI : effect of interconnect requirements on yield improvement of VLSI designs, IEEE trans. on Computers, vol. C-31, no. 7, July 1982

[22] T. E. Mangir, Sources of failures and yield improvement for VLSI and restructurable interconnects for RVLSI and WSI, Proc. of the IEEE, vol. 72, no. 6, 1984

[23] K. K. Saluja, K. Kinoshita, H. Fujiwara, An easily testable design of PLAs for multiple faults, IEEE trans. on Computers, vol. C-32, no. 11, November 1983

[24] F. Somenzi, S. Gai, Fault detection in Programmable Arrays, Proc. of the IEEE, vol. 74, no. 5, May 1986

[25] C.-L. Wey, On the design of a redundant Programmable Logic Array, IEEE Journal of Solid-state Circuits, vol. SC-22, no. 1, 1987

[26] C.-L. Wey, On yield consideration for the design of redundant Programmable Logic Arrays, IEEE trans. on Computer-Aided Design, vol. 7, no. 4, 1988

Wafer Scale Integration, III
M. Sami and F. Distante (Editors)
Elsevier Science Publishers B.V. (North-Holland)

Defect-Tolerant Chip Featuring Multiple Configuration and Built-In Test Approaches[1]

Ulrich Jagau, Hans-Jürgen Iden, and Matthias Grützner

Institut für Theoretische Elektrotechnik
Universität Hannover
Appelstrasse 9A, D-3000 Hannover, West Germany
phone: +49 (511) 762-5049 or ext. -3202
telefax: +49 (511) 762-5051

Working towards a signal processor in monolithic silicon, a demonstration chip was designed and produced. It mainly contains configuration and test techniques as well as extremely long interconnection systems. The configuration techniques are based on laser fuses and anti-fuses as well as on software programming. The test is supported by special test hardware. Measurements on produced chips have been carried out, and the introduced concepts have been found to be suitable for signal processor architectures.

1. INTRODUCTION

In the last years several approaches for the configuration of defect-tolerant systems on wafers via electrical or physical switches have been made [1, 2].

In this work various solutions are provided for substantial problems such as the implementation of test-hardware and the logical and physical reconfiguration of modules and busses. Additionally, we took into account the electrical behavior of long distance wires (e.g. busses) [3].

We have modelled solutions with a monolithic chip (Fig. 1) with redundancy (50 sqmm) fabricated in a 2μm 2-layer Al standard CMOS process (Philips), including full-custom layouts of laser-programmable fuses and anti-fuses (Fig. 3) connecting bus-systems of up to 35 mm in length. This demonstration chip is one of a series of chip designs (test chips for laser anti-fuses, test chips of cache and processing units) which prepares the integration of a video signal-processor realized in monolithic silicon. The final architecture is introduced by Wehberg and Volkers in [4]. Estimates of the yield of systems (including nets) with implemented hierarchical redundancy are proposed by Iden in [5]. The research on the laser anti-fuses is done by Hartmann et.al. [9].

For testing the interconnecting system, additional hardware is implemented using the Echotest [6, 7]. The reconfiguration can be done using laser controlled switches (LCS) for connection and disconnection of system components. Process-compatible laser

---

[1] Supported by Fraunhofer Gesellschaft, West Germany under contract No. T/RF33/D0043/D2203

arrays (to standard CMOS processes) of fuses and anti-fuses are implemented for the repair of bus-systems (Fig. 3).

2. SYSTEM DESCRIPTION

WSI systems presented so far use special architectures as very regular structures or nearly independent signal paths which are well adapted to the implementation of redundancy (e.g. Carlson et.al. in [2] and Lea in [11]). Although this is a good way to restrict overhead, numerous algorithms cannot be fitted effectively to such architectures. We investigated the implementation of redundancy for a proposed architecture [4] which imposes conditions for module size and connectivity.

Within the limited area of the chip (50 sqmm) we designed a model system to verify various configuration techniques for the implementation of these architectures. In contrast to systolic array structures it is composed of a relatively small number of complex modules. To get communication capabilities between most modules, these are connected by bus systems.

The defect-tolerant chip (block diagram, Fig. 2) consists of modules, modelling processing elements (PE), and memory blocks (MEM) which model functional blocks of WSI circuits. The implemented processors (PE) are software programmable allowing user-defined data transfer, modification, and storage operations. For flexibility, programs are loaded via the bus system after system configuration. To achieve this, all modules can be addressed from the busses to load the program.

Intermodule communication is realized by three bus-systems in parallel. For system operation, only one of them is necessary at a time. They are folded to implement various lengths of long on-wafer bus systems (length on designed chip up to 3.5 cm). Bi-directional bus amplifiers (BV) are used to test their behavior for speed enhancement (delay minimization) and crosstalk prevention. Different distances between amplifiers are implemented for the the busses.

The direction of operation of these amplifiers is switched automatically by internal logic depending on the direction of data-flow of the bus. This provides for the transparancy of the system operation. As they partition the busses, defects on bus lines can be isolated to a single bus segment between two amplifiers.

The busses are terminated with blocks for I/O (PORT) and for test support (ECHO). A bi-directional boundary scan-path reaches the interfaces between the modules and the busses thus giving an additional access to the connections between modules and the bus.

The modules are connected by the MX arrays of laser switches. They allow a replacement of single bus-lines by spare lines and a remapping of module I/Os to the bus lines. Additionally, a spare bus can be chosen in case of an irreparable defect. Thus we have a hierarchical redundancy scheme for the busses which are crucial for the yield of a system.

After fabrication, all modules are connected electically (pre-configurated) to the bus-lines. Functionally, they are separated by tri-state buffers after an initial system reset. Thus, at least for a defect-free system, no configuration by laser is needed.

3. TEST OF THE INTERCONNECTION SYSTEM AND TEST OF THE SUCCESSFUL LASER ARRAY CONFIGURATION/RECONFIGURATION

It is the goal of WSI testing to scrutinize wafers as quickly and economically as possible. To achieve this, BIT (i.e. BIST techniques), known from VLSI, are being proposed for use in WSI. The advantages of these techniques over the other usual testing procedures, such as wafer probing, are also valid for WSI. The main difference between WSI and VLSI is, that in addition to fault detection, a fault localization is necessary in order to use the available redundancy effectively.

The demonstration chip contains module redundancy and redundancy in the interconnection systems (busses/configuration networks). These interconnection systems play a significant role, not only for the system architecture of a signal processor, but also for the configuration of modules.

Due to the redundancy in the interconnection systems, an independent test of these systems is necessary to tolerate faults in the lines. Therefore, the Echotest [7] with the Echoregister as a test aid was developed. The Echotest can be applied to interconnection systems of bi-directional use. With the Echoregister, the testing of interconnections (Echotest) is possible from one end. Opens, shorts, and s-a-faults in the interconnections (lines and line drivers) can be detected by this test [6].

Accessing the modules during the test becomes a problem as long as the configuration by laser switches has not been carried out. To overcome this difficulty, the boundary scan-path [8] for bi-directional use was designed and is applied to reach the module boundaries. A testing of the connections between modules is thereby feasible and at the same time supports the testing of the modules. Using bi-directionality, the scan-path can tolerate one fault only.

The chip test in our demonstration chip proceeds in several steps (Fig. 6). First, the interconnection systems are tested by the Echotest, and faulty lines are replaced by the use of laser switches. The scan-paths themselves are then tested (flash-test). If they are fault free, the modules are tested by using the scan-path (it was not the intention to implement self-test techniques for the modules themselves in this demonstration chip).

The laser switches (MX) for the configuration of the chip logic lie between the module boundary and the interconnection systems. If the interconnection systems are found to be fault-free (initially or after repair), and if the scan-paths are also fault-free, the successful configuration can be tested. However, if the scan-path, which is a sequential test bus, is faulty and cannot be used for the test, the test of the whole chip is nevertheless still possible. In such a case, after having tested the interconnection systems, modules are configurated and tested step by step. A faulty scan-path makes a fault localization which distinguishes between a fault in the module itself and the laser switches impossible.

With the described test hardware and by applying the test procedure, the test of the demonstration chip was carried out successfully without utilizing wafer probe techniques.

4. SWITCHES FOR CONFIGURATION

To realize the interconnections of the final system [4], we must choose the right kind of switches (physical or electrical) or combinations of these switches for different kinds of bus-systems or networks. Software programming of multiplexers is reversible on the one hand, but only possible if the interconnection to the switch cells is verified to be fault-free. On the other hand laser-programming is only reversible if spare-paths are designed and can only be programmed right after fabrication. For choosing the correct type of switches, several parameters must be taken into account:

- **programming probability of the switches**
- **point of time of the (re-) configuration (just after fabrication or during life-time, i.e. defect-tolerance or fault-tolerance, respectively).**

Connection of lines can be done by welding specially designed structures [9] (Fig. 4) in the metal layers 1 and 2 using a Q-switched Nd:YAG laser (NEC). On our actual chip, fuses and anti-fuses are implemented in a standard CMOS process (Fig. 3). Our aim is to verify the suitability (yield improvement, optimized number of redundant connection lines) of a universal array of laser switches for WSI-systems.

The laser programming procedure itself is imperfect. Laser programming (of anti-fuses) is done with a probability of less than 100%. Taking this into account, laser programming arrays are designed redundantly and defect tolerant. The amount of redundancy is optimized by a statistics model. By redundancy, non successful shots during system configuration can be repaired.

Additionally, we have designed a new standard cell, the laser controlled switch (LCS, Fig. 5). It consists of a highly reliable laser fuse for the reconfiguration of modules and bus-systems. By means of a single laser shot, configuration is performed. This switch is used in the presence of gross defects, but has no advantages in disconnecting whole modules.

Measurements on produced chips have shown, that the introduced concept is a good method for achieving access to chips with defective bus-systems. Spot defects causing malfunction of single lines are repaired using the above described laser arrays (MX, Fig. 3) and avoid expensive solutions like software programming of multiplexers and fixing or storing of the applied configuration.

5. CONFIGURATION

Various configuration techniques are provided throughout the system. They allow for a reconfiguration with a minimum number of switches to be programmed depending on the location of a defect and the kind of malfunction it causes. It is the objective of this chip to verify a wide range of configuration techniques. That is why they are over-dimensioned in terms of a optimum productivity [5].

Based on the amount of configuration operations done externally, we distinguish the following methods:

0. Initial fault-tolerance:

Since test and configuration must be performed in a system with defects, critical signals (e.g. clock signals) are provided by TMR.

1. Configuration by the loaded program:

When the software is loaded to a module, special registers are set which activate the module. In this way, defective modules can be left deactivated in many cases. As no laser programming has to be applied, a preliminary module activation for test purposes is feasible.

2. Configuration by the programming of single laser programmed cells (LCS):

Modules which are stuck in the "enabled" mode (e.g. because of a defect in the module selection circuitry) can be forced to the "disabled" mode by the programming of single laser controlled switches (LCS). Additionally, a configuration as in (1) can be made permanent in similar fashion.

3. Configuration of laser switches (by MX arrays):

Modules causing bus-conflicts (e.g. by a bus driver, which cannot be switched to the high impedance mode or by a fault in the module selection circuit) can be cut line by line from the bus system within the MX array of laser switches. Additionally, this is an effective way to switch spare lines within a bundle of lines.

These methods are used not only to select functional modules, but also to select one of the three busses for system operation.

The configuration process of the system is done in two steps. First, after testing the busses by the aid of Echoregisters and boundary scan circuitry, defect bus lines are replaced by spare lines. Additionally, a spare bus can be selected. Second, after the interconnections are verified to be fault free, the modules can be tested via the bus and a defect module is replaced by a spare. Alternatively, a module test via the boundary scan is feasible.

The active circuitry of ports, bus amplifiers and Echoregisters is, to a large degree, part of a bus-line and will be replaced with that line.

6. RESULTS

About 2300 chips were fabricated and pre-tested with an ATE (Tektronix 3260/3266) at the Philips factory. The pre-testing operates with standard test steps (input leakage, Idd-leakage, functional test, output voltage) with eliminated break-points in the test program. This enables the monitoring of defects in the functional test, e.g. after detection of a preceding failure.

Our interest was to reconfigurate chips with defects primarily in the interconnection network and modules. Measurements were performed on wafers using a probe-card and on housed chips. Detailed test and reconfiguration was performed on defective chips found from the pre-test procedure. The final test procedure and the proper repair of the defective chips were carried out successfully:

- **defects could be localized in the interconnections and modules**
- **repair was done successfully using laser and software programming**

7. CONCLUSIONS

Techniques for configuration incorporated with new test concepts were verified with the defect-tolerant demonstration chip.

By applying the test and configuration techniques described in this paper, defective chips could be repaired successfully.

The hybrid concept of applying diffent types of switches (electrical and physical, program-controlled or having configurations set by the LCS) at different system levels is useful for repairing chips after fabrication and during their life-time.

Acknowledgement

We would like to thank Mr. Lars Müller for doing the final chip test and the chip reconfiguration.

References

[1] T.E. Mangir, "Sources of Failures and Yield Improvement for VLSI and Restructurable Interconnects for RVLSI and WSI: Part II - Restructurable Interconnects for RVLSI and WSI," *Proceedings of the IEEE*, vol. 72, no. 12, pp. 1687 - 1694, 1984.

[2] R.O. Carlson, C.A. Neugebauer, "Future Trends in Wafer-Scale Integration," *Proceedings of the IEEE*, vol. 74, no. 12, pp. 1741 - 1752, 1986.

[3] H. Grabinski, J.P. Mucha, "A Numerical Approach to the Analysis of Signal Propagation on VLSI Interconnect Systems," *IEEE-ICCD '85*, New York, pp. 683 - 687, October 1985.

[4] T. Wehberg, H. Volkers, "Architecture of a Programmable Real-Time Processor for Digital Video Signals Adapted to Motion Estimation Algorithms," *SPIE-Conf. 'Visual Comm. and Image Processing III,'* Cambridge, pp. 908 - 916, November 1988.

[5] H.J. Iden, "On the Optimization of Hierarchical Redundancies Including Configuration Nets and Switches," *IEEE International Conference on WSI*, San Francisco, pp. 173 - 182, January 1989.

[6] M. Grützner, H. Grabinski, "Simulation and Test of Faults in WSI Interconnect Systems," *IEEE International Conference on WSI*, San Francisco, pp. 345 - 354, January 1989.

[7] M. Grützner, "Design for Testability for Wafer-Scale Integration Interconnect Systems - Design & Test Methodology," *ITC '88*, Washington D.C., pp. 146 - 152, September 1988.

[8] M. Grützner, "A Hierarchical Test Approach for WSI using Boundary Scan Logic," *Wafer-Scale Integration*, Elsevier Science Publishers B.V., North-Holland, pp. 169 - 175, November 1986.

[9] H.D. Hartmann, T. Hillmann, "Investigations of Laser Formed Connections and Disconnections of Standard CMOS Double Level Metallizations," (submitted to the *IEEE International Conference on WSI*, San Francisco, 1990).

[10] U. Ludemann, "Ein Verfahren zur Ausbeutesteigerung regulärer integrierter Schaltungen durch Redundanz," *Ph.D. Dissertation*, Universität Hannover, West Germany, June 1985.

[11] R.M. Lea, "A Reconfigurable WSI Associative String Processor for Highly-Parallel Information Processing," *IFIP International Workshop on WSI*, Brunel University, Egham, England, September 1987.

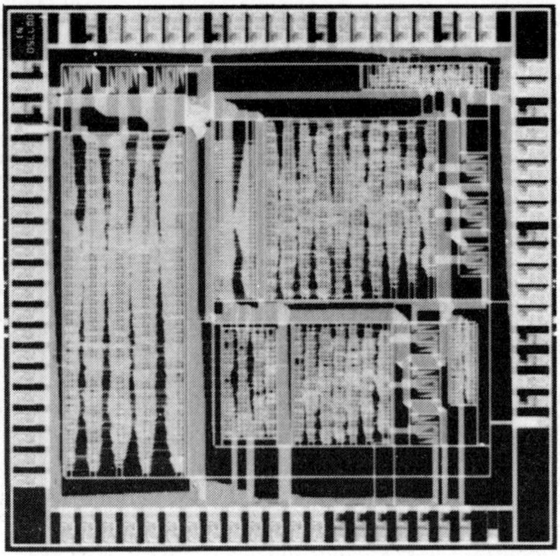

Fig. 1: Micro-Photograph of the Produced Defect-Tolerant Chip

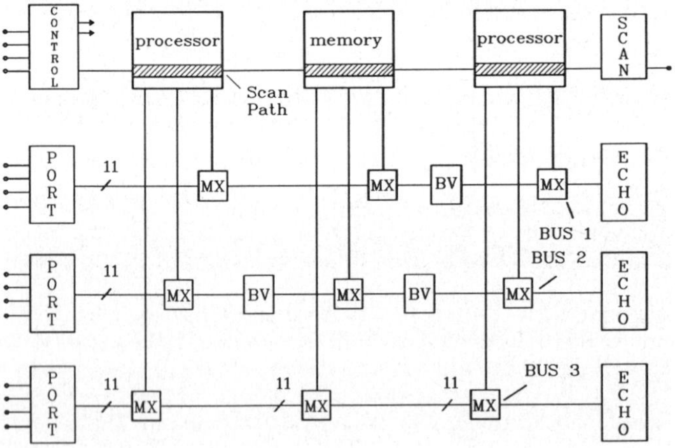

Fig. 2: Block Diagram of the Implemented System

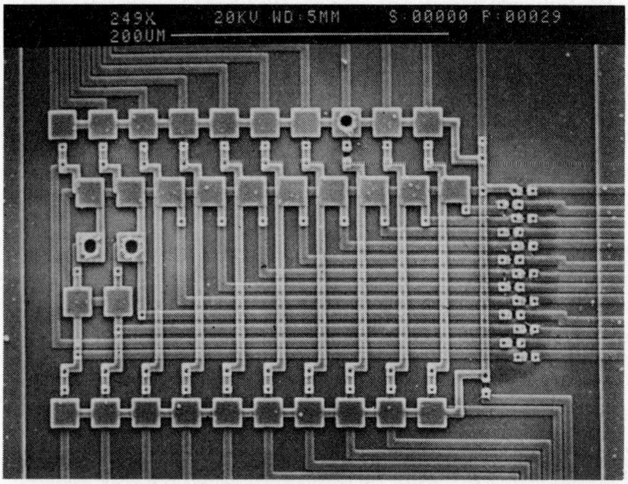

Fig. 3: SEM Micrograph of an Implemented Laser-Array (MX, Fig.2)

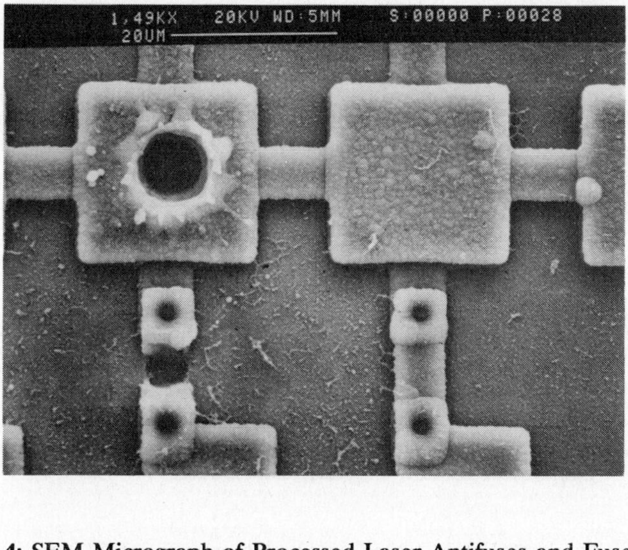

Fig. 4: SEM Micrograph of Processed Laser-Antifuses and Fuses in the Standard CMOS Process (Detail of Fig. 3)

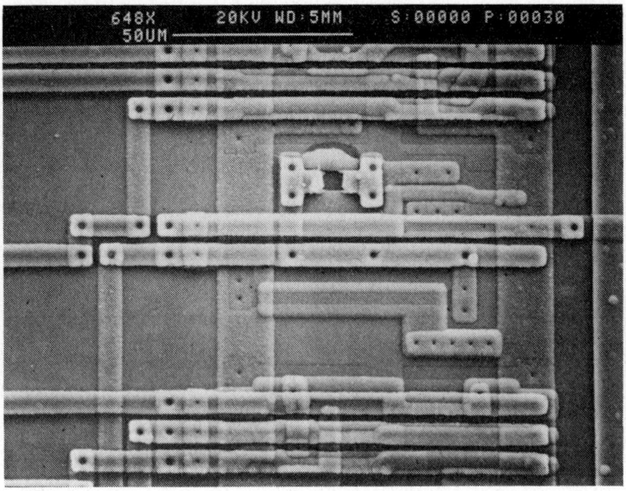

Fig. 5: SEM Micrograph of a Processed Laser Controlled Switch (LCS)

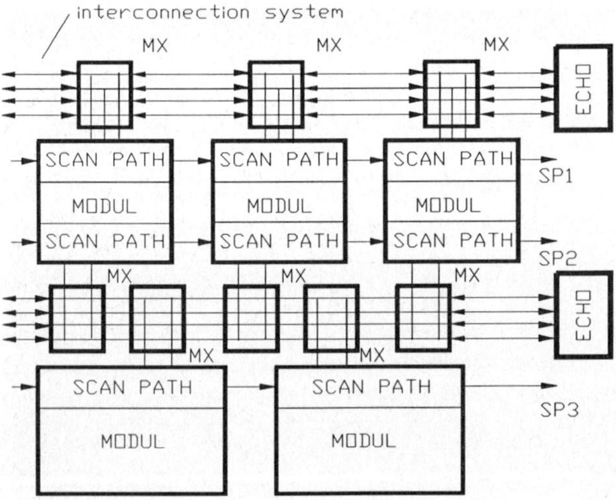

Fig. 6: Schematic Block Diagram of the Demonstration Chip with the Implemented Test Hardware (MX: Laser Array)

Wafer Scale Integration, III
M. Sami and F. Distante (Editors)
Elsevier Science Publishers B.V. (North-Holland)

CONTROLLING UNIT FOR A PIPELINED FLOATING POINT HARDWIRED ENGINE

S. Vassiliadis and E. Schwarz

IBM Corporation, Endicott, N.Y, USA

In the present paper a control unit with maximum absorbance/throughput pipeline is described. The proposed pipelined scheme is incorporated into the Floating Point Unit (FPU) of a computer system. In addition to the FPU, the computer system also includes a cache, a central processing unit (CPU), and a vector processor (VP). The FPU operates on a multiple pipeline structure with dynamically defined instruction streams.

1. INTRODUCTION

A number of computers employ SISD type of architectures with various degrees of piping. A SISD architecture is defined to be a Single Instruction stream Single Data system [1] and it implies that instructions are executed one by one and a single instruction deals with at most a single data operation. The primary reason for the use of pipelined designs is to increase the performance at the expense of complexity of design as well as the addition of extra hardware.

The major difficulty with the design of pipelines lie with the detection and resolution of structural and data dependent hazards[2]. Very often the resolution of the hazards results in performance losses due to underutilization of the hardware.

It is the scope of the present paper to describe a maximum absorbance/throughput pipeline for a floating point unit, and as an extension to other units, that combines the following features:

1. Has a control mechanism that operates on dynamically defined instruction streams and optimizes the stage usage while avoiding collisions.
2. Because of the SISD architecture considered, instruction streams are not defined a priori and therefore a dynamic definition is used to distinguish instruction streams, and also more than one instruction has to be executed in parallel for successful overlap of the instruction streams.
3. Operates on a pipeline structure which is composed of many pipelines. Each pipeline can be of variable length depending on the category of instructions, which will include a minimum number of stages and a maximum number of stages defined a priori.
4. The controlling unit comprises a dynamic history table containing information on the status of the stream in consideration, without having a set position for a stream. It also contains information on all particular pipelines and their status. In addition to this global and external type of information, internal status of each stage in the pipeline is kept to control hazard situations to reinsure the proper operation of each pipeline. In addition the history table controls externally which pipeline

will complete next while the internal to the pipeline controls push the data as far and as quick as possible through to the end of the pipeline.

5. The structure communicates directly with a high speed data cache and other units for data transfers.
6. It frees the CPU of the information needed to control the execution unit (E) thus resulting in a time savings which allows the CPU to execute other instructions, assuming that the CPU is designed to execute out of sequence.
7. The pipeline is controlled not up front, thus, more than one instruction of a stream can be present in the pipeline, and also, at a given time more than one stream may be initiated.
8. The FPU communicates to the CPU the END or ABNORMAL CONDITION of a stream thus the CPU sees a number of instructions as a unique instruction.
9. Not all instructions can be seen as a subset of a stream, this is because in an implementation for hardware savings the dataflow is designed to accommodate certain lengths of operations. Thus it is not possible to initiate instruction streams for all types of mixtures if the dataflow accommodates not all possible length operations. The controlling unit operates on two distinct modes, to accommodate instructions that can or can not be directly accommodated by the dataflow.

2. DETAILED DESCRIPTION OF THE PIPELINE STRUCTURE

By way of background information, the pipeline is incorporated into the Floating Point Unit (FPU) of a 370 or 370/XA computer system [3]-[4]. In addition to the FPU, the computer system also includes a cache, a Central Processing Unit (CPU), and a Vector Processor (VP). The FPU receives data directly from the cache, the CPU, or the VP; and receives instructions from the CPU. The CPU does not control the data coming from the cache. The CPU requests data (from the cache) while sending instructions to the FPU. While the data is being accessed by the CPU from the cache, the CPU continues to send instructions to the FPU without regard to synchronization of the cycle by which data is being accessed from the cache with the cycle by which the corresponding instructions are being sent to the FPU. Therefore, the data arriving at the FPU at cycle N may be data pertaining to an instruction delivered to the Floating point unit in cycle M, where $M < N$. The CPU requests that certain operations be performed by the FPU and other units, such as the cache, via a bus called the CBUS. The CBUS is the only means by which instructions are communicated between the CPU and the FPU. The CBUS conducts handshake control signals and instruction opcodes. When the CPU transmits these requests, the functional units to which these requests are sent are called Processor Bus Units (PBU). The FPU comprises one of the PBUs. When the CPU encounters an instruction which it cannot execute, and a PBU should execute the instruction, the CPU transmits a Processor Bus Operation (PBO) signal to the appropriate PBU. For instance, if the CPU decoded an instruction to be a multiply floating point long, the CPU transmits the PBO signal to the FPU requiring the FPU to perform the operation.

The pipeline of the FPU as illustrated in Figure 1, may be subdivided into two paths: one for instructions and controls (CBUS path), and the other for data flow (DBUS path). The instructions are received via the CBUS and put in the instruction stack, and then decoded via a decoder. The data is introduced into the "MIMD" pipeline of Figure 1 via the DBUS.

The handshakes and global hazards circuit of Figure 1 transmits "handshake" signals to the CPU and detects global hazards.

Depending upon the incoming instruction being decoded, the "MIMD/SISD" switch switches to either "SISD" mode or "MIMD" mode. If an incoming instruction involves operands more than 64 bits in length or if an instruction is determined to be difficult to execute, the "MIMD/SISD" switch selects the "SISD" mode, otherwise it uses the "MIMD" mode.

The specific instructions which are considered to be "difficult" and which invoke "SISD" mode are: The Divides, both floating and fixed point; The Square Roots; and Operations which involve extended operands.

During execution in the "SISD" mode, everything is shutdown (BUSY) except for the execution of the difficult instruction.

The initialization circuit of Figure 1 starts the pipeline, and updates the Dynamic History Table. In connection with the handshake/hazard logic, the initialization circuit determines the beginning and end of an instruction stream and determines if any data dependent hazards exist. After the decode step, the type of instruction, as indicated by the output from the decoder, is compared in the initialization circuit and the global hazard circuit with the completion status of the first cycle status of the appropriate pipeline to use, as indicated by the internal pipeline controls. If there is no global hazards, as indicated by the dynamic history table, and no immediate internal hazards exist, as indicated by the output of the handshakes and global hazards circuit, the instruction is initialized. If the BUSY handshake signal is developed by the handshakes and global hazards circuit, no initialization takes place. Initialization involves starting the status controls of the appropriate pipeline and also entering a

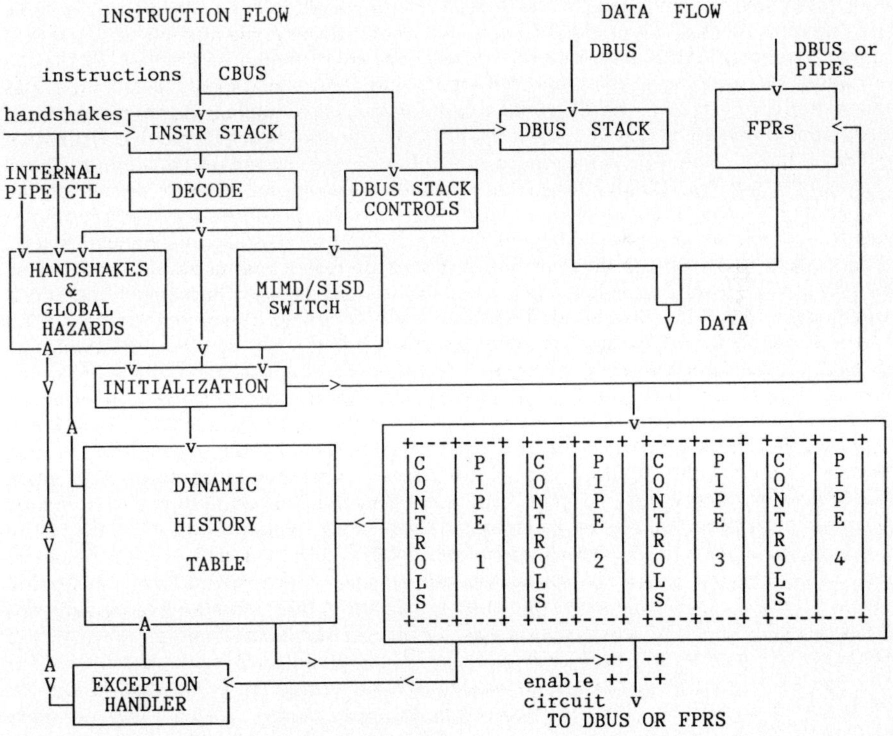

Figure 1. MIMD/SISD Pipeline Floating Point Unit

new line in the Dynamic History Table. Notification of initialization is handled by the handshake controls which sends the acknowledge signal to the CPU indicating that the instruction has been started or by sending a busy signal to the CPU, indicating that the FPU has the instruction but the pipeline of incoming instructions had better be stopped because the FPU cannot handle very many more instructions. The initialization logic and the global hazards logic determine the beginning and end of a stream of instructions. The response "acknowledge" and "not busy" to an instruction not already in a stream indicates the beginning of a stream, and "busy" indicates the end of a stream. The global hazards circuit is used to determine hazards due to data "dependencies". The initialization logic adds new lines to the dynamic history table. Therefore, initialization consists of handshaking, updating the history file, and possibly dealing with data hazards.

The "MIMD" pipeline of Figure 1 includes four pipeline circuits: pipeline1, pipeline2, pipeline3, and pipeline4; for four categories of instructions, one category for each pipeline.

Data on the DBUS is processed by either the FPRs, or the DBUS stack which is controlled by the DBUS stack controls.

The exception handler determines if there is an exception. The types of data exceptions that can occur while executing instructions are: 1.Exponent Overflow Exception, 2.Exponent Underflow Exception, 3.Floating Point Divide Exception, 4.Fixed Point Divide Exception, 5.Significance Exception, and 6.Square Root Exception. When an instruction is detected that causes one of these exceptions, all instructions received after this must be cancelled as if they were never received even though they may be already executing; this is a property of a "SISD" architecture which must be preserved by the dynamic "MIMD" architecture. This is done by changing all the valid bits in the Dynamic History Table to zero after the instruction causing the exception has completed. In addition, the CPU and the other units are notified of the interrupt and they must cancel their instructions until the CPU begins an interrupt handler routine.

The "MIMD" pipeline, disposed in the FPU of the computer system, receives instructions via the CBUS and the FPU responds back, as do the other processor bus units, by transmitting certain "handshake" signals including an ACKnowledge handshake signal, a BUSY handshake signal, and an INTerrupt handshake signal. Since the CPU works in a pipeline mode and sends PBO commands out every cycle, regardless of whether the last PBO was ACKnowledged, the PBUs must determine whether the last PBO was acknowledged before proceeding to execute the next PBO. The PBUs include an interface that checks on the handshakes of other PBUs with the CPU. A PBU is required to send one of the three handshake signals (from the handshakes circuit of Figure 1) to the CPU in the cycle after a PBO was received by the PBU. If hazards are encountered by a PBU, such as the FPU, a BUSY handshake signal is sent to the CPU by a PBU. When the BUSY signal is sent to the CPU, the PBU holds the received instruction and the following instruction in an instruction stack (such as instruction stack of Figure 1 for a FPU) so that the sequence of instructions received from the CPU can be maintained. Thus, implemented on the FPU, as part of the instruction stack, is a CBUS Register and a CBUS STACK which hold the received instruction and the following instruction, respectively. Instructions are not stacked unless hazards, which cause generation of the BUSY handshake signal, are encountered. The FPU accepts as many instructions as it can handle; however, the FPU does not contain as much information as is contained by the CPU, since the CPU can halt an instruction before the instruction is even sent to the bus units if it sees, in its buffer of instructions, that problems may be encountered. When PBOs are sent from the CPU that require execution by the FPU and another bus unit, such as the data cache, the FPU has no power to prevent the data cache from starting the execution of the instruction. Thus, the most efficient method for the FPU pipeline is to go as far as possible until a hazard is encountered.

Referring to Figures 2a-2c, a construction of the instruction stack of Figure 1 is illustrated. Figure 2a illustrates the construction of the instruction stack, Figure 2b illustrates the bits on the CBUS during a hardwired mode, and Figure 2c illustrates the bits on the CBUS during a microcode mode.

In Figure 2a, the instruction stack comprises the CBUS Stack register, and a CBUS register connected to the output of the CBUS stack register. The instruction stack as well as the CBUS consist of 25 bits of information described by the following:

- bit 0 - the PBO bit which indicates whether the FPU is in a hardwired mode or a microcode mode; if in hardwired mode (0), exceptions are reported to the CPU; if in microcode mode (1), exceptions are stored in the status word but are not reported;
- bit 1 - the FPU (FP) request bit which signals the FPU that this instruction must be executed by the FPU;
- bit 2 - the IPU/Cache request bit which signals the cache to decode the instruction;
- bit 3 - the VP request bit;
- bits 4 to 10 - the instruction opcode bits;
- bits 11 to 13 - F1, operand 1's FPR address (encoded);
- bits 14 to 16 - F2, operand 2's FPR address (encoded);
- bits 17 to 19 - in microcode mode, these bits are the SRC, source, identifier bits;
- bits 20 to 22 - in microcode mode, these bits are the DST, destination, identifier bits;
- bits 17 to 19 - in hardwired mode, these bits are the interrupt TAG field which is stored in the status word on an exception and uniquely identifies the instruction causing the exception;
- bit 24 - the parity bit, P, used for checking the validity of the instruction; and
- bits that are reserved are indicated by " /" s.

Thus, the instruction on the CBUS is introduced into the instruction stack via the CBUS bits defined above.

Referring to Figure 3, a construction of the dynamic history table of Figure 1 is illustrated.

In Figure 3, the dynamic history table comprises 17 bits of information for at most 8 instructions at a time. The Dynamic History Table consists of data that is needed when it is necessary for an incoming instruction to enter one of pipelines and to complete from these pipelines. Since the instructions are stacked, the table provides a means of sequencing the completion of execution of instructions of one or more instruction streams. The limitation of the CBUS, to send one instruction at a time, determines the instruction's starting time. Since the execution of the instructions of the one or more instruction streams may take multiple cycles to complete, and since there exists more than one pipeline, it is possible that multiple instructions will be executing at the same time. Due to architectural constraints, the instruction completion sequence must be maintained because of possible unpredictable results if an interrupt occurred and the instructions were not sequential. Therefore, there is a need to maintain and store sequencing information and completion information in the table. The dynamic history table stores the following information:

1. the pipeline number (PIPE NO),
2. the write address (WR ADDR),
3. whether the instruction is a write type (WT) of instruction,

4. a tag (INT TAG) which uniquely identifies it with an instruction in a stack on the CPU,
5. whether the instruction is an "SISD" instruction type (M/S), where "M" implies an "MIMD" instruction type and "S" implies an "SISD" instruction type,
6. the result length (LEN),
7. whether it is a hardwired or microcoded PBO (H),
8. some retry information (PSW PTR), and
9. a valid bit (V).

The pipeline number (PIPE NO) is critical because it sequences the multiple pipelines of pipeline circuits of Figure 1 (four in all). Sequencing is of very little concern in a one pipeline system, but, with multiple pipelines, tracking information must be maintained. The write address (WR ADDR), write type (WT), and result length (LEN) help in completing the instruction. The tag information (INT TAG) is stored away if an exception occurs and helps in identifying the exact instruction which caused the exception. If it is a "SISD" instruction, completion is sensed in a different way than looking to see if valid data is disposed at the end of a pipeline, but instead, is determined by a counter which counts the cycles. The most important bit is the valid bit (V) which indicates whether the instruction in this entry of the stack is valid. The valid bit (V) is cleared when an exception occurs. The valid bit (V) entry is cleared and the stack is shifted upon completion of an instruction. Thus, a quick method is available to cancel all pending instructions in the FPU, that is, by clearing the valid bit (V) in the dynamic history table.

Referring to Figures 4 through 7, a construction of the pipeline of Figure 1 is illustrated. In particular, Figure 4 illustrates the construction of pipeline1 which is used for add type instructions, such as add, subtract, divide, compares, and square roots.
The Figure 4 pipeline functions in three cycles. Figure 5 illustrates the construction of pipeline2, the multiply pipeline, which functions in 5 cycles, and is used for multiply instructions. Figure 6 illustrates the construction of pipeline3, which is used for load

(2.a)

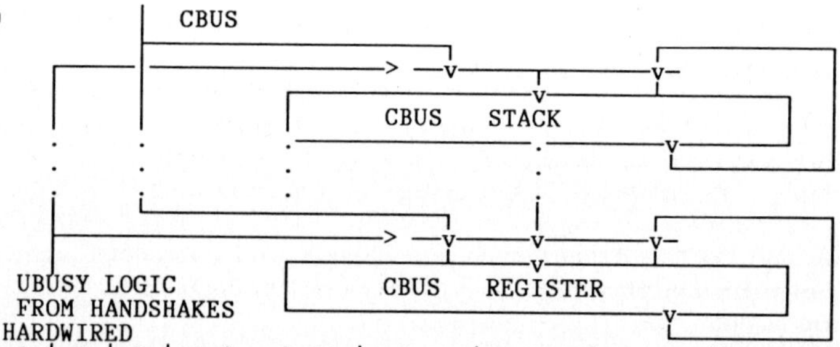

(2.b) HARDWIRED

0	1	2	3	4 |...| 10	11...13	14...16	17...19	20...22	23	24
///	FP	IPU	VP	OP CODE	F1	F2	TAG	////////////		P

(2.c) MICROCODE

0	1	2	3	4 |...| 10	11...13	14...16	17...19	20...22	23	24
PBO	///////////			OP CODE	F1	F2	SRC	DST	////	P

Figure 2. Instruction Stack

RX type instructions and functions in two cycles. Figure 7 illustrates the construction of pipeline4, which is used for all other miscellaneous functions, and which are usually either a write or a read of some auxiliary or status registers.

In Figure 1, the pipeline 1 includes a controls section and a dataflow section. Similarly, pipelines 2 through 4 each include a controls section. The pipeline controls section controls the internal parts of each pipeline by pushing the operations as far as possible through the pipelines and sensing when the FPRs are interlocked and by determining where good data can be found. To better understand these controls, it is best to first understand the layout of each pipeline. In Figure 1 and in Figures 4 7, the pipeline comprises four pipelines: 1 through 4. In "MIMD" mode, these pipelines have different lengths thus creating complexities in controlling them globally. Internal to the pipelines, there are registers, and associated with these registers are status fields.

Referring to Figure 4b, the status fields for the registers in the add pipeline of Figure 4 consist of the operands FPR address (ADDR), whether "SISD" or "MIMD" mode is invoked (M/S), some bypass information (2BY), whether this stage in the pipeline is for a valid instruction (VI), whether the data in the register is valid (VD), and whether the instruction is RX or RR type (RX).

Referring to Figure 5b, the status fields associated with the registers of the multiply pipeline of Figure 5 store this information and further information including the length of the operands (LI), whether it's a floating point or fixed point operand (FLP), whether the data is still valid (VR), even if this stage of the pipeline does not have a valid instruction, and whether Y's operand is interlocked(INTL). The other pipelines do not need status information because they are very short.

Referring again to Figure 1, the status information for each stage of the pipeline flags to the following stage its validity and then, in the following cycle, the next stage becomes valid if there are no contentions. Thus, the flags help in determining contention and push the instruction's data as far as it can through the pipeline. After the pipeline in question locates the data on the DBUS and execution takes place in the pipeline, the pipeline in question waits for the Dynamic History Tables oldest entry to match with the pipeline number (PIPE NO) of the pipeline in question; at this point, the pipeline is allowed to complete thus maintaining instruction completion synchronization.

```
     V  WR ADDR  PIPE NO  H   WT  M/S EXT LEN     INT TAG          PSW PTR
   | 0 |1 |2 |3 | 4 | 5 | 6 | 7 | 8 | 9 | 10 |11 | 12 | 13|14 | 15 | 16|
1  |---|--+--+--|---+---|---|---|---|---|----|---+----+---|---+----+---|
     *    *       *       *   *   *   *   *         *              *
     *    *       *       *   *   *   *   *         *              *
     *    *       *       *   *   *   *   *         *              *
   |---|--+--+--|---+---|---|---|---|---|----|---+----+---|---+----+---|
8
```

```
V       - valid bit
WR ADDR - write address
PIPE NO - pipeline number (00 - add, 01 - mult, 10 - load rx, 11 - misc)
H       - hardwired fpu request if 1, if 0 microcode instruction
WT      - write type
M/S     - if 1 MIMD, if 0 SISD type pipeline
EXT     - extended result must be written
LEN     - if 1 long result, if 0 short result
INT TAG - interrupt tag which uniquely defines instruction in CPU stack
PSW PTR - identifies retry pointer for the instruction
```

Figure 3. Dynamic History Table

Referring to Figure 4, pipeline1 (add) internal control registers are illustrated.

In Figure 4a, pipeline1 includes an alignment register, an FA register, an FB register, an A register, a B register, an adder, an FS register, an S register, and a post normalizer register. In Figure 4b, status fields associated with the FA register, FB register, and FS register are illustrated.

Figure 4 illustrates the add pipeline and its associated internal pipeline control registers. The add pipeline of Figure 4 consists of three cycles. During the first cycle, the data is retrieved from either the FPRs and/or from the DBUS. Alignment is accomplished by the alignment hardware. Operands are latched into A register and B register. In cycle two, the actual add is performed by the adder and the result is stored in S register. In the third and final cycle, the post normalizer shifts out leading zeros if required and the data is sent back to the FPRs. The previously described function reflects the manner in which the pipeline handles add instructions. For other instructions, belonging to the same category, at least some of the registers or some of the internal bypassing controls in the pipeline are used. Thus, the three cycle add pipeline is used for many different instructions. To maintain and control this pipeline, three major control registers are needed: FA register, FB register, and FS register.

Referring to Figure 4b, the status fields for the FA register, FB register, and FS register are illustrated.

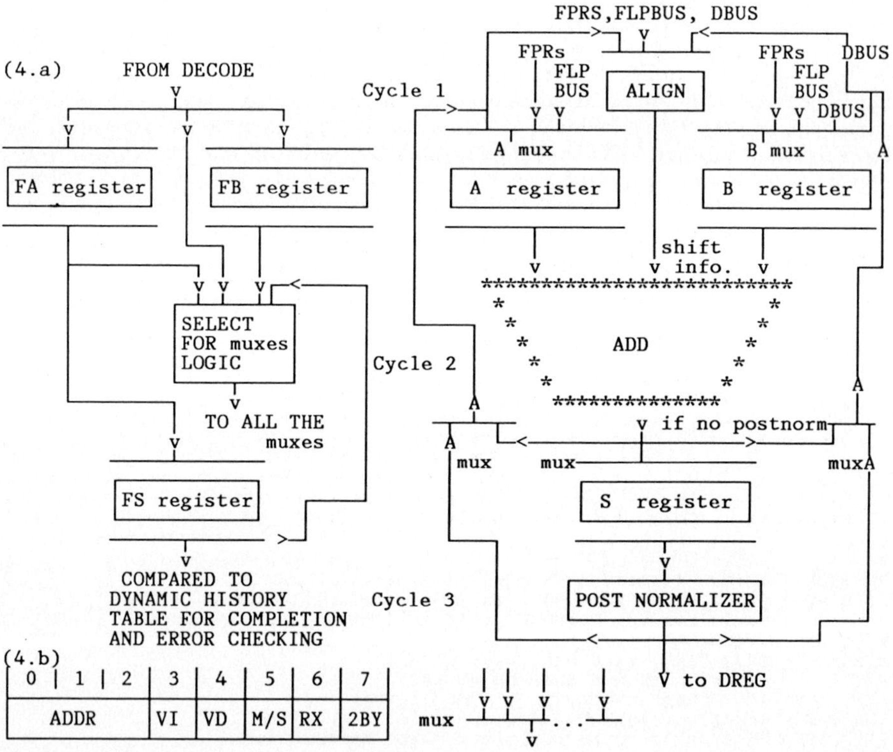

Figure 4. Pipeline 1 (ADD) Internal Control Registers

In Figure 4b, such status fields include the following bits:

1. FPR address bits (ADDR) of the operand which are used to locate operands that may be interlocked
2. A valid instruction bit (VI) which is used to indicate that this stage in the pipeline is valid for an instruction
3. A valid data bit (VD) which indicates that the associated data register is valid
4. A "MIMD/SISD" pipeline indicator (M/S) which signals the instruction end. For "MIMD" mode, its the last stage valid and no contention on completion. For "SISD" mode, its a little more complex because the instruction may loop several times through the pipeline.
5. A bit to indicate that the instruction is an RX type of instruction (RX) which, on the FB register, indicates that its address bits are really invalid and the data bus should be watched for incoming data if not already valid.
6. A bit which indicates the first cycle of a two cycle bypass (2BY). Sometimes it takes two cycles to retrieve interlocked data once it has been located.

(5.a)

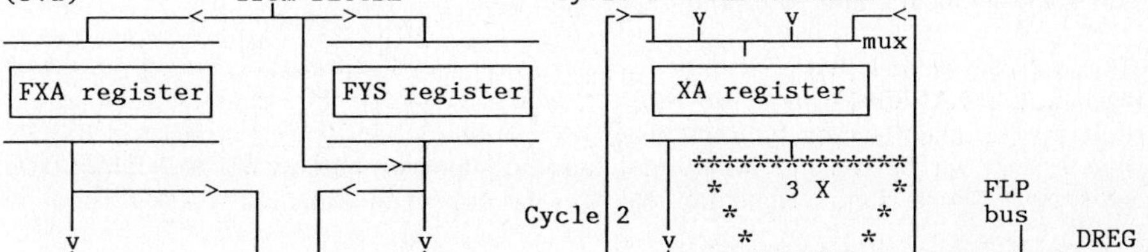

(5.b)

0 | 1 | 2	3	4	5	6	7	8	9	10	11
ADDR	VI	VD	VR	RX	2BY	EXT	LI	FLP	INTL

Figure 5. Pipeline 2 (MULT) Internal Control Registers

Referring to Figure 5a, pipeline2, the multiply pipeline, and the control registers, internal to the pipeline, is illustrated.

In Figure 5a, pipeline2, the multiply pipeline, includes FXA register, FYS register, FXB register, FY register, FP register, XA register, 3X hardware, XB and 3X registers, Y register, M1 hardware, M2 hardware, and P register, details of the multiplier algorithm can be found in [5],[6]. The multiply pipeline consists of 5 cycles if no hazards are encountered:

- Cycle 1 - Operand 1 is loaded into XA register from the FPRs and, if operand 2 is also from the FPRs, it is read and stored in a temporary register because the bus structure limits the loading to one operand at a time.
- Cycle 2 - Operand 2 is loaded from either the temporary register or from the dbus to Y register; concurrently, a 3 times multiple of operand 1 is calculated by the 3X hardware and stored in 3X register. The XA register directly loads XB register.
- Cycle 3 and Cycle 4 - These are the two cycles of actual execution of the multiplier. These cycles, termed the M1 and M2 cycle, use the M1 hardware and the M2 hardware. No registers separate the two cycles of execution. Thus, XB and 3X registers and Y register must be held for these two cycles until the data is latched in P register.
- Cycle 5 - This cycle involves a write from P register to the FPRs. If the result is extended, there is a cycle 6, which is a second write cycle, since the bus structure between chips is limited to one 8 byte data bus.

The control registers that control this pipeline are: the FXA register, which maintains the status of XA register; the FYS register, which maintains the status of a temporary register that initially gets operand 2 on an RR instruction; the FXB register; the FY register; and the FP register, which maintain the status of their associated paired data registers. These status registers maintain 12 bits of information, as described by Figure 5b.

Referring to Figure 5b, a description of each field of the control registers (the FXA register, the FYS register, the FXB register, the FY register, and the FP register), is illustrated.

In Figure 5b, the following is a description of the fields not already described:

1. A valid result bit (VR) which is used to create a local working store on another chip. On an RR instruction, operand 2 remains valid (after the multiply operation) in the Y register until another instruction modifies this FPR. The valid result bit would indicate that the data in the Y register is valid for the address defined. Also, on an RR and a RX instruction, the product register, which contains the result of the multiply, is equivalent to the FPR addressed by operand 1 until another multiply goes through the pipeline or another instruction modifies this address. This is very important in improving performance. Thus, if any of the loads can be eliminated, there usually is an improvement in performance.
2. A bit to indicate that an extended result (EXT) must be written back to the FPRs.
3. A bit to indicate long operands (LI) are in the registers.
4. A bit to indicate the instruction in this stage of the pipeline is a floating point instruction (FLP).
5. A bit to indicate interlock on the Y register's operand (INTL).

The FXA register and FYS register are set in cycle one by the initialization hardware. If there is no "contention" on the XB and 3X registers, the FXB register is set from FXA register. If there is no "contention" on Y register, the FY register is set by the initialization hardware or FYS register. "Contention" can take the form of: M1 valid, or M2 valid and P register contention, or XA already valid for previous multiply and XB contention.

The FP register is set by FXB register when M2 is valid and there is no contention on P register. The valid result bit of the FP register is maintained separately since it is dependent on other writes to the FPRs. Thus, the multiply pipeline is maintained by the internal pipeline controls which push the data as far as possible through the pipeline until internal hazards are encountered or until the P register is valid and no external hazards are encountered which prevent the instructions in this pipeline from completing.

Referring to Figure 6, pipeline3, which is used for load RX type instructions, is illustrated.

In Figure 6, the pipeline3 consists of two cycles, during which there is no execution of data, rather, data merely flows through the pipeline. The instruction is decoded via decoder and the FPU waits for the appropriate DBUS valid signal from the DBUS stack controls. The data, being received and gated into DREG of the DBUS stack, is received during cycle 1. Then, the data is sent to the FPRs during cycle 2. The controls which monitor this are the DREG Valid register of the DBUS stack controls and the rest of the DBUS stack controls.

Referring to Figure 7, pipeline4, which performs miscellaneous functions, is illustrated.

In Figure 7, pipeline4 comprises indirect address registers (IND ADDR REGs) connected to an output of the DREG which forms a part of the DBUS stack of Figure 1, status word register connected to the output of the DREG, retry status register connected to a CBUS register, and an indirect address mode register connected to the CBUS register. The status word register and the retry status register are connected, at their outputs, to the DREG and to the FPRs. Data is loaded from the DBUS into the indirect address registers (IND ADDR REGs) and the status word (STATUS WORD) registers. The indirect address registers are used in microcode mode and contain FPR addresses. The status word registers maintain the status of the FPU, such status consisting of exception status information and the status of the checkers.

In Figure 7, there are five basic groups of instructions which perform the miscellaneous functions in pipeline4:

1. The Load RR group - These are basically one cycle operations where a read (possibly a trivial manipulation of bits) is performed followed by a write to the FPRs.
2. The Load RX group of instructions to other than the FPRs, such as to the indirect address registers and the status word registers.
3. Instructions in this third group involve two cycles: cycle one involves waiting for the DREG Valid register to be valid while loading DREG and cycle two involves loading the DREG valid register from DREG. Instructions which set registers from information off the CBUS are also included in this category, such registers including the retry status register, which maintains retry information and the indirect address mode register, which maintains the mode bit for addressing purposes.
4. The Store RX group of instructions involve reading the FPRs in cycle 1 and storing in DREG. In Cycle 2, the data is driven on the data bus and dbus valid is signalled.
5. The last group of instructions are store type instructions which place the contents of a register, other than the FPRs onto the DBUS, in two cycles. The two registers that can be read in this way are the status word registers and the retry status registers.

There are no control registers associated with the running of pipeline4, except, when applicable, the DBUS stack control registers.

Referring to Figures 8a and 8b, the DBUS stack and DBUS stack controls are illustrated.

In Figure 8a, the DBUS stack controls comprise the DREG valid register, the DCBUS Register, the decoder, the DS5 register, the DS4 register, the DS3 register, the DS2 register, the DS1 register, the S2 valid register, the S1 valid register, and the mux select logic. The valid registers have two bits where P indicates partially valid data (across 8 byte boundary), and F indicates the data is fully valid. The DBUS stack comprises the S2 register connected to the DBUS, the S1 register connected to the output of the S2 register and to the DBUS, and the DREG register connected to the output of the S1 register and the DBUS.

The bits within registers DS1- DS5 include the following:

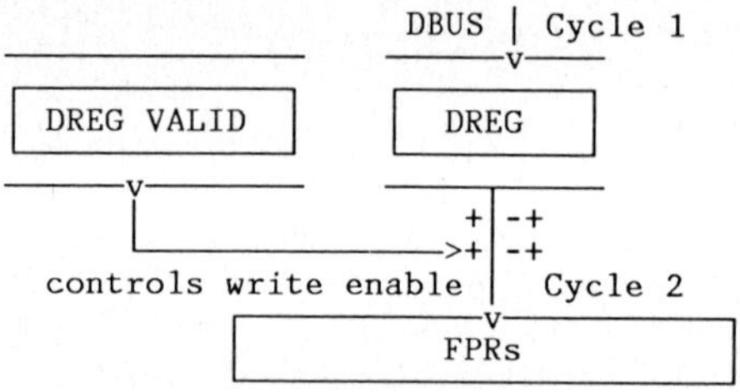

Figure 6. Pipeline 3 (LOAD RX)

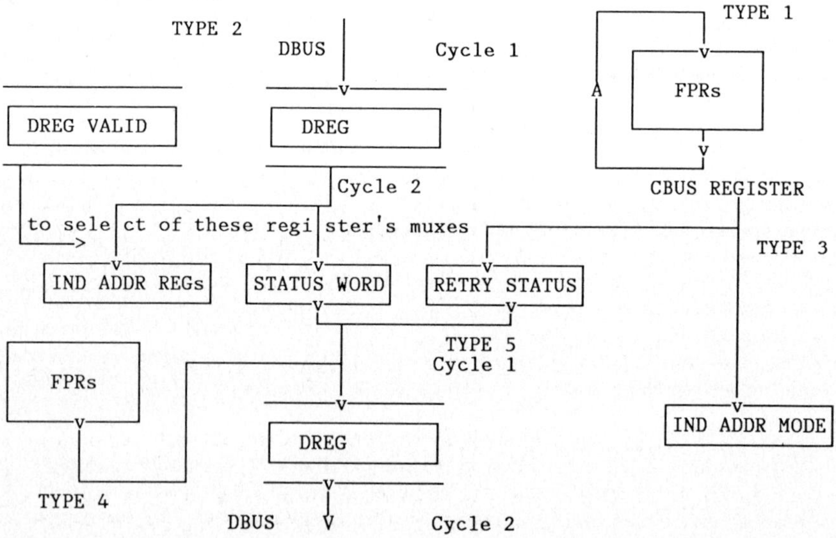

Figure 7. Pipeline 4 (MISCELLANEOUS)

- Bits 0-2 ,IADDR, which indicate the instruction address; the three bits, if non zero, indicate that a cross-boundary load shall take place which requires two DBUS valid signals, one for each part of the data.
- Bits 3 and 4 which are the PIPE number, to uniquely identify the pipeline to which the data should be transmitted.
- Bit 5 is a valid instruction(VI) bit.
- Bit 6 indicates the data for this instruction is in DReg of DBUS stack shown in Figures 1 and 8a.
- Bit 7 indicates the data for this instruction is in S1 register of DBUS stack shown in Figures 1 and 8a.
- Bit 8 indicates the data for this instruction is in S2 register of DBUS stack shown in Figures 1 and 8a.
- Bit 9 indicates that the instruction is EXEcuting and the appropriate pipeline has been initialized.

When data is received, it must find its way to the appropriate instruction, which then transmits through the pipelines as described previously. This creates a chaotic condition. This chaotic condition, involving data finding its way to its correct instruction in the appropriate pipeline is described next.

Data, for an operation involving normal RR operations, comes from the FPRs, unless an interlock exists, in which case, the local pipeline controls find the appropriate data by comparing address fields of status registers of all the pipelines. For RX operations, it is a little more difficult; operand one is found in the same way, but operand two comes from the data bus. Since the data cache or another bus unit is requested to supply the data at the same time that the FPU receives the instruction, and the other bus unit, which is requested to supply the data, may be BUSY, the receipt of the data

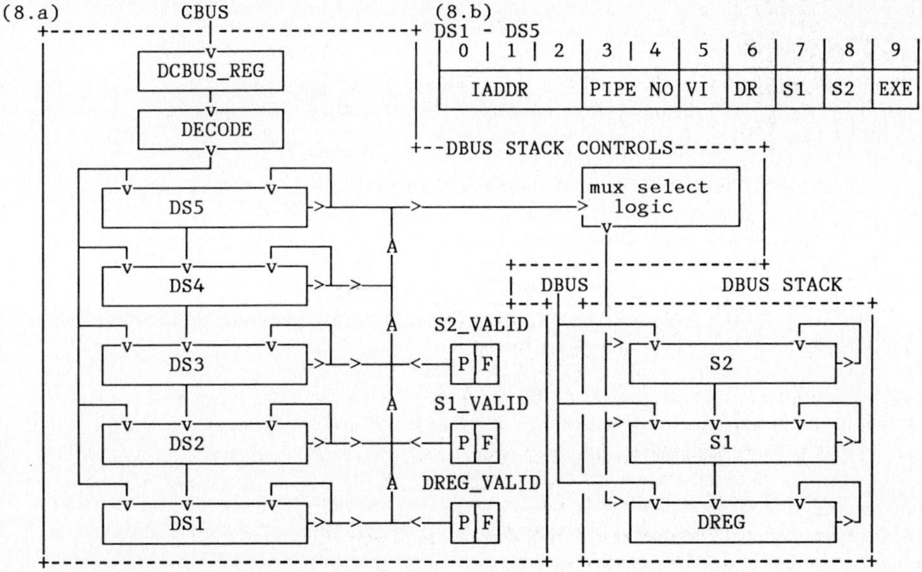

Figure 8. DBUS Stack and Controls

by the FPU may be out of synchronization with the receipt of the instruction by the FPU. As a result, underflow or overflow of the data may exist. The DBUS stack controls inform the local pipeline controls when the DREG is valid for their appropriate pipeline. This is vital in separating the data, especially when several pipelines are starved of data from the data bus and the data bus becomes valid. Therefore, the DBUS stack controls sequences the data with the instructions. After this is accomplished, the pipeline controls move the data through the pipeline and await the approval of the global Dynamic History Table to approve the completion of the instruction. In addition, the data bus stack controls are needed since the memory supplies the data when it can, independently of when the data is requested. Thus, conditions of overflow and underflow of data are possible. Underflow occurs when the memory is slow in supplying the data and several pipelines have been initialized, except for the absence of an operand from memory (RX type). Since several pipelines may be waiting for data, when it finally arrives, controls are needed for sorting out to which pipeline the incoming data should be transmitted. Overflow occurs when a pipeline or several pipelines are full and several instructions, which use the data from memory, are transmitted from the CPU to the floating point unit (FPU) and memory simultaneously; this causes the memory to send the data to the floating point unit, since it cannot immediately put the data into an appropriate pipeline. Therefore, the data must be stacked and identified with the appropriate pipeline number to which the data must eventually be transmitted.

3. CONCLUSION

The present paper described the pipeline structure of a floating point unit hardwired engine. The instruction set considered pertains to the floating point 370 and 370/XA architecture. The same structure can be also extended to other execution units as the fixed point and with modifications to other architectures with "SISD" type instruction sets.

REFERENCES

[1] K. Hwang and F.A. Briggs, **Computer Architecture and Parallel Processing,** McGraw-Hill, 1984.

[2] P.M. Kogge, **The Architecture of Pipelined Computers,** McGraw-Hill, 1981.

[3] **IBM System/370 Principles of Operation,** GA22-7000-10, IBM Corporation, Sept. 1987, available through IBM branch offices.

[4] **IBM System/370 Extended Architecture Principles of Operation**, SA22-7085-1, IBM Corporation, available through IBM branch offices.

[5] S. Vassiliadis, E.M. Schwarz, and D.J. Hanrahan, "A General Proof For Overlapped Multiple-bit Scanning Multiplications" **IEEE Transactions on Computers,** Vol.38, No.2, pp. 172-183, February 1989.

[6] S. Vassiliadis, E.M. Schwarz, B.M. Sung, "Hard-Wired Sign-Magnitude Multipliers with Encoded Partial Products," in preparation for **IEEE Transactions on Computers.**

Wafer Scale Integration, III
M. Sami and F. Distante (Editors)
Elsevier Science Publishers B.V. (North-Holland)

A CONFIGURABLE SINGLE WAFER BIT-SLICED CONVOLVER: EFFECT OF THE INTERCONNECTION NETWORK ON THE YIELD

Sylvie LASSERRE

Dipartimento di Elettronica
Politecnico di Milano
Milano (Italy)

This paper is a statistical study on the way to implement fault-tolerance on wafer integrated repetitive systems so as to optimize the fabrication yield. Fault-tolerance is introduced at two different levels: wafer- and chip level. Then, the optimum amount of redundancy at both levels in terms of area is analytically determined. Taking also into account the existence of the configuration structure, we show that it determines completely the optimum solution. Moreover, two different configuration approaches at wafer level are compared as regards to their effect on the yield.

INTRODUCTION

Highly parallel and regular architectures are particularly well adapted to the wafer scale integration approach. An architecture has been envisioned in [1], that may be interesting for a monolithic integration on a single wafer. It consists of an important array made of simple and identical cells. The length of the convolutions, and their number, is an application-dependent parameter, that commands the configuration of the convolver. This completely iterative and programmable convolver, well suited to our application, and consisting of a unidirectional systolic linear array of identical bit-slices, has been defined in [1]. For reasons of timing, space requirements and power limitation, a monolithic integration of the convolver on a wafer is envisioned. At this point, our problem is to define a fault- tolerance politic, which is necessary for yield enhancement. Fault-tolerance for wafer scale integration of our system may be implemented at two different levels [4,5]: at chip level, where the chip itself is fault-tolerant, and at wafer level, where a larger number of chips than nominally necessary are integrated on the wafer. At chip level, a Diogenes approach [2] is envisioned, that is, a simple bypass of the faulty slices will be done. At wafer level, two different approaches are envisioned. The first one is a Diogenes approach too. In the second one, the chips are embedded into an interconnection network made of buses and switches. The bad chips are disconnected, while the good chips are connected in order to form a linear array.

The array is then partitioned into chips, each one containing a given number of slices. Since in the architecture here envisioned different levels for introduction of redundancy -and for the consequent fault-tolerance strategies- may be chosen, the present paper focuses on this specific problem, aiming at identifying through statistical methods the optimum solution, that is, the best repartition of both wafer- and chip level amount of redundancy. While this model takes into account the existence of the configuration structure at both levels, both wafer level configurations strategies are compared.

In the first section, the repetitive structure of the convolver upon which the work is based will be briefly described. In the second section, analytical models of the yield at chip- and wafer level are recalled. Finally, in section 3, two different wafer level configuration approaches are compared as regards to the yield.

1 THE BASIC FUNCTIONAL ARCHITECTURE

The basic architecture has been presented in details in [1]. We briefly recall in this section the most important points that concern the regularity of the architecture. The basic architecture is given in figure 1 [1].

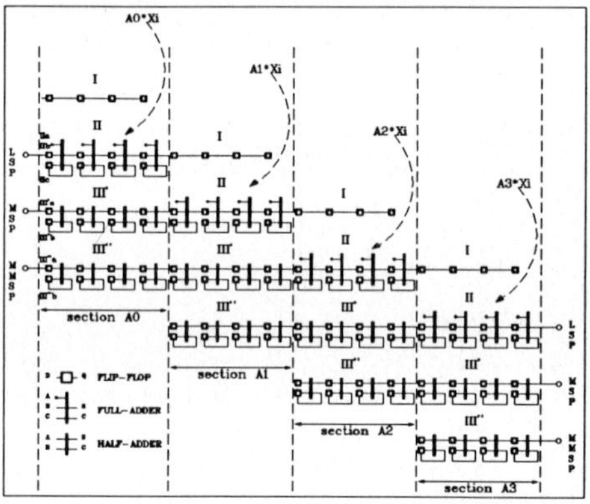

Figure 1 *Bit-sliced convolver*

Each slice contains three serial carry-save adders, a flip-flop to store one bit of a coefficient, and an AND-gate that computes the partial product between the stored bit of the coefficient and the incoming bit of the sample. The adders are used to sum this partial product to the part of the convolution already computed by the preceding slice, and to pass the result to the subsequent slice. The samples (inputs) and results (outputs) are both in serial form. Coefficients are stored in internal registers. The sample bits are broadcast to all subunits: each subunit executes the multiplication with the stored coefficients and accumulates the product with the result received from the preceding unit. The number of slices is an application-dependent parameter. A stack of slices can be programmed as a convolver for prescribed sample and coefficient bit-length, and the number of slices may vary from tens to thousands, depending on the required application. The bit-rate does not depend on the number of slices, that is, the structure is fully iterative. Each slice is bypassable for fault-tolerance requirements. The configuration into an operational or non-operational state is achieved through the use of a control bit contained into each slice. The convolver is divided into sections of equal length. The length (or number of slices) of a section is equal to the length (or number of bits) of the coefficient. The length of a convolution is defined as the number of terms involved into the sum and is equal to the number of sections (see figure 1). Each slice contains a functional configuration bit that establishes the connection with the preceding slice: parallel from slice to slice into the same section, and shifted from a section to another one.

2 FAULT-TOLERANCE STRATEGIES AND YIELD CONSIDERATIONS

The overall fault-tolerance approach adopted here is a hierarchical one, such as the one proposed in [2]. For this purpose, the wafer is subdivided into chips and each chip contains a given number of slices so as shown in figure 2. Fault- tolerance is then envisioned at two different levels: chip- and wafer level.

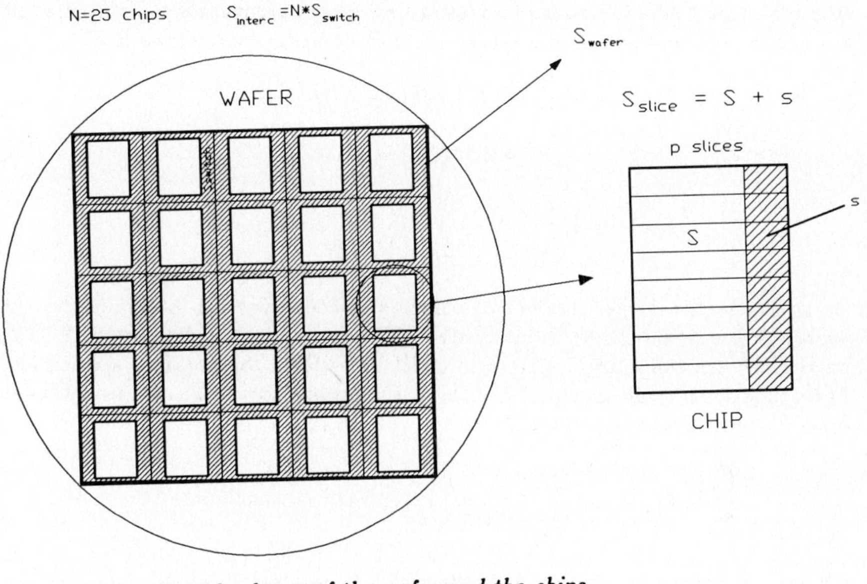

Figure 2 *Topological scheme of the wafer and the chips*

In that previous paper, both chip- and wafer level redundancy are described with the evaluation of the corresponding yields, that we briefly recall:

(1) At chip level, a Diogenes approach is adopted [3]. That is, more slices than nominally necessary are implemented on the chip, and, in case of a faulty slice, a simple bypass of this faulty slice is done. To this aim, a bypass structure has been added to each slice. For a chip consisting of p slices, among which k are redundant, the yield R of a redundant chip is the probability to obtain at least $(p - k)$ working slices, and no faults on the global bypass structure. Then, the bypass structure area is a critical area, that is, the occurrence of only one fault on it makes the whole chip faulty. The yield is expressed as follows:

$$R = [P_o(s)]^p \sum_{i=0}^{k} [\binom{p}{i} - f(i,p)][P_o(S)]^{(p-i)}[1 - P_o(S)]^i$$

where:

p is the number of slices per chip.

k is the number of redundant slices per chip.

S is the functional area of the slice.

s is the area of the bypass structure of one slice.

$P_o(S)$ is the probability that no fault occurs on the area S. For a Poisson distribution, and a fault density D: $P_o(S) = e^{-DS}$.

$f(i,p)$ is the number of combinations where two or more faulty slices are adjacent. It is subtracted to $\binom{p}{i}$, because in our case, we do not recover chips with adjacent faulty slices. The function f has been recursively defined as follows in [2]:

$$i = 0, 1 : f(i,p) = 0$$

$$i = 2 : f(i,p) = p - 1$$

$$i > 2 : f(i,p) = 1 + \sum_{j=i-1}^{p-2} [\binom{j}{i-2} + f(i-1,j)]$$

(2) At wafer level, a larger number of chips than necessary are integrated on the wafer. A linear array has to be created. The bad chips are disconnected. The good chips are connected together in order to perform the required application. For a functional requirement of N_R chips, the probability $P(N_R, N)$ to obtain at least N_R working chips among N is:

$$P(N_R, N) = \sum_{i=0}^{N-N_R} \binom{N}{N_R + i} R^{(N_R+i)}(1 - R)^{N-(N_R+i)}$$

where R is hereabove defined. It is the yield of a redundant chip.

As proved in [2], the existence of the interconnection network has to be taken into account in the evaluation of the yield for the both following reasons:

(1) First, for the area S_{interc} it occupies, that makes diminish the remaining area for the functionality, that is, for the spare chips.

(2) Second, for the effect of the occurrence of faults on this area. This effect depends on the configuration strategy used. This point is studied in details in section 3.

Both levels redundancy are closely linked together. In effect, for a N chips wafer, and for a global functional requirement of S_R slices which are distributed among these N chips, parameters are evaluated as follows:

Interconnection area (S_{interc}): to each chip is associated an area S_{switch} which is used for the wafer level configuration. For a N chips wafer, the global interconnection area is then:

$$S_{interc} = NS_{switch}$$

Chip area (S_{chip}): given an integrable area S_{wafer} of the wafer, the remaining area for the functionality after the interconnection area is subtracted, is divided by the number of chips per wafer N, and this area is assumed to be the chip area:

$$S_{chip} = (S_{wafer} - S_{interc})/N$$

Number of slices per chip (p): it is calculated as the chip area divided by the slice area S_{slice}:

$$p = S_{chip}/S_{slice}$$

and $S_{slice} = s + S$ for a redundant chip, $S_{slice} = S$ for a non-redundant chip.

Number of required chips (N_R): for a global functional requirement of S_R slices distributed among N chips, and for $(p - k)$ functional slices per chip, the number of chips necessary to achieve the application is:

$$N_R = S_R/(p - k)$$

In our study, we take too into account the harvesting, that is, the percentage of the good chips that may be connected together as a linear array through the use of a configuration algorithm. The harvesting depends mostly on the choosen algorithm, on the number of good chips, and on their disposition on the wafer, that is, on the yield R of a redundant chip.

We assume an elementary chip yield R, and a harvesting h. h depends on R: $h(R)$. Taking into account the harvesting is equivalent to assuming a 100% harvesting with an equivalent elementary yield $R' = R * h(R)$. With this assumption, the probability $P_h(N_R, N)$ to obtain at least N_R chips becomes:

$$P_h(N_R, N) = \sum_{i=0}^{N-N_R} \binom{N}{N_R + i} [Rh(R)]^{(N_R+i)} [1 - Rh(R)]^{N-(N_R+i)} \qquad (1)$$

The effect of the occurrence of faults in the interconnection network depends on the configuration strategy used, and this is the purpose of the next section.

We showed in [2] that there is a balance between wafer- and chip level amount of redundancy. It was proved that the wafer level interconnection network determines completely the optimum solution of this compromise. In the next section, two configuration approaches at wafer level are envisioned. Then, the effect of the interconnection network is studied with much details.

3 WAFER LEVEL INTERCONNECTION NETWORK INFLUENCE

The effect of the configuration at chip level has been studied in [2]. Configuration at wafer level is now our purpose. A rectangular array of identical chips has to be configured into a linear array. As it was said hereabove, considering the existence of the configuration is necessary (as regards to the yield) for the following reasons: the first one is that the interconnection network area S_{interc} necessary to the configuration makes decrease the remaining area for the functional elements. Second, the configuration area is susceptible to defects too, and a configuration area with faults has obviously a negative effect on the yield too. Third, a given interconnection network will allow specific configuration algorithms; through the use of these algorithms, a certain amount of chips will be collected: in that case, this is the influence of the harvesting on the yield.

Yield is evaluated for two distinct configurations strategies. The first strategy is based on a Diogenes approach, which principle is the bypass of a bad element: the data path is determined in advance and will be concretized through the interconnection structure. The principle of the second strategy (let's call it strategy B) is the deturn from the bad elements so as to evit them. Then, a convenient data path is created, which was not first predetermined.

To implement these two strategies, two different interconnection networks are used. In the Diogenes approach, the linear array will be trivialy configured, without the use of any algorithm, and 100% of the good chips will be recuperated. On the other way, one fault on this type of interconnection network is *globally* fatal: only one fault is sufficient to make the whole wafer faulty. The strategy B allows the use of different configuration algorithms, which efficiency depends on the number of bad chips, and on their distribution. The impact of a fault on a B interconnection network is determined further. In that case, a fault is only *locally* fatal.

In the following, for each strategy and each configuration algorithm, we evaluate the effect of the interconnection network on the yield of the wafer, and we exhibit the results with the help of some appropriate curves.

3.1 Diogenes strategy

3.1.1 Interconnection network

To each chip is added a switch (which area is S_{switch}) able to connect or disconnect the chip from the linear array. Chips are connected to each others on the wafer by the mean of these switches so as shown in figure 3.

The main disadvantages of this method are the following:

(1) First, a fault on the interconnection structure is fatal to the whole wafer. In effect, a bad switch cuts the global linear array. We say that a failure in the interconnection network is globally fatal.

(2) Second, assuming a good interconnection network, the bypass of some chips introduces irregular propagation delays in the transmission of the datas along the global array.

On the other hand, the advantage of such a method is that, assuming a working interconnection network, the value of the harvesting is always 100%. Then, while $h(R) = 1$, (1) may be rewritten as:

$$P_{100\%}(N_R, N) = \sum_{i=0}^{N-N_R} \binom{N}{N_R + i} R^{(N_R+i)} (1-R)^{N-(N_R+i)}$$

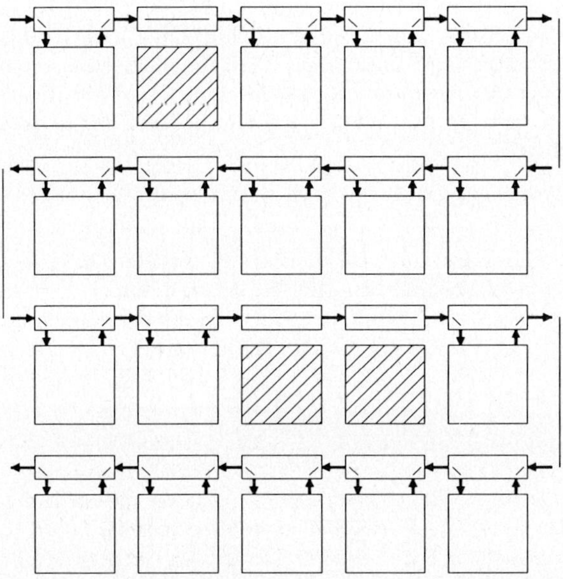

Figure 3 *Diogenes configuration strategy*

3.1.2 Yield considerations

Taking into account the globally nefast effect of the interconnection network, the yield is the probability to obtain at least N_R working chips among N, and to obtain a working interconnection area S_{interc}. That is:

$$Y = P_o(S_{interc}) \sum_{i=0}^{N-N_R} \binom{N}{N_R + i} R^{(N_R+i)} (1-R)^{N-(N_R+i)} \tag{2}$$

where $S_{interc} = NS_{switch}$. Then, (2) becomes:

$$Y = [P_o(S_{switch})]^N \sum_{i=0}^{N-N_R} \binom{N}{N_R + i} R^{(N_R+i)} (1-R)^{N-(N_R+i)} \tag{3}$$

where we recall the values of the parameters:

$N_R = S_R/(p-k)$ is the number of chips necessary to the application

$p = S_{chip}/S_{slice}$ is the total number of slices per chip

$S_{chip} = (S_{wafer} - S_{interc})/N$ is the chip area

$S_{interc} = NS_{switch}$ is the global interconnection network area

3.2 Strategy type B

3.2.1 Interconnection network

A rather efficient interconnection network, which is well suited to many configuration algorithms has been defined and is represented in figure 4. The direction of the chips is given by the direction of the arrows. Adjacent chips have opposite directions because of the necessity of configuring a linear array with the most little connection length as possible. In effect, a same orientation for all the chips would introduce some very long connections, for example, at the end of a column, because of the necessity of linking together two columns.

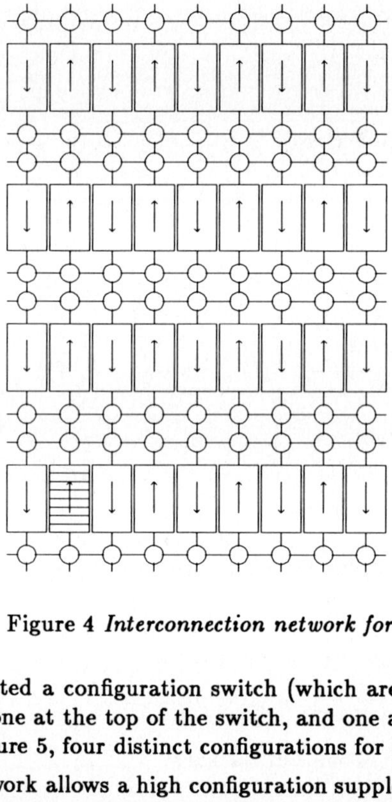

Figure 4 *Interconnection network for the strategy B*

To each chip is associated a configuration switch (which area is S_{switch}), consisting in two little switches: one at the top of the switch, and one at the bottom (see figure 13). So as shown in figure 5, four distinct configurations for each switch are possible. This configuration network allows a high configuration suppleness. In effect, a pair of two adjacent chips may be connected so as in figure 6, depending on the configuration of the switches. Connections between chips are respectively symbolized so as shown in figure 6.

An example of the configuration of a rectangular array into a linear array using this interconnection network is shown in figure 7. The hatched chips are the bad chips that have to be discarded from the linear array. Because we do not afford the direct bypass (in order to maintain a constant crossing of two switches), the chips that are adjacent to the bad ones may remain unused. The connection rules and the interconnection network we have defined, make possible a configuration such that in any case (faults

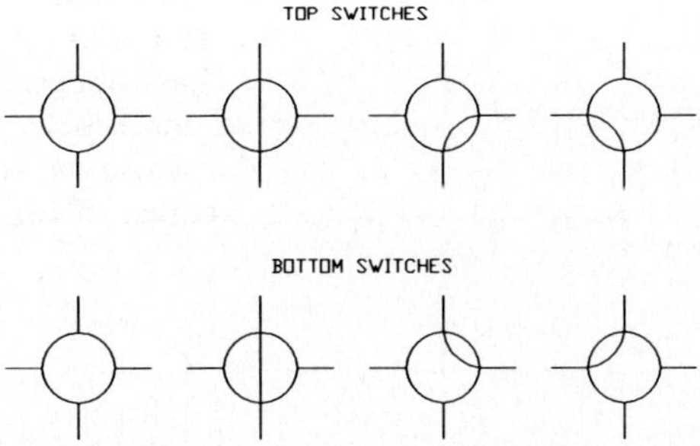

Figure 5 ***Positions of the configuration switches***

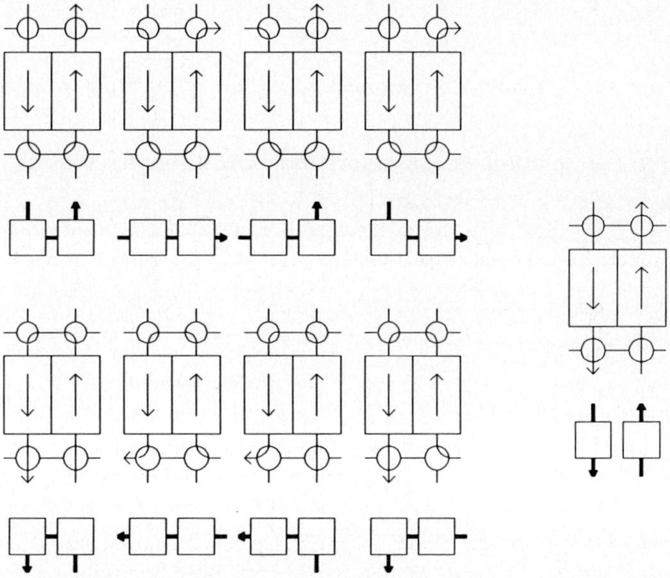

Figure 6 ***Data propagations allowed through a pair of chips***

or no faults), the datas always cross 2 switches.

The advantage of such an interconnection network is that a fault is only locally fatal. But the disadvantage is that the harvesting is often much inferior to 100%.

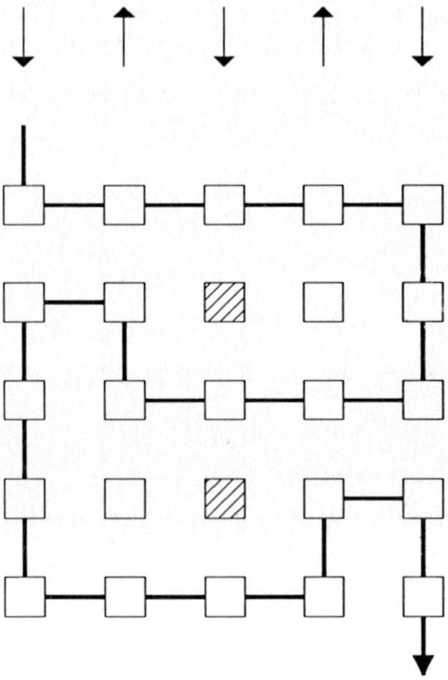

Figure 7 *An example of the use of the B interconnection network*

3.2.2 Two configuration algorithms and their harvesting

Two algorithms for the configuration of a rectangular array into a linear array are described. They both use the interconnection network hereabove defined, where the only connections between chips that are permitted are shown in figure 8.

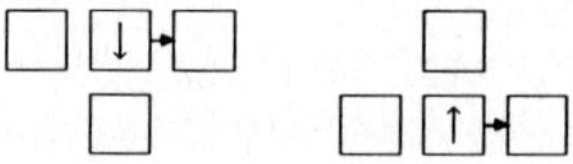

Figure 8 *Permitted connections between adjacent chips*

Algorithm 1: As an example, one of its manifestations is shown in figure 9. Its principles are the following:

(1) The linear array begins at the top of the array and ends at the bottom.

(2) The snake goes from left to right, then from right to left, and must remain the highest as possible, so as to collect the most chips as possible.

(3) Case where the snake does not finish at the bottom of the physical array, other solutions are tried until the snake ends at the bottom.

(4) Once the linear array is created, it may happen that some pairs of chips remain unused. Then the snake is augmented so as indicated in figure 9 by the dashed lines.

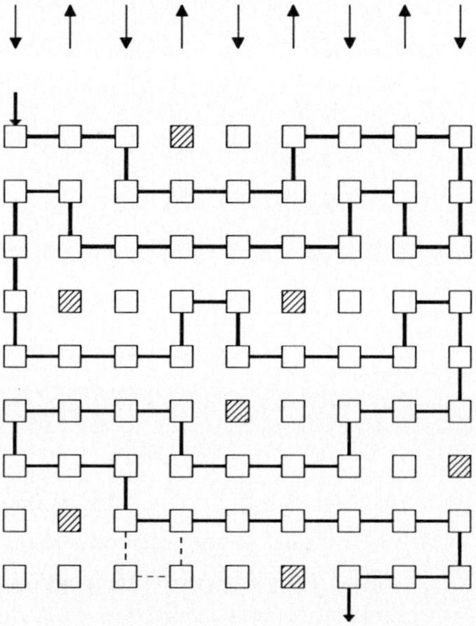

Figure 9 *Example of the algorithm 1*

Simulations were done for a 15 X 15 array, that is $N = 225$, and the harvesting versus the individual yield R of one chip is plotted in figure 10.

Algorithm 2 or spanning tree algorithm: An example of its application is shown in figure 11. The chips are grouped into sets of four chips. Then, the Koren's spanning tree algorithm is applied [3].

Simulations were done for a 16 X 16 array, and the harvesting versus the yield is plotted in figure 12.

3.2.3 Yield considerations

The effect of faults in the interconnection network is now evaluated. Because the connection rules (see figure 8) and the connection network are the same for both algorithms, the effect of faults on the interconnection network is the same. To each chip is associated an interconnection structure which is shown in figure 13. Then, the global array is a cascade of such cells (see figure 13).

We assume that a fault on a switch or on a bus propagates untill the adjacent switches. The connection rules herabove defined allow us to make the following assumption: a

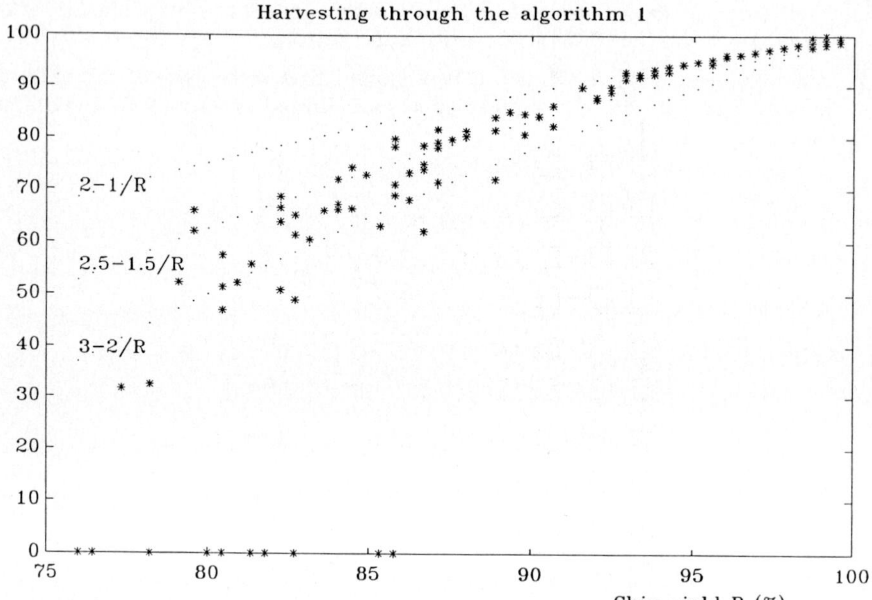

Figure 10

fault on a bus or on a switch of the part of the interconnection network associated to a given chip is equivalent to obtaining a bad cell such as the one defined in figure 13. If r is the yield of a configuration element associated to a chip (let's call it switch now), the yield of the cell is $R" = Rr$. This means that the cell works if: first, the chip works, second, if its configuration structure works too. We assume then that the yield of the wafer that takes the interconnection network into account is equal to the yield of the same wafer without taking into account the interconnection structure, but assuming bigger chips with an equivalent yield $R"$. The global yield (yield of the wafer) is then:

$$Y = \sum_{i=0}^{N-N_R} \binom{N}{N_R + i} R"^{(N_R+i)} (1 - R")^{N-(N_R+i)} \tag{4}$$

where $R" = Rr$, and $r = e^{-D_{switch} \cdot S_{switch}} = P_o(S_{switch})$.

D_{switch} is the fault density on the switch, which may be different than the fault density on the chip: they have not necessarly the same density of integration, then, the occurrence of a defect may not have the same impact.

Introducing the harvesting, $R"$ becomes $h(R") * R"$. That is:

$$Y = \sum_{i=0}^{N-N_R} \binom{N}{N_R + i} [R" h(R")]^{(N_R+i)} [1 - R" h(R")]^{N-(N_R+i)}$$

It is important to note that it is assumed that no fault occurs in the structure supporting the command of the configuration switches.

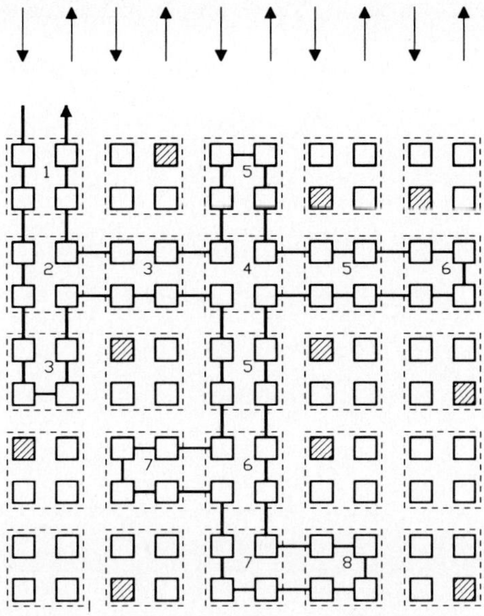

Figure 11 *An example of the algorithm 2*

3.3 Comparison of both strategies

As it was shown in [2], the interconnection network necessary to configure fault-tolerant systems has a negative effect on the enhanced yield. We have distinguished hereabove two distinct classes of interconnection networks:

- The class of the interconnection networks on which a fault is *globally* fatal (used in the Diogenes approach)
- The class of the interconnection networks on which a fault is *locally* fatal (used in the strategy B)

Our purpose is now to evaluate the advantages and the disadvantages of both methods. In this paragraph we compare for both strategies the effect of the configuration switch area (that is, the percentage of the interconnection network area), the effect of the harvesting, and finally, the effect of the partitioning of the wafer.

All the curves plotted in the following are plotted for the optimum corresponding values of k (the number of redundant slices per chip); for reasons of clarity, these values do not appear on the curves.

3.3.1 Effect of the switch area

The yield of the wafer is plotted in figure 14 versus the switch area S_{switch}, for 3 different partitionings of the wafer into $N = 225$, $N = 144$, and $N = 100$ chips, and for both approaches. The dashed curves concern the Diogenes approach, while the continuous curves concern the second approach. In a first step, we assume a 100%

harvesting for both strategies. This is true for the Diogenes approach, but not for the second one.

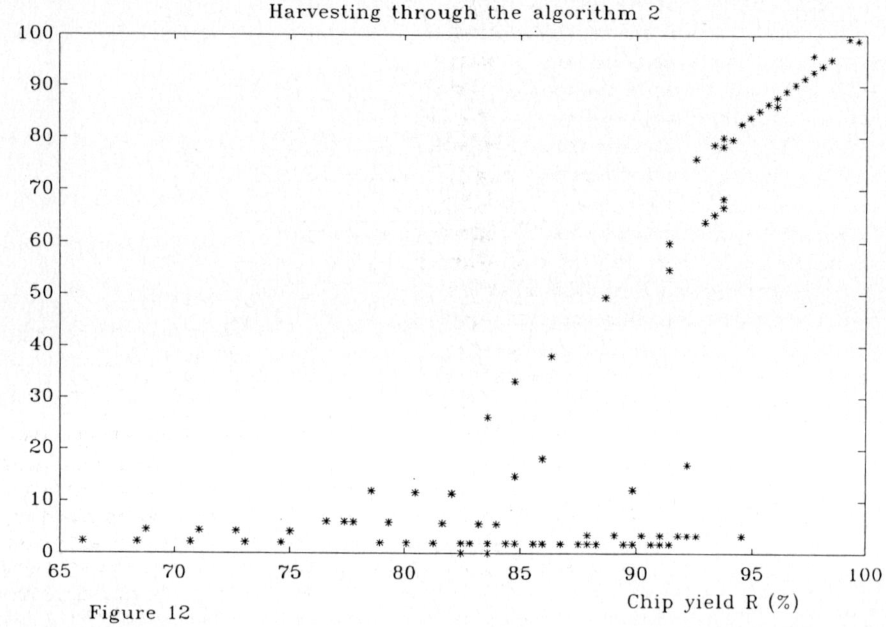

Figure 12

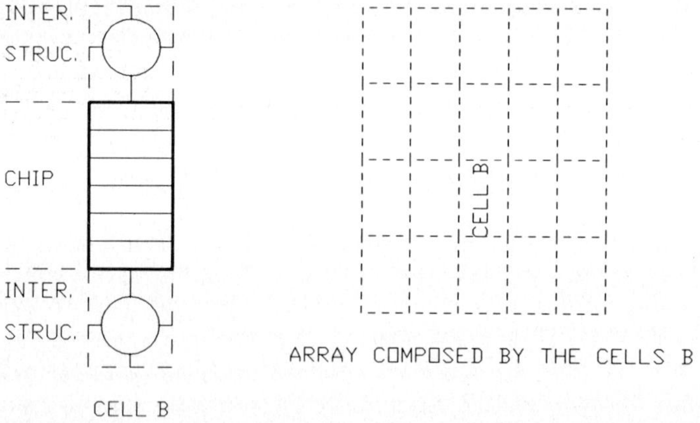

Figure 13 ***Chip with the associated interconnection structure for the B interconnection network***

In both cases, we assume a 5 inch wafer, a global functional requirement of 15000 slices ($S_R = 15000$), a 0.003 cm^2 slice area, a 3 faults/cm^2 slice fault density, a 1 fault/cm^2 switch fault density, and a 15% bypass area (that is the percentage of each slice dedicated to the configuration at chip level which effect is implicitly considered in the computation of the individual yield R of a chip).

The reason why the curves decrease with the increase of the switch area S_{switch} is obvious: as it was said before, this area is susceptible to defects, and is a functional area consumer too. It is more important to note the difference of both approaches as regards to their reaction to the configuration switch area: while the continuous curves (approach B) decrease quasi-linearly, the dashed ones (approach B) decrease exponentially. The explanation is simple: in the Diogenes approach, the effect of a fault on the interconnection network is *globally* fatal, and for a Poisson distribution, the decrease of the global yield (see (3)) is in $[P_o(S_{switch})]^N = e^{-ND_{switch}.S_{switch}}$, while in the approach B, the effect of a fault is *locally* fatal (see (4)), then, the decrease of the yield is much smoother. In other words, the influence of faults in the configuration network is weak in the B approach while it is very pronounced in the Diogenes approach. Obviously, this difference is as much pronounced as the partitioning is high: for a 100 chips wafer, the difference is almost negligeable, while it is very amplified for a 225 chips wafer. In effect, increasing the partitioning makes increase the global interconnection network area $S_{interc} = NS_{switch}$, then, the difference between both approaches appears more distinctly. Lets note that the harvesting has not been taken into account. We consider its influence on the B approach in the following.

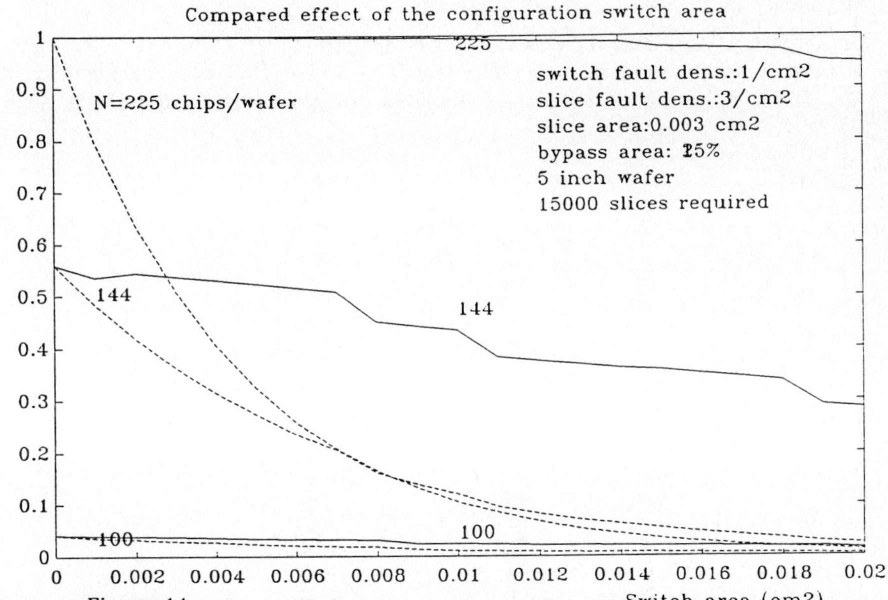

Figure 14

3.3.2 Effect of the harvesting

The disadvantage of the Diogenes approach as regards to the global fatality of a fault in the interconnection network may be balanced now by the fact that the value of the harvesting it allows is 100%, while the harvesting offered by the B algorithms is lower. As it was said hereabove, the harvesting depends on R: $h(R)$. Determining theoretically (and precisely) this function $h(R)$ is a very complex problem of statistics (see figures 10 and 12) that is still unsolved [6] because of the occurrence of isolated islands of good chips in the array, that remain unconnected. We restrict our study to the first algorithm, and we limit ourselves in the present paper to an empiric evaluation of $h(R)$, that gives (see figure 10):

$$93\% < R < 100\% : h(R) = 2 - 1/R$$

$$86\% < R < 93\% : h(R) = 2.5 - 1.5/R$$

$$67\% < R < 86\% : h(R) = 3 - 2/R$$

This means that for high yields ($93\% < R < 100\%$), no clusters are present (or no good chips are isolated): thus, as shown in figure 7, only one good chip is lost for each bad chip. When the yield decreases, good chips may remain isolated from the linear array, due to their position in the physical array: we assume that this is equivalent to lose 1.5 chip per bad chip or 2 chips per bad chip, or more. It depends on the yield. The dashed lines in figure 10 plot the harvestings for 1 chip lost per bad chip ($2 - 1/R$), 1.5 chip lost ($2.5 - 1.5/R$), and 2 chips lost ($3 - 2/R$). With the help of these curves, and comparing with the experimental values of the harvesting, we have determined the function $h(R)$ as hereabove.

It is important to note that this method is not rigourous at all, but however, it is much more precise than assuming constant harvestings, that do not depend on the yield. In the following, we use this approach for the global yield computation.

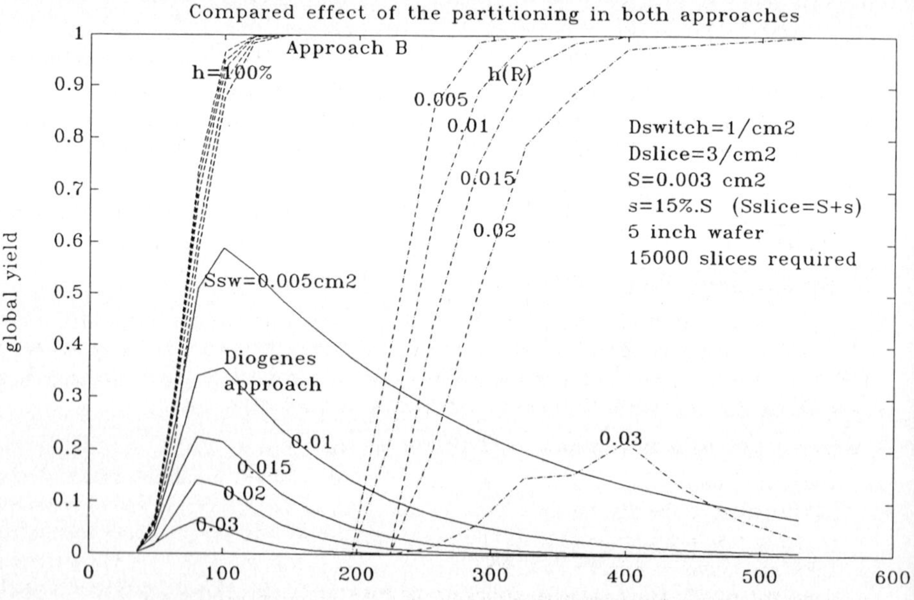

Figure 15

As shown in figure 15 by the two classes of curves of the approach B, the effect of the harvesting has to be considered too. In the first class, a 100% harvesting is assumed, while in the second class, the harvesting depends on R. The curves are plotted versus the partitioning of the wafer (number of chips per wafer). Each class of curves contains 5 curves, each one corresponding to a given configuration switch area (0.005, 0.010, 0.015, 0.020, and 0.030 cm^2). In both cases, the partitioning makes obviously increase the yield (see 3.3.3). But the increase appears much later in the $h(R)$ case. In effect, taking into account the harvesting is equivalent to decreasing the individual yield. This diminution is unfortunately very important: as an example, in figure 15, and for a 0.005 cm^2 switch area, before 196 chips per wafer, the yield $R' = Rh(R)$ is not sufficient anymore to allow a global yield different from zero. Lets examine now in more details the effect of the partitioning for both approaches.

3.3.3 Effect of the partitioning

It is obvious that the partitioning makes increase the yield [2]. We exhibit here the fact that this effect is unfortunately limited by the interconnection network effect. In figure 15 are plotted 3 different types of curves: the first one concerns the Diogenes approach, the second one concerns the approach B without taking into account the harvesting (it is assumed to be 100%), while the third concerns the approach B that takes into account the harvesting defined in 3.3.2. All the curves are plotted for a 5 inch wafer, a global functional requirement of 15000 slices ($S_R = 15000$), a 0.003 cm^2 slice area, a 3 faults/cm^2 slice fault density, a 1 fault/cm^2 switch fault density, and a 15% bypass area (at chip level). All the curves are plotted for the optimum value of k, the number of redundant slices per chip. Each class of curves contains 5 curves, each one corresponding to a given configuration switch area: $S_{switch} = 0.005$ cm^2, 0.010 cm^2, 0.015 cm^2, 0.020 cm^2, and 0.030 cm^2.

Two aspects fondamentaly different appear immediately: in the Diogenes case, vertexes appear while they do not in the approach B. This can be easily explained: the partitioning makes increase the global interconnection network area: $S_{interc} = NS_{switch}$. As it was said before, the interconnection network area is nefast for two reasons:

(1) For the area it occupies, that takes place for the remaining functional area (spare chips)

(2) For its susceptibility to defects.

Both approaches react the same way to the first point. At the contrary, they react differently to the second point: the Diogenes configuration network is much more sensible to defects than the B configuration network: in effect, a fault is *globally* fatal in the Diogenes case, while it is *locally* fatal in the second case.

Lets now examine in more details both classes of curves:

Diogenes class: the yield increases rapidly, then, after a critical value of the partitioning, it begins to decrease. In effect, after the interconnection network area has reached a critical value, the positive effect of redundancy becomes weaker than the negative effect of the configuration. Besides, the values of the vertexes decrease significantly with the increase of the switch area: for a 0.005 cm^2 switch area, the yield reaches 60%, while for a 0.003 cm^2, it reaches 7%. For the 0.005

cm^2 switch area, the optimum partitioning is 100 chips per wafer, that allows a 60% yield. In effect, for a 64 chips wafer, the yield reaches 40%; for a 81 chips wafer, it reaches 52%. Then, after the optimum, the yield decreases rapidly; for example, a 200 chips wafer involves a 40% yield, and the yield of a 400% chips wafer reaches only 13%.

B class: two sets of curves are plotted: the first one on the left corresponds to a 100% harvesting, while the second one on the right takes the harvesting into account, considering it as R dependent (as defined in 3.3.2): $h(R)$. Both sets of curves have the same aspect, that is, the partitioning does not impose any limit to the yield. This is, of course, due to the local effect of a fault in the interconnection network. The essential difference between both sets of curves is that the curves $h(R)$ are shifted on the right: a higher partitioning is necessary to enhance the yield. In effect, while all the chips may be collected in the case where $h = 100\%$, only a percentage $h(R)$ may be harvested in the other case. There is a particular curve which does not reach 100%, but begins to decrease after having reached 20%: the 0.03 cm^2 curve with $h(R)$: this is due to the fact that the global interconnection network area S_{interc} is to high to allow a sufficient number of remaining chips. It is important to note that for an ideal infinite wafer area, the yield would always reach 100%.

CONCLUSION

After having brielfly recorded the yield models used at chip- and wafer level in this paper, we examine the effect of the configuration network influence on the enhanced yield (through redundancy) at wafer level for two different configuration strategies. The first configuration strategy (Diogenes strategy) allows a 100% harvesting, but the effect of a fault in the interconnection network used to implement it is globally fatal. At the contrary, the effect of a fault in the interconnection network used for the second strategy is only locally fatal, but this method involves a harvesting lower than 100%. The effects of the interconnection network on the yield for both configuration approaches were examined (at wafer level configuration), as regards to their reaction to the partitioning of the wafer. We highlight the existence of an optimum value of the partitioning (that is, the optimum size of the chip) that depends entirely on the switch area and on the strategy used to implement the configuration network. In other words, we show that a given way to configure determines the optimum size of the chip to implement on the wafer, and at the same time, the optimum amount of redundancy to implement at chip level.

AKNOWLEDGMENTS

The author wish to thank Pr. M.G.Sami for her advices and encouragements and is also gratefull to Prs L.Dadda and R.Stefanelli for their help.

REFERENCES

[1] L.Dadda, L.Breveglieri: "A serial-input serial-output bit-sliced convolver", ICCD 88, p 490.

[2] S.Lasserre: "A single wafer bit-sliced convolver: an optimum statistical solution for the implementation of redundancy", International conference on systolic arrays, 1989.

[3] R.Negrini, M.G.Sami, R.Stefanelli: "Fault-tolerance through reconfiguration of VLSI and WSI arrays", MIT Press-Cambridge.

[4] C.Jesshope, L.Bentley, "Techniques for a wafer scale RPA", IFIP Workshop on WSI, 86.

[5] P.Ivey, M.Huch, T.Midwinter, P.Hurat, M.Glesner, "Design of a large SIMD array in wafer scale technology", IFIP Workshop on WSI, 87.

[6] J.W.Essam, "Percolation theory", Rep. Prog. Phys., Vol. 43, 1980.

Wafer Scale Integration, III
M. Sami and F. Distante (Editors)
Elsevier Science Publishers B.V. (North-Holland)

Design of an universal switching network for reconfigurable 2D-arrays

J-Luc Patry, G.Saucier
INPG/CSI
46, Avenue Felix Viallet 38000 Grenoble

Introduction

A lot of papers in the literature deal with the reconfiguration of 2D-arrays and very attractive results are available ([1] [2] [3] [4] [5] [6] [7] [8] [9]). Nevertheless no information is given about the actual implementation of switches and even not about their architectures and practical feasibility (area, speed, overhead, etc). In [21], a switch is proposed but the overhead is unacceptable with the yield increase expected for reconfigurable arrays. The second problem is that defect tolerance of the connection themselves is not clearly outpointed and, even more, in most of the approaches, they are supposed to be defect-free.

We present here an universal switching network associated with a defect tolerant interconnection scheme. The originality of the switching network is the ability to cope with hierarchical design. Switching nodes are able to steer n bits wires and can therefore be associated with basic chips having n bits connections with their neighbours. A flexible floorplan allows to "stick" these switching nodes on each side of the basic chip fitting exactly with its width. The second originality is the ability to be programmed externally and serially from the I/O pads. The architectural and logical design, the electrical characterization and related buffering problem and the layout will be show in this paper.

1) Architecture of the switching array

1.1) Global architecture

The general architecture is a 2D-array of one-bit processing elements for SIMD computing. The PEs are grouped in elementary arrays called chips. Each chip contains a redundant column for defect tolerance. A double rail of connections and switching nodes surround the chips. The switching nodes allow to bypass the faulty chips and to build the final array. Eight switching nodes surround a chip (figure 1), four of them are vital because they connect the chip to the network.

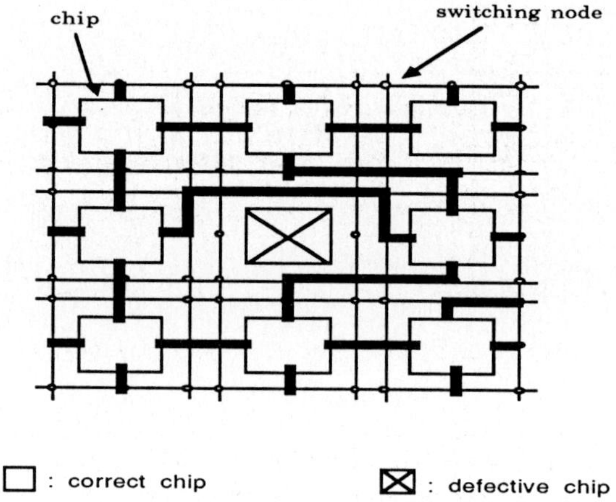

Figure 1: A 2D-array connected by a double rail switching network

A switching node can be crossed in 4 different ways as shown in fig 2.

Figure 2: The four possible configurations of a switching node

To choose the connection network and the crossing capabilities of the switching nodes, typical defect patterns have been considered. They correspond for instance to 1, 2, 3, 4 consecutive defective chips with and without surrounding defective switching nodes. For each case, we look for the deformation of the array and the number of chips lost. As an example we show 2 defect patterns and the corresponding paths created in the network (figure 3 and 4). For each case, we loose 2 chips and pratical experience has shown that the number of unused chips and the deformation of the arrays are acceptable four our array.

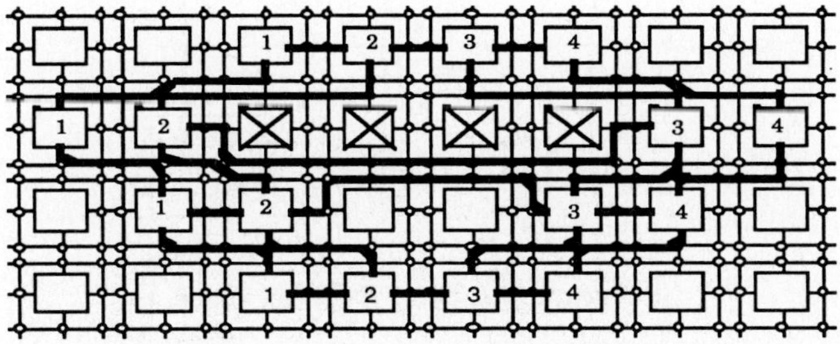

Figure 3: Four consecutive defective chips

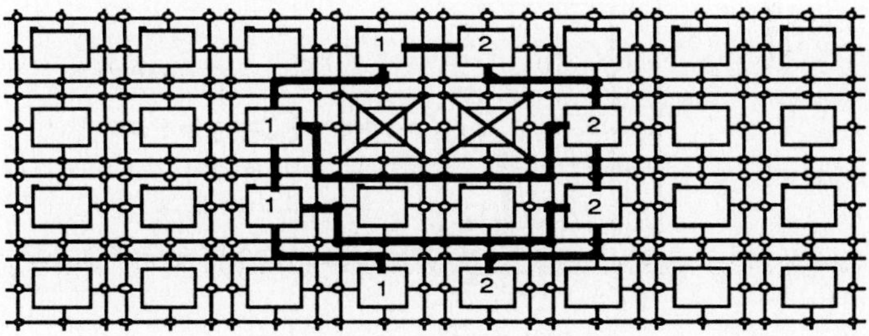

Figure 4: Two consecutive defective chips with defective neighbour switching nodes

1.2) Architecture of the switching node

1.2.1) The basic switch

The architecture presented in ([8], [22]) will be recalled briefly here. The basic switch is made up of two parts. The first part shown in figure 5 implements the set of paths and is called "basic transfer block". The second part shown in figure 6 is the control part in which two steering flip-flops A and B control the transfer block through random logic. Figure 7 shows the correspondence between the values of the variables A and B and the crossing of the switch.

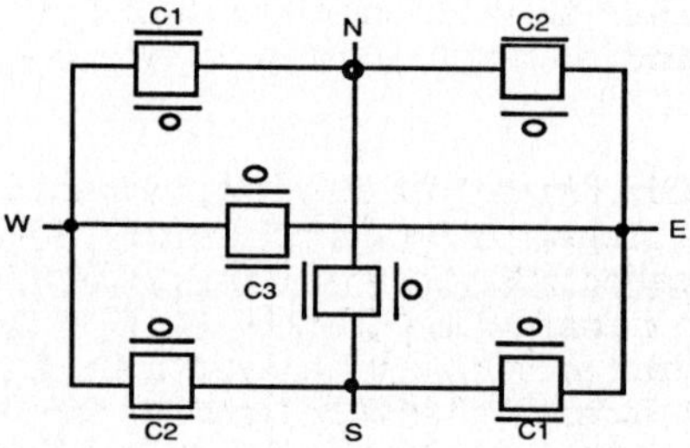

Figure 5: A basic transfer block

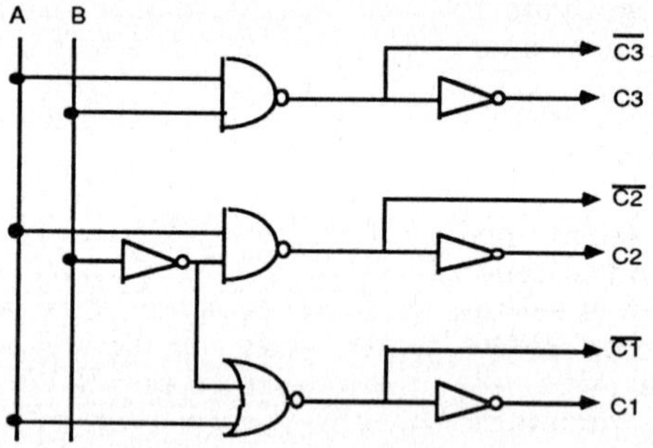

Figure 6: The control logic

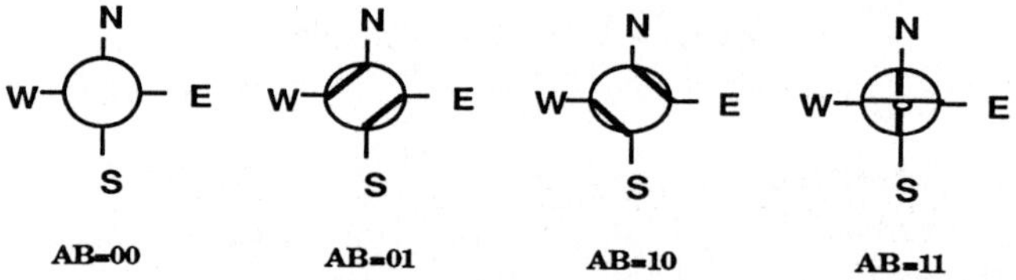

Figure 7: Correspondence between the values and the paths

1.2.2) Paths programming principle.

To program a path in the switching network from the I/O pads, a two-bit shift register is implemented. To each switching node are associated two bits (X,Y) of the shift register. During the path programming, information is sent serially all over the wafer. If at a given instant, the steering information concerns a switching node, it will be sampled in its steering flips-flops (A,B); otherwise it will circulate further on. Control wires CTRPROGn at the north input, CTRPROGs at the south input, CTRPROGe at the east input and CTRPROGw at the west input are used to indicate whether this steering information is just circulating (CTRPROGx =1) (x=n,s,e,w) or whether this information has now reached the target switching nodes and must be sampled in the steering flip-flops (CTRPROGx =0).

Wires called PROGDATAn, PROGDATAs, PROGDATAw, PROGDATAe are used to send the programming data from the four direction to the shift register. The wires called DATAn, DATAs, DATAw, DATAe carry the data between the chips in normal functioning.
Let us illustrate this and show the creation of a path crossing two successive switching nodes (SW1,SW2) (figures 8 and 9), and the associated chronograms (figure 10). Only the CTRPROGx, PROGDATAx and DATAx of the switching nodes at the border of the wafer are connected to pads. First, a signal CTRPROGn is set to 1 during two clock cycles; at the same time the two bits 11 are sent on PROGDATAn which cross the shift register. Then the signal CTRPROGn is set to 0 during one clock cycle to allow the PROGDATAn to go from the shift register (X,Y) to the steering flip-flops (A,B). After these two operations, the first switching node is programmed and PROGDATAs, CTRRPROGs and DATAs are connected to PROGDATAn, CTRRPROGn and DATAn of the south neighbouring switching node (SW2).

To program the second switching node (SW2), we need to cross the first switching node and to send the PROGDATAn serially. For this purpose, the signal CTRPROGn is set to 1 during four clock cycles; at the same time four bits 0111 are sent on PROGDATAn. Then, the signal CTRPROGn is set to 0 during one clock cycle. The two first bits of PROGDATAn program the second switching node and the two last bits program the first switching node. The first switching node is programmed during the two steps, so we need to send the same PROGDATAn at each step.

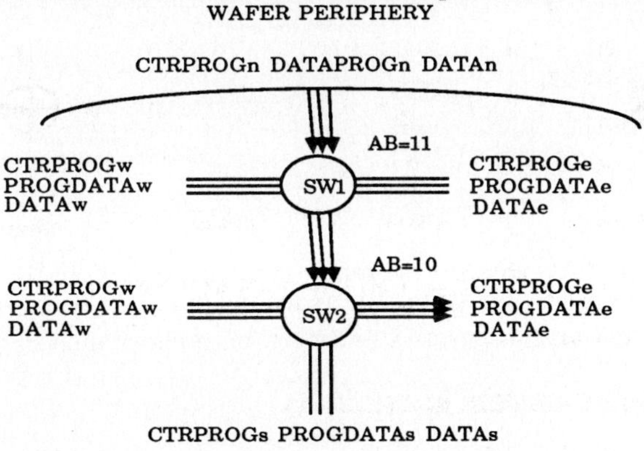

Figure 8: Path programming information

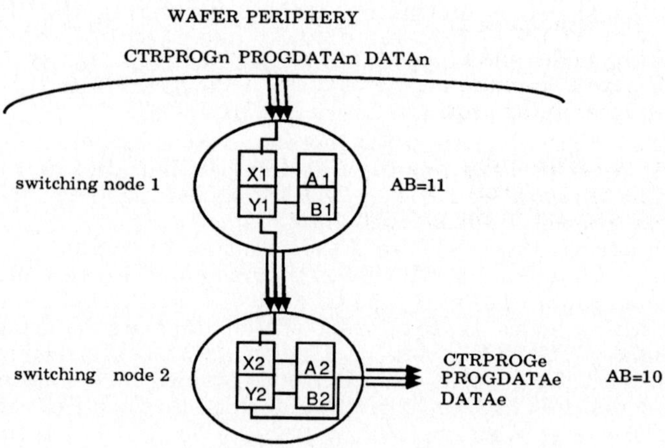

Figure 9: Shift register and steering flip-flops values associated with the two switching nodes.

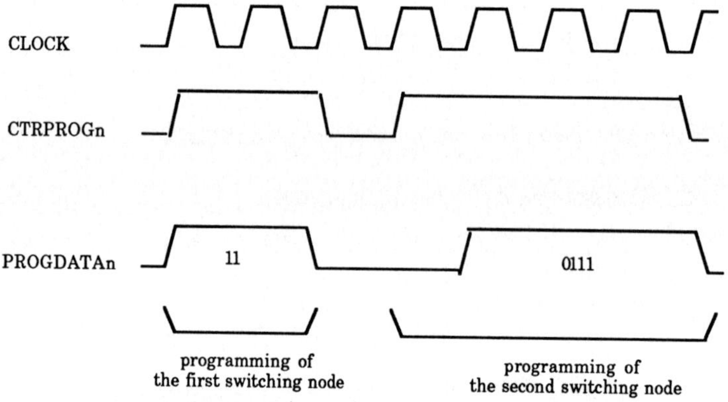

Figure 10: Programming the paths

The programming of several paths can be done serially or in parallel. Suppose a set (S_i) of the switching nodes belong to a path I; to create a new path J we must only program the switching nodes which belong to :

$$S_j - (S_j \cap (\bigcup_{i=0 \text{ to } j-1} S_i)) = S_j - \bigcup_{i=0,j-1} (S_i)$$

Let us illustrate these paths on figure 11.

To create the path I, the programming data flow must be :

(11)+(1110)+(111010)+(11101011)+(1110101111)+ .....

As shown below, the path 2 can be added in programming SW_{21} and SW_{22}. The switching node SW11 does not need to be programmed during the creation of path 2, because it belongs to the previous path 1. The beginning of path 2 programming sequence is then (01)+(0101) + ...

During the programming phase, the first programming of a switching node can only be performed at a given instant in one path. This restriction is taken into account in the path program plan.

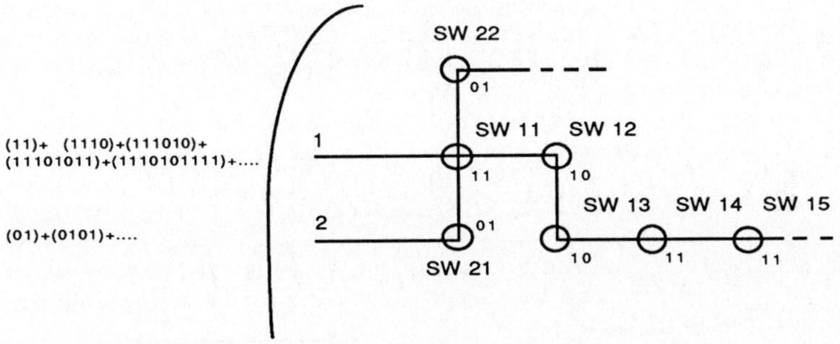

Figure 11: Two paths with their programming sequence

1.2.3) Detailed architecture of the switching node.

The switching node is made up of two parts, the CONTROL block and the TRANSFER block (figure 12). Figure 13 shows the general architecture of the switching node.

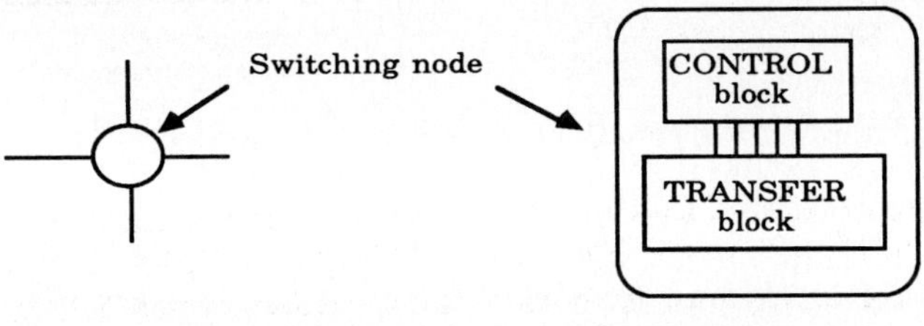

Figure 12: Description of a switching node

i) The TRANSFER block

The TRANSFER block is made up of (n+2) basic transfer blocks (cf figure 3), where n is the number of data wires which cross a switch. One basic transfer block carries the CTRPROGx signal and one basic transfer block carries the PROGDATAx signal.The n others carry the data (figure 13). All the different transfer blocks are in the same configuration.

ii) The CONTROL block

A block called "SELECT" selects the inputs/outputs of the "CONTROL" block. It receives the programming control signal CTRPROG (ctrprogn, ctrprogs, ctrproge, ctrprogw) and the programming data signal PROGDATA (progdatan, progdatas, progdatae, progdataw) from all its four neighbours. It is made up of a XOR gate, a block called "FLAG" and a block called "TFG". "FLAG" is made up of 4 flags (flagn, flags, flage, flagw) and "TFG" is made up of transfer gates.

At the beginning, all the basic transfer blocks are "open" or cut and all the flip-flops are at 0. The programming is detected by the XOR when one CTRPROG are high. Then, the flags store the source of the programming by storing the CTRPROG wire status into the four flags (flagn, flage, flagw, flags). The flags control the "TFG", which selects the PROGDATA wires from which the programming data flow arrives. The status 1 of a flagn opens a transfer gate which connects the datan to the shift register. The CTRPROGx is set at 0 to allow the data to go from the shift register to the steering flip-flop (figure 9). These flip-flops are connected to a decoder (figure 6), which controls the TRANSFER block.

After the programming of the switching nodes, the paths are fixed with fuses and floating gates. The fuses and floating gates are implemented in the steering flip-flop (A,B).

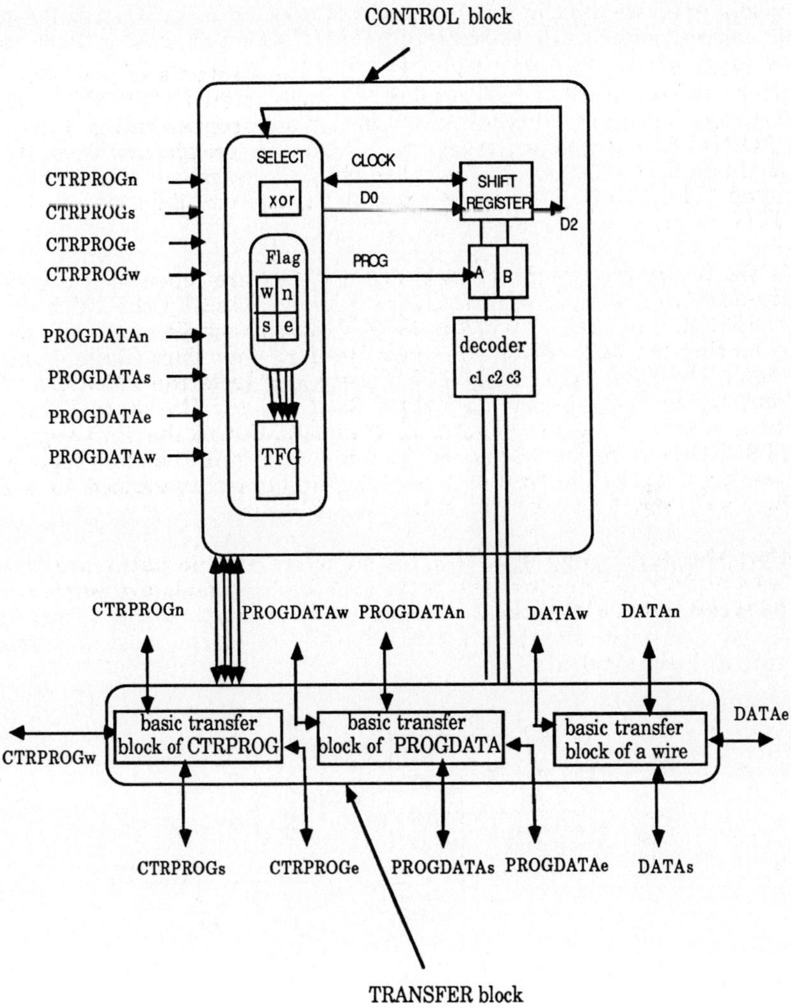

Figure 13: Architecture of the switch

2) Electrical design

2.1) Location of the buffers

As shown in figures 3 and 4, a path between two PEs may cross 2 to 15 switching nodes. As the target array had to work at 20 MHz, different types of bidirectional buffers were studied. The number of buffers must be minimized and this must take into account the requirement of a repetitive layout, due to a step and repeat process. Therefore the location of the buffers has been chosen as shown in figure 14.

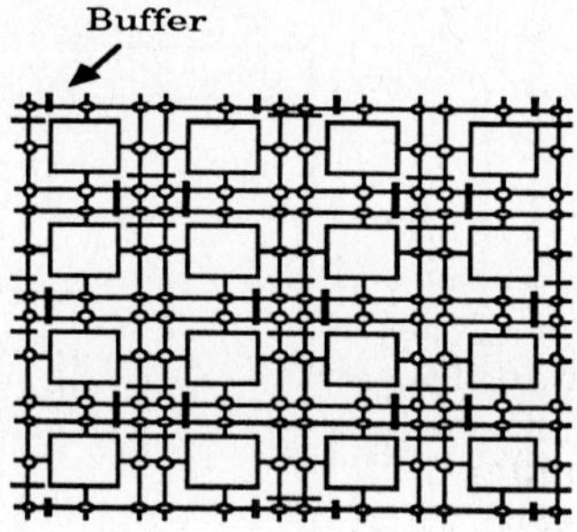

Figure 14: Location of the buffers inside a reticle

2.2) Choice of the buffer.

Different types of buffers were simulated, to compare and choose the best design. The type shown in figure 14 seems to be the most adequate, after electrical simulation.

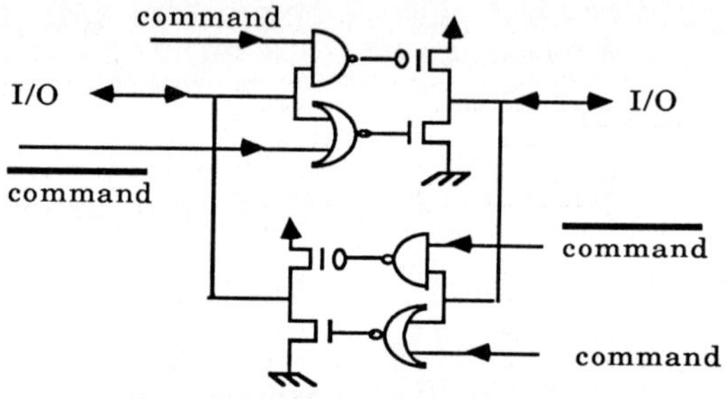

Figure 15: A two-level bidirectionnal buffer

2.3) Programming the buffers.

Data crossing a buffer may either connect the west side of a PE to the east side of a another PE or connect the north side of a PE to the south side of a PE. The two types of connections are shown in figure 15.

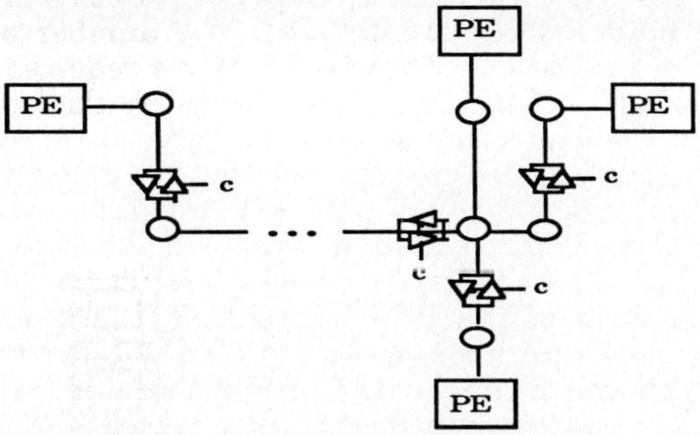

Figure 16: Types of data which cross a buffer

The command of a buffer may be a command west-east or a command north-south, and for each type of command we have two directions. Four commands of the bidirectionnal buffer are implemented (ie: Cns, notCns, Cwe ,notCwe) and after the reconfiguration, one is chosen through a multiplexer.

To program the buffer, one solution is to connect every buffer to a switching node, and use the programming of the switching node to program the buffer. We added two bits to the shift register and two steering flip-flops. These bits control the multiplexer which selects one buffer command among the four (figure 17).

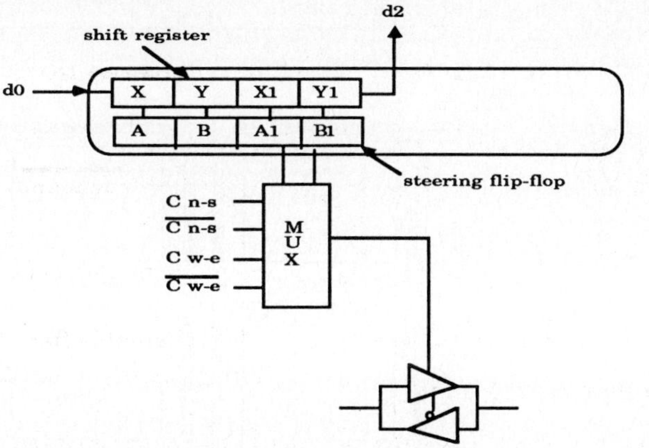

Figure 17: Programming of the command buffer with the switching node

3) Layout

3.1) Floorplan of the reticle

This section reports on a switching network implemented on a 2D-SIMD array designed in an ESPRIT project. Due to step and repeat process, three levels of hierarchy existed at the layout level: the PE, the chip, the reticle. The figure 18 shows the different architectural possibilities. The computation of the pourcentage of area devoted to defect tolerance is reported in figure 18. These results have led to implement 2x2 chips in a reticle and a 12x6 logical array of PEs with one redundant column (12 x 7) in a chip. In this case, the redundant area and the total area of a chip are shown to be acceptable with respect to the yield. The yield at the wafer level depends on the yield of the PE, the yield of a chip and the number of redundant elements and has been calculated as shown in (8). Figure 19 shows the floorplan of a reticle. The reticle contains four chips with their eight surrounding switching nodes. On the whole wafer, 36 reticles have been implemented on a 6 inch wafer. Then the size of the physical array is 144 x 72 PEs.

	Number of chips per reticle: 2	2x2	3x3
Area of a chip	6.25 x 11 sq mm	6.33x5.29 sq mm	4.2 x3.54 sq mm
Number of rows and columns in a chip	12 x 15	12 x 7	8x5
ratio = area devoted to defect tolerance / global area:	1 redundant column 7%	18%	25%
	2 redundant columns 15%	Too large	Too large
	3 redundant columns 25%	Too large	Too large

Figure 18. Table of the different possibilities

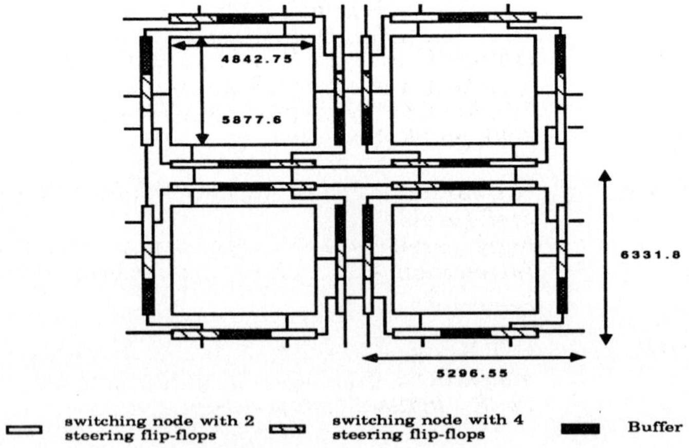

Figure 19: Reticle floor plan

3.2) Layout of the switching node

On the layout, in figure 20 we identify at the left side the 14 basic transfer bocks (12 for the data of the PE, one for CTRPROG and one for PROGDATA), in the middle the CONTROL block and at the right side the 14 buffers (figure 20).

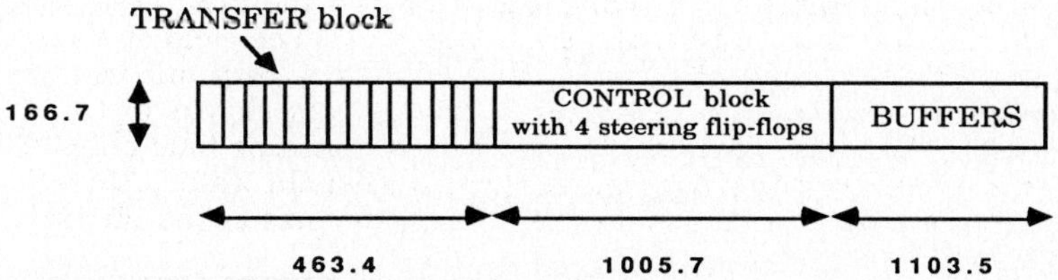

Figure 20: The layout of the switching node and buffer

Conclusion.

An universel switching network has been designed and implemented in HCMOS3C Technology. In an ESPRIT project, this network has been used succesfully within the Wafer Scale European Large SIMD array (ELSA) dedicated to image processing designed in cooperation with BRITISH Telecom and T.H.Darmstadt.

References

[FUS82] D.Fusell and P.Varman. "Fault-tolerant wafer-scale architectures for VLSI". *Proceeding of the 9th annual IEEE/ACM Symposium on Computer Architecture, April 1982, pp.190-198.*

[HED82] K.Hedlung and L.Snyder. "Wafer-scale integration of configurable,
higly parallel (CHIP) processors". *Proceeding of the IEEE International Conference on parallel Processing, 1982 , pp.262.264*

[HUN82] H.T.Kung. "Why systolic architectures". *Computer magazines,*
IEEE , January 1982,pp.21-27.

[CHU83] F.R.K Chung, F.T.Leighton, and A.L.Rosenberg. "Diogenes: a methodology for designing fault-tolerant VLSI processor arrays". *Proceeding of the IEEE Symposium on fault-tolerant Computing, June 1983, pp.26-31.*

[WILL83] G. J. William. "Configuration of VLSI Arrays in the Presence of Defects". *Ph. D dissertation. Stanford University. December 1983.*

[KOR84] I.Koren and M.Breuer. "On area and yield considerations for fault tolerant VLSI processor arrays ". *IEEE Transactions on computers, Vol.C33, No 1,1984, pp. 21-27*

[CHE85] G.Chevalier, G.Saucier. "A programmable switch for fault tolerant wafer scale integration of processor arrays". *International Workshop On Wafer-Scale Integration 1985.Southampton, U-K.*

[EVA85] R. A. Evans, J. V. M. Mc Canny, K. W. Wood. "Wafer Scale Integration based on Self Organization". *Int. Workshop on WSI, Southampton, 1985.*

[LEI85] F.T Leighton and C.E Leiserson. "Wafer-scale integration of systolic arrays". *IEEE Transactions on computers, vol. c-34, No.5 May 1985, pp.448-461*

[NEG85] R. Negrini, R. Stefanelli. "Algorithms for Self Reconfiguration of Wafer Scale Regular Arrays". *Proc. of Int. Conference on Circuits and Sytems, IEEE, Beijing 1985.*

[SAM85] M. Sami and R.Stefanelli. "Fault tolerance approaches for VLSI/WSI arrays". *Proc.IEEE Pheonix Conf. on Comp and Comm; 1985, pp.460-468.*

[LEI86] T. Leighton, C. E. Leiserson. "A Survey of Algorithms for Integrating Wafer Scale Systolic Systolic Arrays". *Saucier & Trihle eds., North Holland, 1986.*

[NEG86] R.Negrini and R.Stefanelli. "Comparative evaluation of space and time redundancy approaches for WSI processing arrays".*IFIP Workshop on Wafer-scale integration, Grenoble , 1986.*

[SAM86] M. G. Sami, R. Stefanelli. "Reconfigurable Architectures for VLSI Processing Arrays". *Proceedings of the IEEE, vol. 74, n° 5, pp. 712-722, May 1986.*

[KUO87] S. Y. Kuo, W. K. Fuchs. "Efficient Spare allocation for Reconfiguring Arrays". *IEEE Design and Test", V.4, N°1, Feb. 87, pp. 24.*

[LOM87] F. Lombardi, R. Negrini, M. G. Sami, R. Stefanelli. "Reconfiguration of VLSI Arrays : A Covering Approach". *Proceedings of the 17th FTCS conference. pp. 251-256. Pittsburgh, July 1987.*

[HAR88] Jim C. Harden and Noel L R.Strader II."Architectural Yield Optimization for WSI". *IEEE trans. on computers, vol.37,pp. 88-110,1988*

[SAU88] G.Saucier, J-L Patry, E.F.M Kouka. "Defect Tolerance in a Wafer Scale Array for Image Processing". *International Workshop on Defect and Fault Tolerance in VLSI systems. Massasuchett, October 1988.*

[SAU89] G.Saucier, J-Luc Patry, E-F Kouka, "A Reconfigurable Wafer Scale Array for Image Processing". *Proc. International Conference on Wafer Scale Integration, San Francisco US, p.277-288.*

Wafer Scale Integration, III
M. Sami and F. Distante (Editors)
Elsevier Science Publishers B.V. (North-Holland)

Low Temperature Fabrication to Enhance Discretionary WSI and WSHP Wiring Yield

S. Dabral and J. F. McDonald

Center for Integrated Electronics
Rensselaer Polytechnic Institute
Troy, NY, 12181, USA

The need for low temperature processing for WSI and WSHP reliable interconnects is shown. A complete low temperature fabrication process is discussed. The importance of low temperature vapor depositable dielectrics is brought out. A repair and testing strategy is also examined.

1. Introduction

1.1 Reasons for low temperature processing

Wafer Scale Integration involves interconnection of working dies on a wafer to produce the desired circuit. The implicit assumption in making such a circuit work is that processing subsequent to die testing will not cause the dies that passed to fail. Temperature has been shown to contribute significantly towards chip failures. These failures arise mainly due to mechanical stress resulting from the temperature differences between fabrication and the operation. The high temperature processing during CVD oxide deposition, anneal of deposited materials, curing of polymides and passivation by nitride can all disturb the tested die sites.

When the thermal coefficient of materials is not the same, then temperature deviations away from the fabrication conditions will lead to stress. The stress can cause the wafer to curve. This curvature becomes larger as the new layer gets thicker. Larger TCE differences also lead to larger curvature. This stress can cause wires on the surface to break, leading to failure [1]. It has been shown that narrower interconnect lines fail at a faster rate than broader ones [2]. The curvature can become larger than the depth of field of conventional optical exposure tools causing problems with photolithography. Stress has been also shown to cause threshold voltage changes in MOSFETs [3]. This could lead to reduced noise margins and to improper circuit functioning. Figure 1 shows the two possible curvatures of a wafer and the associated mechanical stress.

There are two ways to reduce mechanical stresses, namely reduction of differences between TCE and thermal excursions. It may not be possible match the TCEs for all the materials used. This is shown by the choice of interconnect materials in use today, aluminum and polymides on silicon. The dielectric materials have thermal coefficients of expansion in the range of 25-50 ppm/deg centigrade, which matches the TCE of aluminum. Both these then are not very compatible with the underlying silicon substrate which has a TCE of 3 ppm/deg. centigrade. One could choose to have the substrate as aluminum itself to get rid of the problem, but it would cause a loss of available sites for termination resistors, diodes or chip sites.

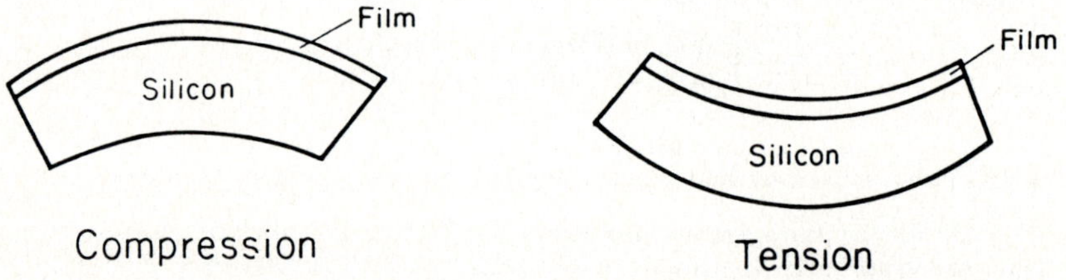

Fig. 1: Wafer curvature and associated mechanical stress caused by difference in TCE.

These problems have prompted us to look at fabrication options which use low temperatures. The low temperature fabrication should be lower than 125 degrees centigrade. The choice of upper temperature limit is set by the maximum operating conditions of military chips. It is also a practical limit because most of the photoresist in use today require these temperatures for curing. Fabrication at these temperatures would ensure low thermal deviations causing a reduction in stress.

2. Fabrication Processes.

2.1 Choice of Interconnect Dielectric.

The electromagnetics of the problem show that the interconnect metal has to be thick to reduce the resistance[4], [5]. Additionally the dielectric has to be sufficiently thick to decrease the capacitance to such a point that the transmission line characteristic impedance is much greater than the line resistance. Further more the use of power planes is essential for a controlled impedance environment and coupling reduction [6].

The requirement of thick dielectric can not be met with glass or silicon dioxide which has a tendency to crack when deposited in thick layers. The dielectric constant for these are high, increasing the capacitance. Thus a thicker dielectric and metal would be required to obtain transmission line like qualities. Polymides were found to be suitable as they were pliable, had lower dielectric constant and required a lower deposition temperature. These polymides solved some problems but created some of their own, namely:

1. the polymides had a tendency to take up water and expand.
2. The polymides could not be completely cured resulting in outgassing at a later time. This led to providing escape holes in the metal planes, thus compromising the electromagnetics and interconnect densities.
3. The temperature which was lower than most semiconductor processes was still higher than required to achieve small wafer curvatures.

4. The spin on polymides used solvents as a base . The solvents themselves added contamination and particles.

5. The polymides normally required adhesion promoters for satisfactory adhesion.

Most of these problems can be overcome by a vapor depositable polymer [7]. One such suitable polymer is parylene. It is available in four forms, normal (n), chlorinated (c), dichlorinated (d) and aliphatic flourinated versions (f). The n,c and d forms are unstable in the simultaneous presence of radiation and oxygen. The flourinated version is stable. The main advantages of this material are listed :

1. The polymer is vapor depositable in slight vacuum. It uses no solvents and the particulates in the raw material are left behind in the sublimation boat. This leads to a cleaner surface.

2. The deposition is at room temperature.

3. the water take up is negligible (less than 0.01 percent by weight), and it has low gas diffusion.

4. There are no pinholes formed and the film is conformal and uniform.

5. the dielectric constant is low (2.14) and the loss tangent lower than those of polymides.

6. Stable dimer form of the material, the batch processing of hundreds of wafers and no curing requirements allow for better processing economics.

The low temperature deposition and other additional benefits over polymide were reasons enough to use parylene as the chief dielectric material. Figure 2 shows the depositing equipment.

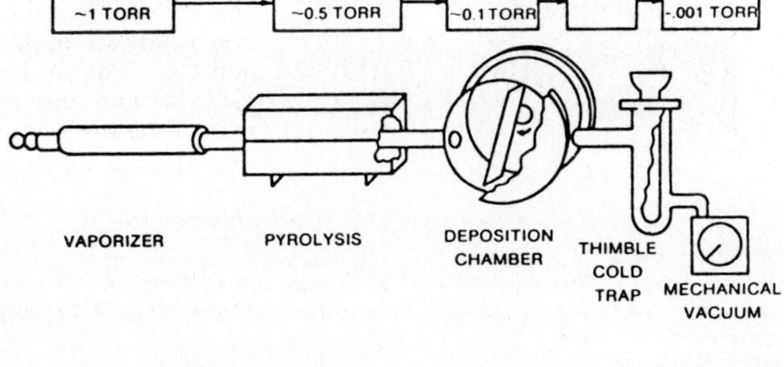

Fig. 2: Parylene deposition chamber.

2.2 Metal Deposition.

In conventional metal deposition there is an annealing step. This high temperature step causes alloying between semiconductor and metal. It creates a better bond at the metal to metal interface. This step significantly reduces the contact resistances. Radiation damage to devices is another consideration during metallization or reactive ion etching. This can arise due to X-ray generation in electron beam metal sources or plasmas. Process condition during these operations must be optimized to reduce these since the only way repair the damage is to subject the wafer to a high temperature rapid thermal anneal. Unfortunately the temperatures required for such a step (400 degrees and higher) are not permitted in the process under consideration.

The partially ionized beam method was found to be a good alternative. Partially ionized beam processing is a simplified form of the ionized cluster beam method . In this process aluminum is evaporated from a boat. Part of the atoms are then ionized and accelerated towards the substrate. The kinetic energy of these ions gets converted to thermal energy on contact with the substrate. This thermal spike is sufficient to cause enough mobility among depositing atoms to enter suitable sites in a lattice. The substrate temperature can be maintained low. Further this ionic bombardment helps in cleaning the interface yielding better contacts. The high directionality of the atoms make it suitable for a liftoff scheme [8]. The thickness of the metal can be very well controlled and varied. With the same technique it is possible to fill in vias using high ionization and energy. The one problem with this approach is the low throughput. Figure 3 shows aluminum deposited by the PIB technique.

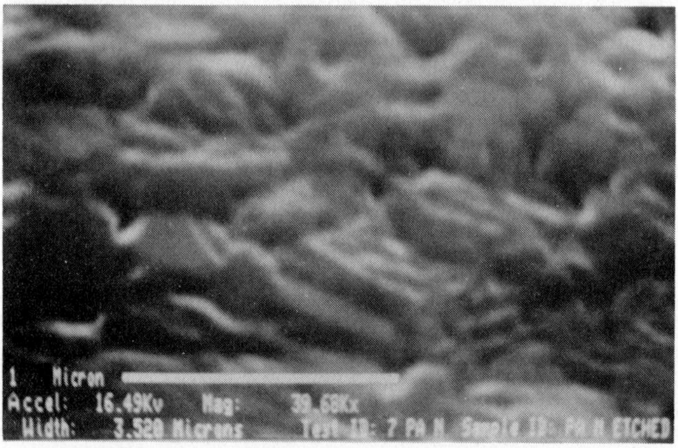

Fig. 3: Aluminum grown on a plasma etched parylene surface using PIB.

2.3 Exposure Scheme.

Though a low temperature process is being employed to deposit the dielectric there is some intrinsic stress at the interface. The stress when compounded several times will cause some wafer curvature. This could lead to exposure difficulty because of the limited depth of field of optical exposure machines. Clearly different exposure tools or small wafer curvature is a requirement.

An electron beam lithography system has a greater depth of field. It can also be dynamically focussed for each writing site. These properties can be effectively used to overcome lithographical problems due to curvature. The e-beam has additional advantages over conventional optical photolithography. As there is no mask, faults caused by damage or particulates on the mask do not get transferred to the substrate. Presently it is possible to route and process this information in a short duration. This makes it possible to have customized routing for different wafers, only the appropriate database has to be accessed by the e-beam driving computer [9].

2.4 Etching and Liftoff Schemes.

The dielectric and the metal involved in this process are thick. To use wet etching would unnecessarily waste valuable area caused by the undercutting. It would lead to jagged edges due to nonuniform etching. This would increase the high frequency resistance of the microtransmission line. Further wet etching causes more particulates and bubbles resulting in more faulty wires.

To avoid the above mentioned problems dry etching was found favorable. This avoids particulate and gas bubbles problem. The anisotropy ensured better area utilization and smoother vertical sidewalls. As shown in the next section the vertical etching is essential to preserve planarity. The plasma etching is more controllable in terms of etch rates and anisotropy.

The excess aluminum was lifted off. This process saves a lithography step, thus enhancing yield, registration problems and saving in time. Release layers, organic or metal could be used. Molybdenum has been examined because it has volatile flourides, thus is capable of a dry lift off process. Organic release layers could be photoresist or polysulfone [10]. As these can be etched using oxygen plasma only, which also etches the organic dielectric, dry liftoff would not be possible. A wet liftoff process is however possible by using orthogonal solvents which do not etch the metal or parylene. Figure 4 demonstrates the lift off technique.

2.5 Planarized Processing.

One important factor in this thick layer interconnect scheme is that the planarity has to be maintained through out the wafer. Lack of planarity can cause problems with lithography. The wires at steps are considerably thinner and are thus likely to fail. The other alternative to planarity is to have sloped vias. This via scheme would require a large area at the top of the via due to the thickness of the dielectric. Further vias could not be stacked on top of each other to avoid gross departures from planarity. These would cause design rule complications and reduced routing areas. The planar alternative was found to be simpler and possible using lift off techniques.

The planarizing could be achieved in a number of ways. To obtain an initial plane surface a spin on polymide would have to be coated on the non-planar substrate. The spin on polymers have a tendency to planarize surfaces when sufficiently thick layers are put down. Multiple coatings have been found to be more effective in this respect. Once

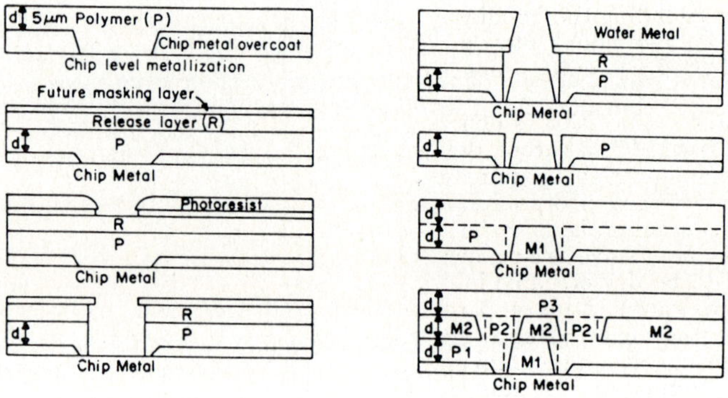

Fig. 4: Liftoff sequence preserving planarity.

the surface has been planarized further dielectric material would be parylene which coats conformally. Using metal stud vias and liftoff process the planarity of the top surface could be maintained. If the initial substrate was planar there is no need to have the planarizing step, and the fabrication starts using parylene itself. Figure 5 shows the resultant structure after all these fabricating steps.

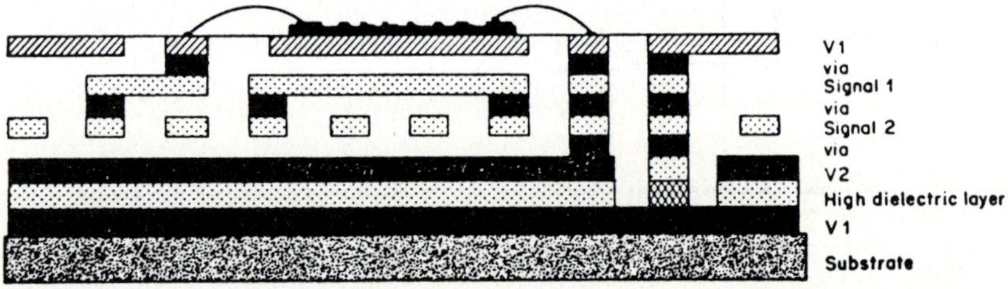

Fig 5.: The wafer transmission module.

2.5. Encapsulation of Wires and Package.

Parylene is known to be a good diffusion barrier for gas and ions. However the n, c and d versions are unstable in the presence of both oxygen and radiation. Thus it is necessary to exclude one of the two when using these materials. This problem does not arise with the flourinated version.

The water takeup of parylene is extremely small. Water takeup, however could cause failure of the interconnect lines. To provide additional sealing the use of plasma deposited carbon films is considered. These carbon films could be used to encapsulate the whole wafer including the bonding wires. It could also be used in extremely thin layers to encapsulate individual interconnect wires [11] [12]. This would lead to minimizing corrosion related reliability problems.

The carbon deposition is done using methane in a chamber as shown in Figure 6. This gas is cracked and carbon deposited on the substrate. The bias voltage, temperature are the main variable parameters. At higher bias voltages the cracking is better and there is less hydrogen incorporation in the deposited carbon film. It has been shown that smooth films of amorphous carbon are depositable with the right conditions. Uniform good quality films have been grown on silicon at room temperature. These films have been found to be dry etchable in oxygen plasma, although at a very slow rate. Using such a plasma carbon films on parylene have been patterned. These films can be grown on the sidewalls thus making it possible to encapsulate the wire completely. Figure 7 shows an amorphous carbon film grown on a parylene surface at 250 degrees centigrade. A similar patterned carbon film is shown in Figure 8.

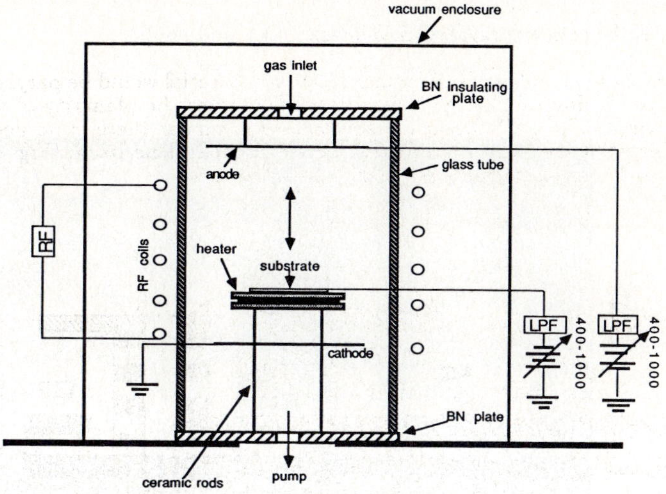

Fig 6: Plasma deposition chamber

3. Testing and Repair

3.1 Focussed Ion Beam

This process uses low temperatures, tries to minimize lithography and wet etches. Even with these precautions there is bound to be some defects introduced in the process. It has been shown using Stapper model of particulate faults that it may be possible to

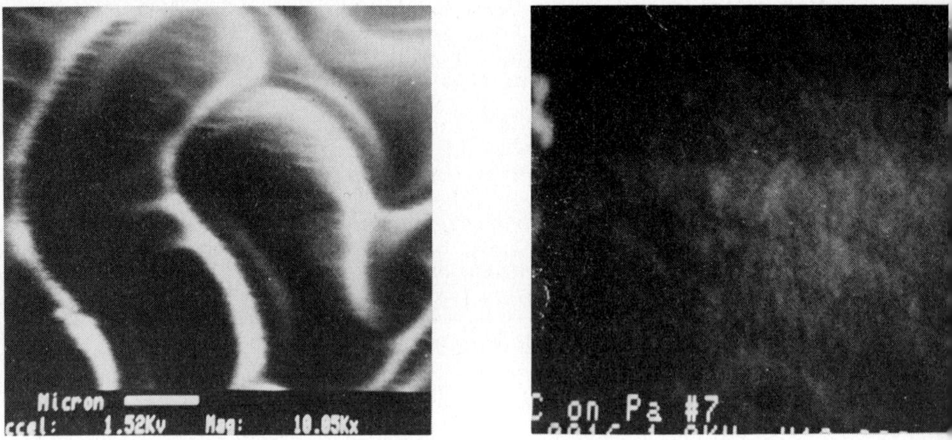

Fig 7: A-carbon films on a parylene surface; poor and good

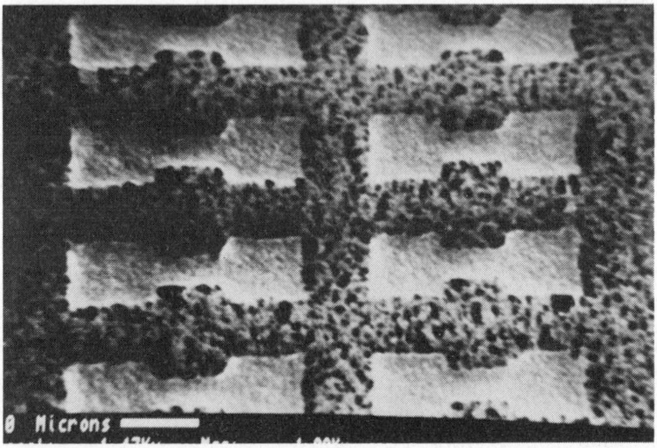

Fig 8: A plasma patterned a-carbon on Pa film.

fabricate 100 meters of 10 micron wire with a 90 percent or greater yield, in a class 100/10 clean room. This model does not take into account stress related and other nonparticulate faults which may occur. In general wires will have to be tested and the faulty ones repaired. As mechanical testing would damage the substrate, it is not possible to do conventional testing.

The focused ion beam is the tool required for testing. The testing is done using voltage contrast testing [13]. In this process one end of the wire is charged and the presence of this charge is tested for at the other end. If the wire is healthy the charge will appear at the other end. This makes it possible to identify the faulty wires. Further this same tool can be used to cut/repair wires In-Situ. Ion milling can be used to isolate bad nets from the wiring. Also new wires could be added by allowing small quantities of a suitable gas at the place the beam is sweeping which will deposit metal to form the right connections.

4. Conclusion

It has been shown that a complete low temperature process is possible. This reduction in stress will ensure that the working chip sites on a wafer are not damaged. This process would reduce mechanical stress significantly thus providing higher yield and reliable interconnect. Further by combining WSHP and WSI it may be possible to have two planes, one on the substrate and the other mounted as a hybrid, of working chips connected by short microtransmission line interconnects.

Acknowledgement

The assistance provided by Eastman Kodak, IBM, Raytheon and DEC over the years is acknowledged.

References

[1] Hinode, K. et al, Stress-induced Grain Boundary Fractures in Al-Si Interconnects, J. Vac. Sci. Technol., b 5(2) Mar/Apr 87, pp 518.

[2] Li, Cheu-Yu et al, Analysis of Thermal Stress-induced grain boundary cavitation and notching in narrow Al-Si metallizations, Appl. Phys. Lett., Vol 53, No.1, July 1988.

[3] Stienbaugh, W. H. et al, Correlation of Gm degradation of submicrometer MOSFETs with refractive index and mechanical stress of encapsulation materials, IEEE Tran. on ED, Vol 36, No. 3 March 89.

[4] Jaesok, Kim et al, Transient and crosstalk Analysis of slightly lossy interconnect lines for WSI and WSHP - weak coupling case, IEEE tran on Circuits and Systems, Vol 35, No.11, Nov 1988, pp 1369.

[5] Capt. Donlan, B. J. et al, The Wafer Transmission Module, VLSI System Design, Jan 1986, pp 54.

[6] Frankel, S., Multiconductor Transmission Line Analysis, Artech House Press, 1977.

[7] Majid, N. et al, The Parylene- Al Multilayer Interconnection System, IEEE JEM vol-18(#2), March 89, p301-311.

[8] Selvaraj, R. et al , Ion Cluster Beam Metallized interconnections for Wafer Scale Integration. Proc of 1987 VMIC, santa Clara, CA, June 1987, pp 440-448.

[9] McDonald, J. F. et al, Adaptive discretionary Wiring for WSI using e-beam lithography, SPIE, Vol 773, March 87, pp 140.

[10] Rothman, L. B., Process for Forming Passivated Metal Interconnection System with a Planar Surface, J. of Electrochemical Soc., Vol 130, No. 5, May 83.

[11] Tsai, Hsia-chu et al, Characterization of Diamond like Film and Their Applications as Overcoats on Thin Film Media for Magnetic Storage, J. Vac. Sci. Technol., A 5(6), Nov/Dec 87.

[12] McDonald, J. F. et al, Low Temperature Plasma Amorphous Carbon Encapsulation for Reliable Multilevel Interconnections - with Applications to Wafer Scale Multichip Packaging, VMIC , June 89 (to be published).

[13] Lin, H. T. et al, Focused, electron and Ion Beam Repair Strategies for Wafer Scale Interconnections in Thin Film Packaging, Proc. Int. Conf. on Metallurgical Coatings, 1988.